生态建筑实验与实践

International Ecological Architecture

《设计家》编

FOREWORD
序言

已建成环境的可持续设计

可持续性，是建成环境领域的建筑师及其他设计师在为提升其作品品质的努力过程中广泛使用的术语。关于建筑对景观环境的影响，我们的了解微乎其微，导致我们所赖以生存的自然环境遭受污染、破坏、恶化、停滞。可持续设计的目标在于创造一个平衡且具敏感性的成果——低影响度，保护水资源，减少能源消耗，使用当地现有的材料。可持续性关乎如何确保设计的可持续性，把对未来环境的破坏降至最低程度，同时使它具有功能上的灵活性、可调整性，以持续支持社会和经济的需求。

可持续性设计及建造不仅要考虑项目基地的环境因素，也包括对影响项目设计及施工的社会及经济因素的理解。 通过对基地现状的认识和与周边元素的协作，可持续性设计的过程就开始了。

在设计之初，考量基地状况往往至关重要。在建筑、公园、城镇的设计规划中，人们倾向于利用自然景观来满足使用者、基地内部及周边区域社区的需求。而在此过程中，可能会出现阻碍设计及施工的负面作用力，其中部分因素包括景观地形学、土壤条件、现有植被、水路及野生生物的分布状况。通过最初的基地调研，人们往往会对其特性进行评估，确保最具可持续性的方法得到落实，确保在项目开发的过程中，基地得到细致的对待。

在城镇规划（再规划）、设计（再设计）这个更大的尺度上，可持续性最重要的价值在于它能确保该地长久的生命力与繁荣。这不仅需要考虑环境因素，还要考虑到社会及经济因素。 公共空间的改善为城市的可持续性创造出大范围的效益，从而大大影响城市的宜居性和安全性，有助于城市吸引购房者和观光客，保持经济繁荣。规划师和建筑师倾注心力研究新开发项目中的现有条件，密切关注和利用各种因素，如地形，确保道路系统得到良好的连接，易于识别，布局恰当，来保护现有的动植物生长和栖息之地。山地、山谷、平原等景观的自然构造也对特定功能区域的分布有决定性的影响。

更具体地说，在确保可持续性设计的原理被整合到特定的、具体的建筑中时，仍需作一些思考。在建筑设计中，许多措施都可以整合到物质结构中去。建造高效节能建筑是建筑师优先考虑的主要目标。有许多种技术可以用来降低能源消耗，然而也有一些途径可用于获取和生成更多的能源。高效供暖、降温及通风系统的利用将有助于降低成本，从而确保更低的能源消耗，以及能源循环使用的可能性。利用太阳能板有助于为建筑供暖、降温和提供热水。在炎热的季节，恰当的建筑朝向使其最大程度地实现通风，有助于降低室内气温，从而减少对空调的需求。

使用可循环的建材，是有助于降低对环境负面影响（如砍伐森林）的又一可持续性措施。尤其是在改造项目中，可以对旧窗户、旧地板进行再利用来保留建筑原先的风格。如今，在大尺度的建筑中，也可使用再生钢来构建框架。

为了确保建筑的可持续性，也需要优先考虑如何把室内空间和功能的灵活性最大化。使用者的需要对应着对空间的一些要求，在设计初期进行空间规划，则有助于使两者相匹配，相得益彰。创新型的室内设计解决方案可能包括可移动的墙，及具有一定适应性及便捷性的储藏空间。从更大的尺度来看，富于智慧、巧妙的空间规划通过防止城市扩张、将城市广场和公共空间最大化，将有助于有助于在建成环境中保障其可持续性。

公共交通运输导向型设计（TOD）是一种规划和设计方式，可强化城市的连接性，集中活动区域并提供交通模式的可选性及更大区域间的到达性。这一易于使用、高度便捷的系统被设计用于连连通中央区域的不同节点，并实现城镇内不同节点与区域间的连通。在这些交通枢纽周围往往有高密度的住宅区，它们拥有高品质的人行及自行车流线。该系统支持建成环境内部的可持续性设计，它旨在通过相互连接、便捷且舒适的高效公共交通将城市串联起来。

建成环境中的可持续性设计看似复杂和十分缜密，然而考量之后，可以发现许多设计都采用了简练、合乎逻辑的方式来应对建筑及环境的演变。只要全面考虑关键的环境、社会及经济因素，创造可持续城市和场所是可以实现的。从细节来说，这些因素正是通过建造和设计的技术而得到支持与强化的。在满足各社区的功能需求的同时，这些具有响应度、平衡的建成环境实现了对自然景观影响的最小化，创造出自然与建成环境之间的和谐关系，促进了社会的长期发展与经济层面上的成功。

范铁
英国皇家建筑协会注册建筑师(RIBa)，
福斯特建筑事务所(Foster and Partners)助理合伙人

SUSTAINABLE DESIGN IN THE BUILT ENVIRONMENT

Sustainability is a popular term amongst architects and other designers in the built environment industry as they strive to improve the quality of their work. The natural environment which within we live has for a long time been polluted, demolished, degraded and hindered because of our inability to understand the affects of building on the landscape. The aim of designing sustainably is to create a balanced and responsive outcome that is low impact, conserves water, reduces energy usage and requirements and utilizes materials that are locally available. Sustainability is about ensuring there is longevity in the design to minimize future disruptions to the environment while being flexible and adaptable in function to continually support the needs of society and the economy.

Designing and building sustainably is about not only considering the environmental factors on the project site, it also involves understanding the social and economic factors associated with the impact of the design and construction of the project. Through recognizing the current conditions of the site, and working with the surrounding elements, the process of sustainable design has started.

Considering the site conditions at the start of the design process is always important. When designing a building, park, town or city, the natural landscape is likely to be manipulated to the needs of user or community within and around the site boundary. It is possible that in this manipulation, adverse affects may arise which hinder the design and constructability. Some of these elements include the landscape typology, soil conditions, position of existing trees and waterways and wildlife. Through the initial site survey, an assessment is typically made which identifies these and ensures that the most sustainable measures are put in place so that the ground is treated sensitively in the development of the project.

At a larger scale where towns and cities are planned / re-planned and designed / re-designed, sustainability is of upmost importance to ensure the longevity and prosperity of place. Not only is the environment considered but also the social and economic factors. Improving public space creates a wide range of benefits to the sustainability of a city. It can greatly affect the livability and safety of a city, making it desirable for home buyers and tourists alike and may also boost businesses. Planners and architects take great care in researching the existing conditions and in the case of new developments, work closely with factors such as landform to ensure that roads systems are well connected, easy to comprehend and appropriately placed to conserve existing flora and fauna habitats. Working with the natural formation of landscape such as hills, valleys and plains can also dictate the location of particular zones of use and function.

At a more detailed level, there are also considerations to be made when ensuring sustainable design principles are integrated particularly to architecture. In building design there are a range of measures that can be integrated into the physicality of the structure. Creating energy efficient buildings are one of the main, prioritized objectives of architects. There are various techniques used to reduce the energy consumption of a building however there are also ways to capture and generate more energy. Efficient heating, cooling and ventilation systems are all able to be utilized to support cost cutting measures ensuring energy consumption becomes less and also may be recycled. Harnessing the sun using solar panels helps to warm and cool buildings and also heat water. The orientation of a building, particularly in hot climates allows wind flow to be maxamised through a building helping to cool the interior and thus reducing the need for air conditioning.

Recycling materials for building is another sustainable measure that helps to reduce the need for negative impacts on the environment such as deforestation. Particularly in renovations, old windows and floor boards may be re-used to retain the same style of the house. In larger scale buildings, it is now possible to use recycle steel for framework.

To ensure sustainability of a building, maximizing the flexibility of interior space and function must be a priority as well. In the early stages of design, spatial planning helps to ensure that the needs of the user compliment the space required. Innovative design solutions for interiors may include movable walls and adaptable, convenient storage spaces. Smart spatial planning at a larger scale helps to ensure sustainable built environments through preventing urban sprawl and maximizing city squares and public space.

Transport Orientated Design is a form of design and planning that strengthens the connectivity of a city, centralizing activities and providing choice in transport modes and accessibility to a wide range of areas. This easily to use, highly convenient system is designed to connect different nodes of transport in central locations and to various nodes and places within towns and cities. Around these transport hubs are typically a higher density of housing where there is a high quality of pedestrian and cycle routes. This system supports sustainable design in the built environment as it aims to bring together cities through efficient public transport in an interconnected, convenient and comfortable manner.

Sustainable design in the built environment may appear to be complex and meticulous however when considered, much of the design is a simple and logical approach to the evolution of architecture and the environment. Creating sustainable cities and places is very achievable as long as key environmental, social and economical aspects are considered. In the detail, these elements are supported and strengthened through building and design techniques. These responsive and balanced built environments minimize impact on the natural landscape while meeting the needs and functions of the communities, creating a harmony the natural and built elements, promoting the longevity of society and success of the economy.

CONTENTS
目录

1 规划

2 商业|办公

3 公共

Planning

Office | Commercial

Pubic

4 文化|教育

5 酒店|住宅

Culture|Education

Hotel|Residential

Planning

规划

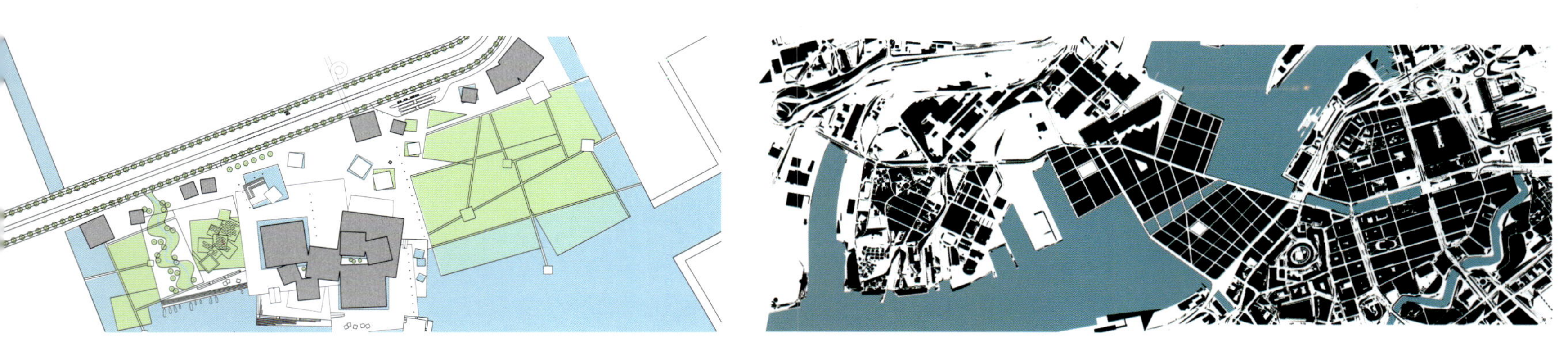

Caofeidian Sustainability Center

曹妃甸国际生态城可持续发展中心

Sweco

项目地点	中国，唐山
项目时间	2008年—2009年
建筑主创设计师	Johannes Tüll, Anna Hessle
建筑设计主创团队	Johannes Tüll, Anna Hessle, Alessio Boco, Anna Markström, August Wiklund
景观设计主创团队	Shira Jacobs, Staffan Sundström, Anna Gustafsson, Helena Eriksson
室内设计团队	Johannes Tüll, Barbro Sjöfall, Bo Lundberg, Sonia Garbajosa, August Wiklund
建筑设计配合团队	Lis Kjaer, Elsa Dahlman, Elin Mossberg Johan Svanholm, Elsa Dahlman, David Essinger, Petter Appelfeldt, Bosse Lundberg
专家配合团队	Ulf Ranhagen, Gunnar Nordberg, Lars Olof Matsson, Thomas Nordh, Henrik Berg von Linde,David Essinger, Peter Wiss
3D效果设计	August Wiklund, Petter Appelfeldt, Fredrik Ericsson
当地设计研究院建筑工程师	中国建筑设计研究院，张通

Project Location	Tangshan, China
Project Milestones	2008-2009
Chief Architects	Johannes Tüll, Anna Hessle
Architecture Main Design Team	Johannes Tüll, Anna Hessle, Alessio Boco, Anna Markström, August Wiklund
Landscape Main Design Team	Shira Jacobs, Staffan Sundström, Anna Gustafsson, Helena Eriksson
Interior Design Team	Johannes Tüll, Barbro Sjöfall, Bo Lundberg, Sonia Garbajosa, August Wiklund
Architecture Support Design	Lis Kjaer, Elsa Dahlman, Elin Mossberg Johan Svanholm, Elsa Dahlman, David Essinger, Petter Appelfeldt, Bosse Lundberg
Supporting Expert Team	Ulf Ranhagen, Gunnar Nordberg, Lars Olof Matsson, Thomas Nordh, Henrik Berg von Linde, David Essinger, Peter Wiss
3D -images	August Wiklund, Petter Appelfeldt, Fredrik Ericsson
Local Architects and Engineers	CADG, Zhang Tong

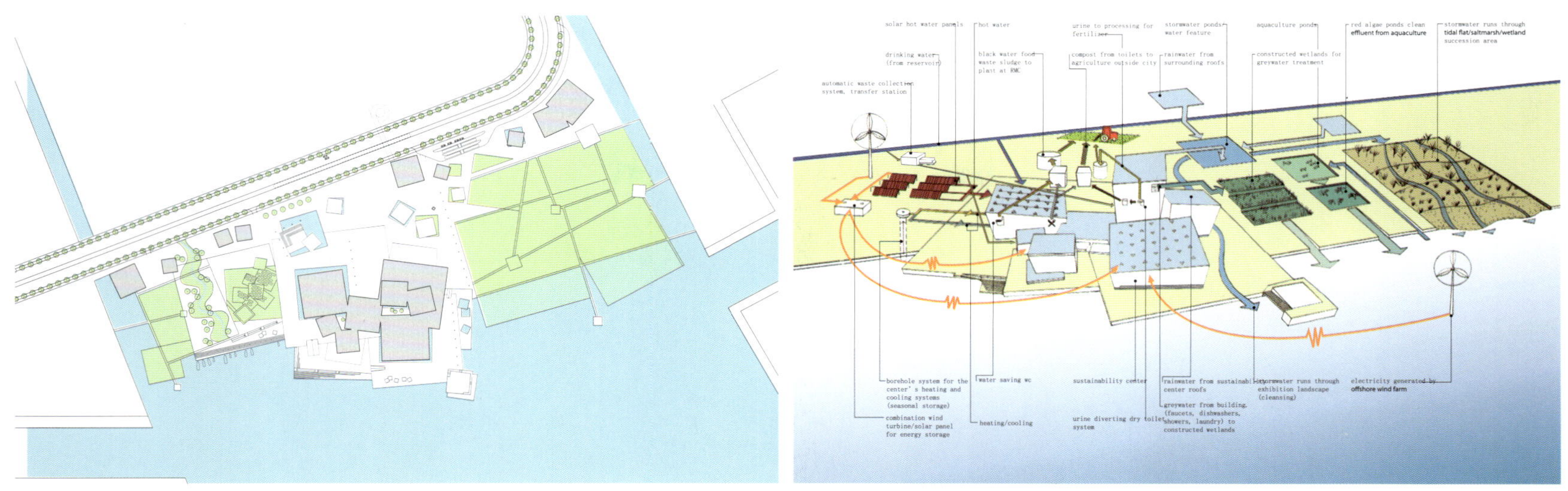

唐山曹妃甸国际生态城由瑞典Sweco集团规划设计，可持续发展中心作为生态城的一部分，是生态城的展览和信息建筑。这座占地2万平方米的中心将展示有关这座生态城的开发信息以及瑞典的环境技术和智慧。

可持续发展中心的建筑体型受到盐粒结晶的形状和聚集方式的启发。盐从咸水中析出，不断扩大，这就是晶体方案建筑和景观设计的背景。可持续的盐的生产是这个可持续发展的建筑的创作灵感。晶体的体型组合考虑了当地日照和风向的条件，创造出有庇护的空间、宜人的尺度和对可持续发展未来的美好憧憬。

场地分为三个不同的区域，各自的特点稍有不同。景观设计的目的围绕着三个原则：节约用水，展示可持续技术和曹妃甸地区不同的生态系统，并提供一个世界级的公共空间。

功能组织

可持续发展中心是曹妃甸生态城的标志。建筑将展示技术、交流总体的远景和传递唤起公众意识的信息。建筑中的常设展厅、多媒体视频大厅和用于临时聚会的活动厅，都会展示并促进可持续发展的系统解决方案和城市的远景发展。中心还向公众展示了一个功能良好的可持续发展的建筑，包括所有内部系统和一个新系统及新设备的展示区域。

节能环保

下列系统将应用在本建筑中或以展览方式展示：

良好的外围护系统；

节能型照明系统；

用于外立面和室内装饰照明的LED照明系统；

节能型通风和空调系统；；

节能型设备；

太阳能电池、燃料电池和风车的组合，用来展示一个应急电源的解决方案的新系统；

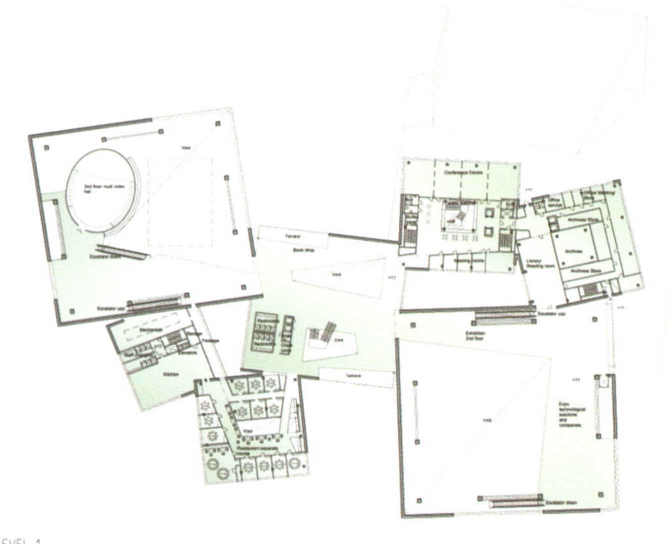

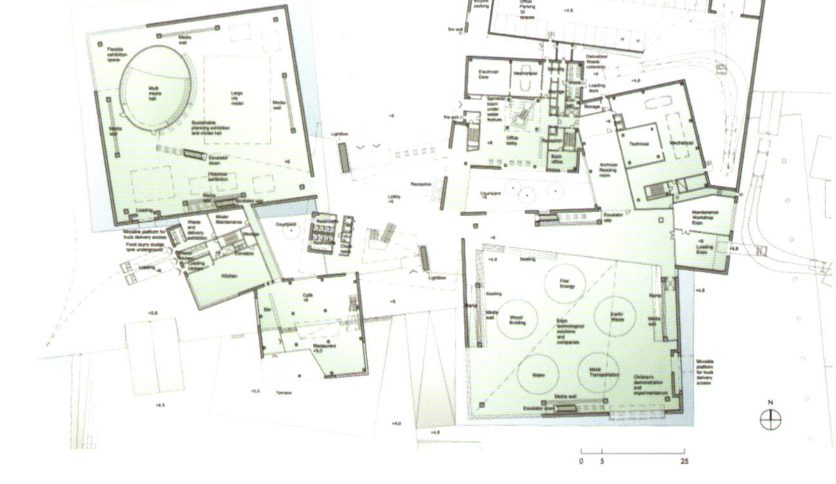

平面布局图

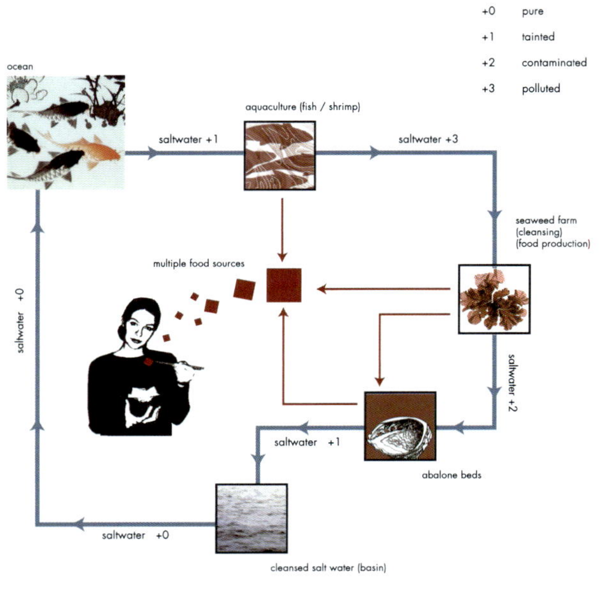

海水养殖系统

0 greywater from building

1 wetlands for greywater treatment

cleansed greywater

excess water to landscape

2 riparian area

3 estuarine area

3.1 tidal wetland

3.2 brackish marsh

3.3 salt marsh

inner sea lagoon

循环再造系统

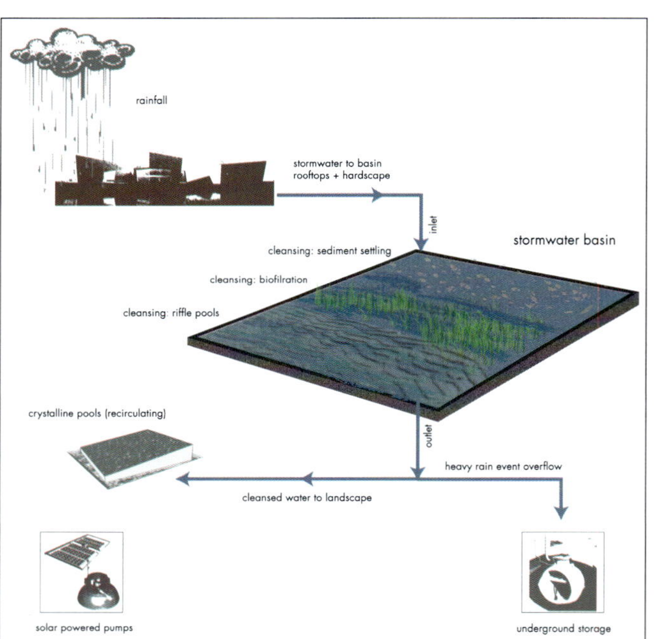

雨水循环系统

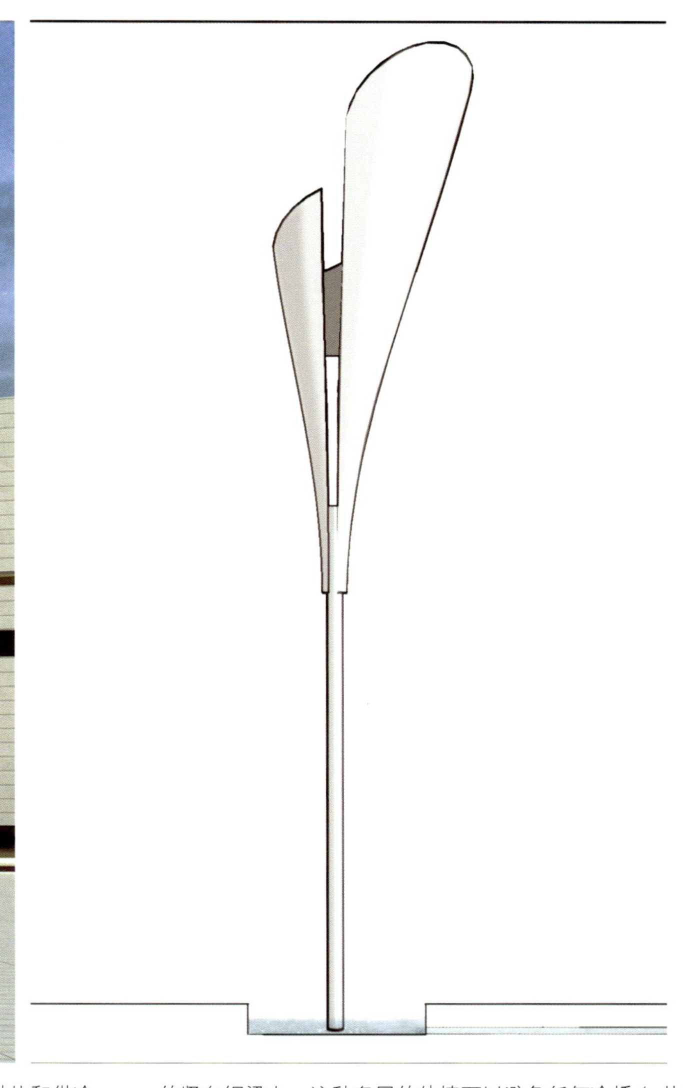

采用深井进行季节性能量储藏，用于供热和供冷；

环保的垃圾收集和处理系统；

连接城市和农村的生态循环的可持续发展的解决方案（食物垃圾粉碎和黑水系统）；

有利于节水的系统；

不同类型的可持续发展的卫生设备和污水处理系统，包括一个人工湿地。

建筑特点

晶体建筑考虑给建筑内部的不同功能赋予不同的空间体量。每一个体量的大小、形状和特性又受到不同内部功能的影响，比如是否需要自然光、对视野的要求、对内部小气候的要求、空间净高要求以及对灵活性的要求。

建筑布局时重点考虑组织清晰，使人在建筑中能容易找到方向。通过不同体量的组合，创造出自然的人流、物流路线。建筑成为一个讨论可持续发展理念的聚会场所，包括随意使用的聚会空间，从而鼓励可持续发展理念的讨论和发展。

立面系统

该立面系统基本上是一个微通风立面，由硬质保温材料和石材饰面板组成。用螺丝把不锈钢构件固定在合适的框架上构成的点状固定系统（外面看不见），将石材（切成板状）固定在外墙上。下层结构固定在外墙内的竖向钢梁上。这种多层的外墙可以避免任何冷桥/ 热桥点，并达到最好的内部气候效果。

景观设计概念

可持续技术包括了尽量减少能源和水的消耗；利用无毒，回收，循环再造以及当地最好的材料；以及利用现有的废水作为进水，寻求“闭环回路”。可持续发展中心的景观计划能满足所有这些标准。通过储存雨水，废水回用和日益增长的食物与盐水，全部进行计算废水、雨水和海水系统，以减少淡水的需求。露水收集器演示了如何从空中收集水和从当地获得铺路材料。

室内设计概念

可持续发展中心的室内设计概念从唐山当地的文化和自然环境中获取灵感，并融合了北欧斯堪的纳维亚的设计哲学，从而创造其独特的风格。设计中的一些元素的灵感来自当地三角洲的自然环境、盐湖湿地和石楠丛生的田野。当地海边的鸟类以及海洋生物也是图案装饰和织物的灵感来源。

室内设计概念的另一部分特征元素来自唐山当地文化的形象、京剧的脸谱和色彩、中国织物和壁纸的图案以及当地悠久的陶瓷生产历史。我们以这些图案、色彩和图像为基础，把它们以斯堪的纳维亚的方式进行解构和诠释。建筑的不同功能区域有不同的主题。

Background

Tangshan Caofeidian International Eco-City is planned and designed by Sweco. The Sustainability Center being part of the new Eco-City, is the building for exhibition and information exchanging in the Eco-City. The Sustainability Center will be a symbol of the International Caofeidian Eco- City's founding principle of economically, socially, and environmentally sustainable urbanism. The Center's mission is to demonstrate cutting edge environmental technologies through temporary and permanent exhibitions, communicate sustainable principles through public education, and showcase successful climate-neutral construction through integrated building design.

This climate-neutral Center will showcase the eco-cycle model of the Caofeidian Eco-City; the building's construction and function will demonstrate both applied and experimental technical solutions including small-scaled renewable energy production. Event spaces accommodate both permanent and temporary exhibitions while a business arena and expo will showcase innovative companies and products within the field of environmental technology and serve as a venue for the presentation and promotion of the Eco-City. Some of the municipality's administrative functions responsible for monitoring the ongoing construction of the Eco-City will also be housed in the building. This establishes the Center as a

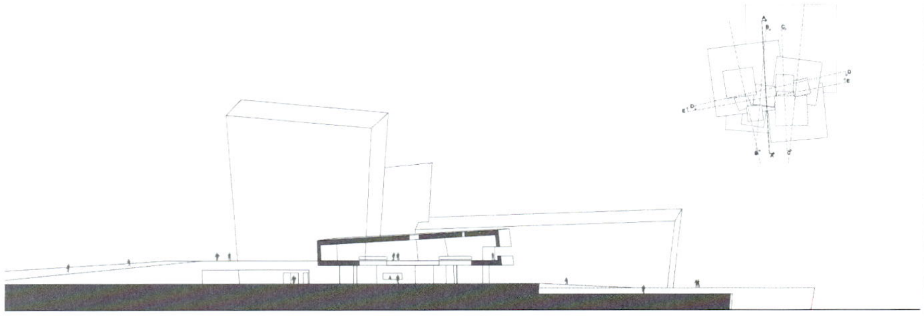

Section A-A

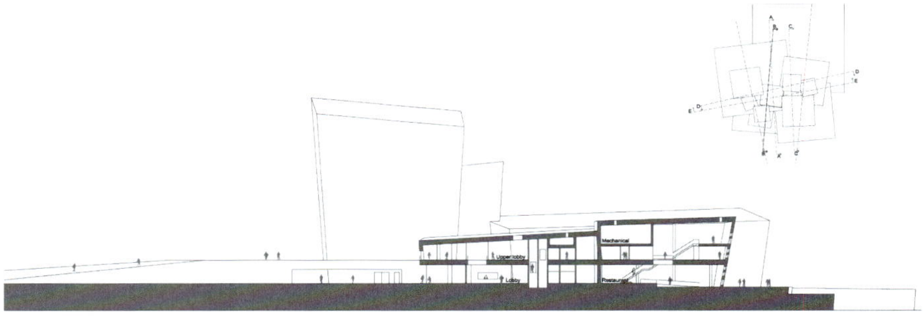

Section B-B

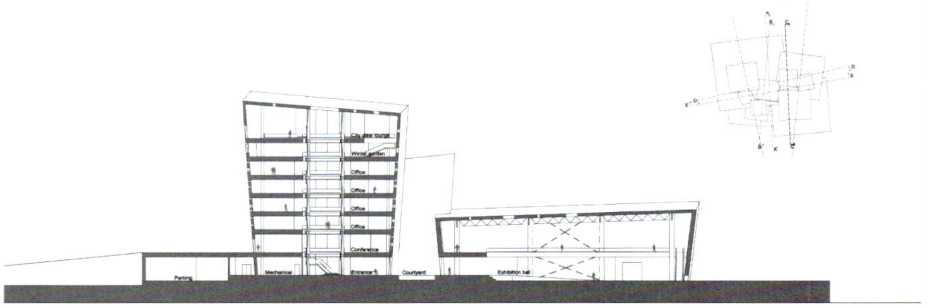

Section C-C

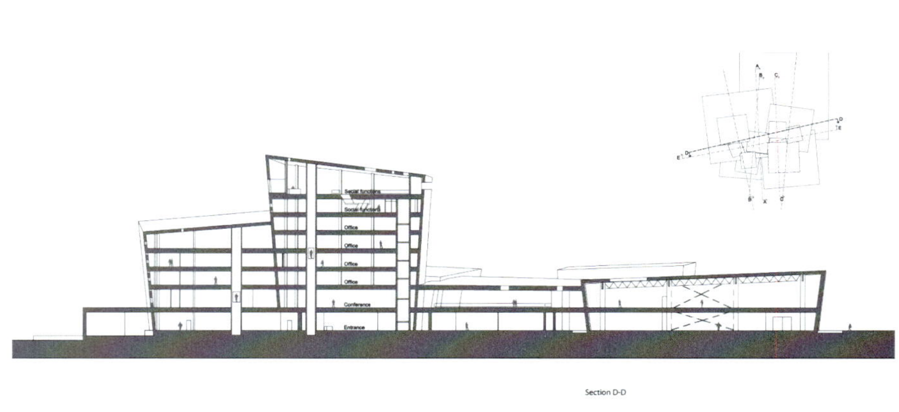

Section D-D

main node in a valuable feedback loop, where information, research and assessment of completed phases informs the City's future development.

Sustainability goals and objectives

The goal for the Center is to address all social, economic, and ecological factors impacting human health, safety, and quality of life, to create an arena for a continuous dialogue between international developments in sustainable practices and the Caofeidian community, and to embody the soul and character of the Caofeidian heritage.

Urban context

The Center is strategically located along the green diagonal axis in the Intermediate Area, along with other significant venues such as the administrative center, the cultural center and the civic center.

Set on a peninsula in the lagoon area of the Eco-City, the Sustainability Center engages both the diagonal city grid and the coast. The restaurant, café, public landscape, and shops will draw visitors day and night, encouraging activity along the quays and public parks. Pedestrian, bicycle and public transportation methods are encouraged through the strategic location of monorail, bus station and bicycle parking adjacent to the Center.

Conceptual idea

Local salt production, with its references to both the history of the site and the physical location of the Center on the edge of the sea, provides inspiration. The geometric shape, and clustered formation of salt crystals also underlie the conceptual foundation for the Sustainability Center and indeed the entire Caofeidian Eco-City. Just as salt crystals join and divide, gradually spreading and growing, the Eco-City will develop outwards from the Sustainability Center and the core area around it. Just as salt is traditionally formed into mounds, rising from pools of seawater, the context for the Center and its surrounding landscape forms. The building volumes are clustered to adapt to the local conditions of wind and sun, creating sheltered spaces and a human scale.

Architectural character

The unique silhouette and sculptural appearance of the Crystalline Building, inspired by the behaviour and geometry

of salt crystals, becomes a focal point of the harbour with its open, inviting and architecturally distinct form.

Crystalline's clustered volumes rest on clean horizontal planes. Some clusters appear to rise from planes of shallow water, their volumes and tilted roof —planes mirrored on the reflective surface. Their facades and skylights evoke a light and crystalline character with the use of light natural stone cladding and irregularly placed horizontal openings.

Clustered, varying spatial volumes create dynamic, sheltered and human-scale spaces experienced both inside and outside the building. Two courtyards formed by the juxtaposition of the volumes are designed to help regulate the temperature of the building.

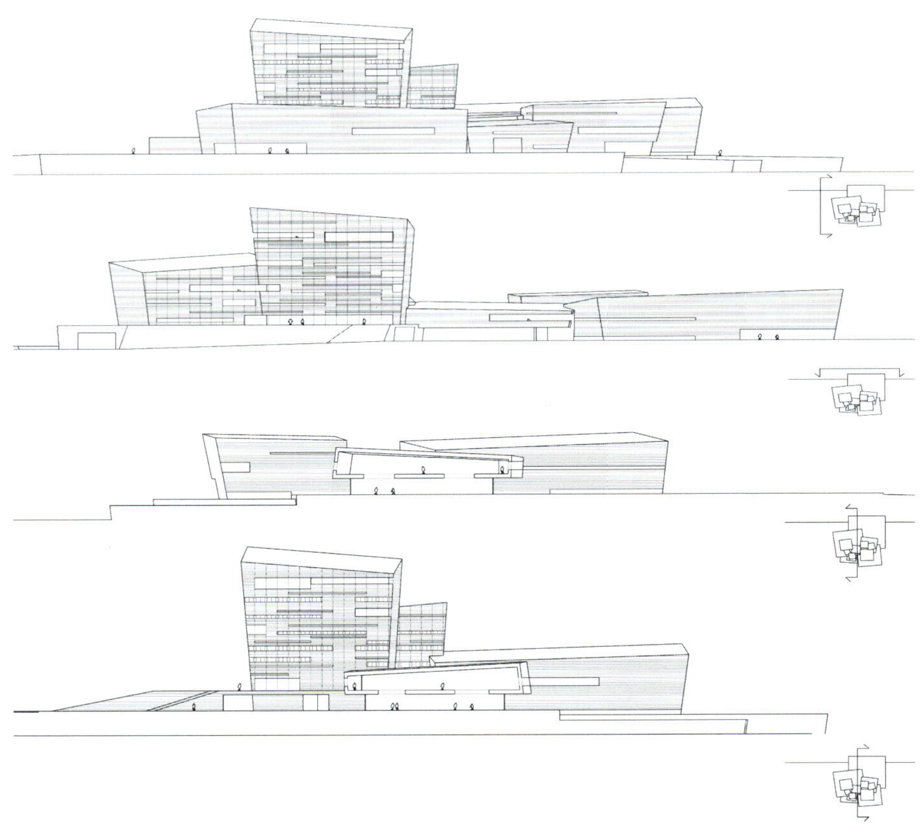

Organisation/ Function

Each interior function is represented by a unique building volume and composition. The size, shape and character of the volumes adapt according to considerations of daylight, views, microclimate, roof height, and use.

The building's entry plaza extends beneath a lifted crystal volume into the main lobby, capturing sea views. The lobby is approachable from both the entry plaza and the quayside.

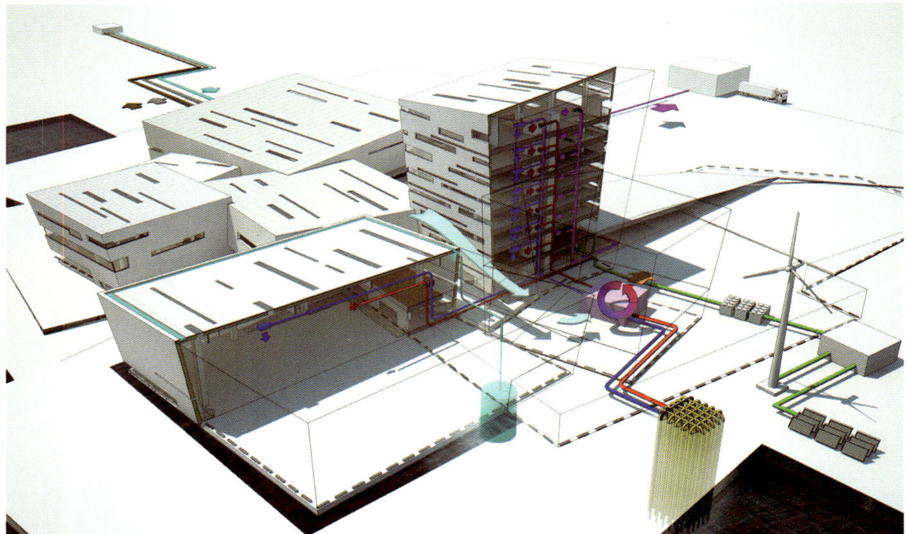

The two main exhibition halls are accessed directly from the main entrance lobby. The first contains the Multi-Media Hall and models of the Eco-City. The second contains an exhibition of sustainable technologies. Both exhibition spaces are large, flexible, and lit by skylights. Horizontal openings in the facades afford views to the landscape.

The restaurant, easily accessed from both the lobby and the outdoor public entrance, offers an open space ground floor with café and bar.

The higher building volumes containing offices and archives are accessed to the east of the main entrance lobby. Central light wells ensure adequate daylight. Geological conditions prevent basement structures; as an alternative, office and archive volumes sit on a common podium containing parking, mechanical facilities and delivery access. This allows the building to maintain a close connection to the boardwalk, thus avoiding high quay edges at the sea.

The building is organized to encourage intuitive navigation and spontaneous interaction, creating multi-use interior spaces to allow for various scales of use. The dynamic, modular form of the building allows for new volumes to be constructed in the future, maximizing the flexibility of the complex.

Interior design

The interior design is inspired by the Tangshan culture and local natural ecosystems: delta, salt wetlands, and heather fields. Bird and sea life lend inspiration to patterns and textiles, as do the masks and colours of the Chinese opera, traditional Chinese textiles and wallpapers, and a long local history of fine porcelain production. Filtering these local patterns and colours through a Scandinavian design philosophy marries simplicity with the richness and beauty of the Chinese heritage, creating a unique interior atmosphere.

Landscape architecture

Three main goals inspire the landscape architectural design for the twelve hectares surrounding the Sustainability Center: prioritize water conservation, showcase sustainable technologies appropriate to Caofeidian ecosystems, and create a world-class public space.

Fresh water is a finite resource. Caofeidian, a coastal city in a relatively arid region, will be designed with systems in place to enable the conservation and recycling of fresh water. The Sustainability Center will lead by example, capitalizing on the value of grey water, stormwater, and saltwater to alleviate fresh water demands. Dew-catchers collect water from the air, edible plants are nourished with salt water, and grey water from the building is rerouted and cleansed for use in the landscape.

There are three special ecosystems along the coast of Caofeidian; the Sustainability Center's landscape strives to demonstrate, through stylized forms, their function and appearance.

The first is the intertidal zone where molluscs and other aquatic animals live in shallow waters, mudflats attract migratory birds and salt marsh species sculpt the sand into protective dunes. The second is the river itself, with riparian flora and fauna. The third is the individual and collective chain of tidally influenced wetlands upstream, brackish marshes, and saltwater estuarine habitat. Each of these is represented in the design, making the landscape a teaching tool rooted in the context of place.

While the landscape supports exhibition space, demonstration areas, and the Caofeidian ecosystems, it also showcases the building and provides visitors with a world-class public space. The large space in which the building rests functions as an exterior extension of the building, allowing growing and colliding "crystals"to flow between inside and outside. The esplanade, crystalline pools, overlook, water stairs, and marina provide a welcome respite from the bustle of the city.

A climate neutral building

An objective of the Center is to demonstrate how a climate

neutral building can be built with existing techniques, and to test and demonstrate possible future technical solutions. The building must meet existing principles for an energy efficient building, and also produce its own renewable energy supply.

The following objectives produce an energy efficient building:

- Plan the site and the building with regard to the sun and wind.
- Minimize energy demand through architectural and constructional design.
- Minimize energy demand through energy efficient installations, lighting, and water saving appliances.
- Provide a building envelope to control solar heat gain and control daylight.

Advanced project management is required throughout the life of the project to regulate quality assurance, environmental assurance, and energy assurance.

The energy demand for the Sustainability Center is expected to be lower than the target levels for the Caofeidian International Eco-City. The overall energy demand for the Sustainability Center is estimated to be less than 1,000 MWh electr city per year, 300 MWh heat per year, and 400 MWh comfort cooling per year.

All electrical equipment and appliances will have the lowest electrical demand available. Other requirements include: energy efficient ventilation and cooling systems; energy efficient fluorescent lamps with timed daylight control systems; wiring, power distribution and communication network sectioned to facilitate energy savings, service and reconstruction; and installation of energy meters to monitor demand, providing a quantitative tool to encourage reduction in use.

Renewable energy Supply

Energy is partially supplied by demonstration units within the Center, securing complete energy delivery all year round. Though the site's geotechnical conditions are not ideal for seasonal storage, the cold winter and warm summer of the Tangshan Region make it an ideal situation for a demonstration unit. Several boreholes provide regulating heating and cooling at competitive costs, with a very low environmental impact. The demonstration unit calls for 50 boreholes, each at 50 meters deep. Seasonal storage of this size is estimated to provide180 MWh of heat or comfort cooling.

Also proposed are solar cells, fuel cells and a small windmill. The combined system can "store" electricity from the windmill and solar cells by using an electrolyser that produces hydrogen. This provides energy to the fuel cell at times of low wind or sun. The proposed system has an average capacity of 5 kW of electricity or heat, and demonstrates a flexible power supply with the capacity to supply electricity to a telecom station within the Center.

Energy production from seasonal energy storage would supply 50% of the demand for heat and cooling, and the fuel cell system would supply roughly 5% of electricity demand and 10% of heat demand. Additional energy is needed to cover the remaining demand, including a combined heat pump and electrical chiller which, during the summer, will produce cooling for the buildings and regulate temperatures via air-cooled condensers. In the winter ,the system is reversed and the outdoor air is used as a heat source. Solar collectors installed adjacent to the solar cells produce heat for tap water.

The electrical systems also connect to the main grid, which supplies energy as needed; this power is produced in the

Caofeidian International Eco-City wind farm or in the waste incineration plant at the Resources Management Center.

Principle system design for main comfort systems

The building is heated and cooled through a mechanical ventilation system with heat recovered from exhaust air. The inlet air is preheated from either the seasonal energy storage or the combined heat pump. The building envelope has been designed so that no additional radiator systems are needed. The building is air-conditioned in summer months with a "warm" cooling beam and cooling of the inlet air. The ventilation systems provide an energy efficient and reliable design with minimal maintenance requirements. Large air handling units and low air velocities help maintain steady pressure in the building.

Office and archive areas will have constant air volume systems, and all other areas will have a demand-controlled variable air volume system.

Conference rooms and areas with changeable internal loads are designed so that an extra forced ventilation air diffuser may be installed.

Sanitation and water

The preferred technical sanitation solution in an area with a limited supply of water is a urine diverting toilet with dry handling of feces, consuming little to no water. As this solution is not yet well adapted to multi storey or public buildings, it will be tested as a demonstration unit.

Water-saving faucets will be installed to reduce consumption of hot and cold water, and water-saving WC with low consuming black water handling will be incorporated. The black water is pumped to an anaerobic digester at the Resource Management Area to be treated together with kitchen residuals. There is a separate system for grey water, which will be treated in situ and reused for landscape irrigation. Grey water recycling begins with pre-treatment using sieves or sedimentation. Grease removal produces fat from cooking o ls, which is transported to the RMC to be treated with the black water in the anaerobic digester.

After the pre-treatment, grey water is diverted to a constructed sub-surface flow wetland for natural cleaning. In order to discourage evaporation, the wetland has no open water surface and is constructed with several basins or lines depending on the load. The wetland removes oxygen-consuming organic material, as well as some nutrients. Pathogens will also be eliminated in the natural cleansing process.

Storm water

Though summer rains provide plentiful fresh water in the Tangshan Area, cold winters can be very dry, requiring storm water retention, purification, and storage. The Sustainability Center aims to function as a pedagogic and educational resource for the Caofeidian community, highlighting the importance of a visible storm water system connecting the building and landscape.

Waste management and recycling

The Center proposes an applied waste management system of source-separation, a demonstration of organic waste treatment, wastewater treatment with anaerobic digestion, and biogas production.

To enable recovery of resources, including building materials, plant nutrients and energy, waste is source-separated by designated waste generators:

1) Hazardous components: chemicals, motor oil, small batteries, small electronics, pesticides, drugs etc.

2) Bulky/inert waste/broken glass

3) Mixed recyclables, excluding glass

4) Food waste (not wooden materials, hard shells)

5) Remaining waste

The Sustainability Center and neighbouring office and housing are connected to an Automatic Waste Collection (AWC) system. The AWC container terminal, located on the site of the Sustainability Center, serves several surrounding city blocks. Its proximity to the Center enables it to be part of the exhibition of sustainable technologies. In the office and archives building, waste chutes on each floor connected to the AWC system dispose of remaining waste and paper for recycling.

The lunch room service area is equipped with a food waste disposal, which is then connected to the black water pipe. The restaurant is equipped with inlets to a food waste grinder connected to a tank, which is to be located underground, saving valuable space on the ground floor. Food waste slurry is to be collected by a sludge vehicle as necessary.

Conclusion

The Sustainability Center highlights principles of economically, socially, and environmentally sustainable urbanism through research, education, and integrated design. Sensitive to its cultural and ecological surroundings, the Center is a model for the successful marriage of local character with applied sustainable building techniques. This holistic approach to place-making constructs a forum to tend to the cultural, economic, and ecological needs of the community while positioning the Caofeidian Eco-City as a world leader in the promotion of sustainable building strategies and technology.

Summary

The Sustainability Center, named Crystalline, is a key component in the realization of the environmental goals of the Caofeidian Eco-City. Through its construction and program, the Center demonstrates future sustainable technologies and communicates a vision consistent with the principles of the Eco-City, the construction and expansion of which will be monitored from the building. Located on a peninsula with close proximity to Caofeidian's administrative core, the Center will be an iconic symbol for the growing Eco-City and for China's sustainable urbanism initiatives. Crystalline is integrated with a surrounding landscape showcasing sustainable technologies and ecosystems. The Center is a model for an energy efficient building, projecting low energy demand and demonstrating working examples of renewable energy technologies. With its unique architectural style and demonstrations of innovative environmentally efficient technologies, the Center will embody and promote the sustainable principles central to the development of the Caofeidian Eco-City.

Afsluitdijk
世界可持续中心

Francis-Jones Morehen Thorp

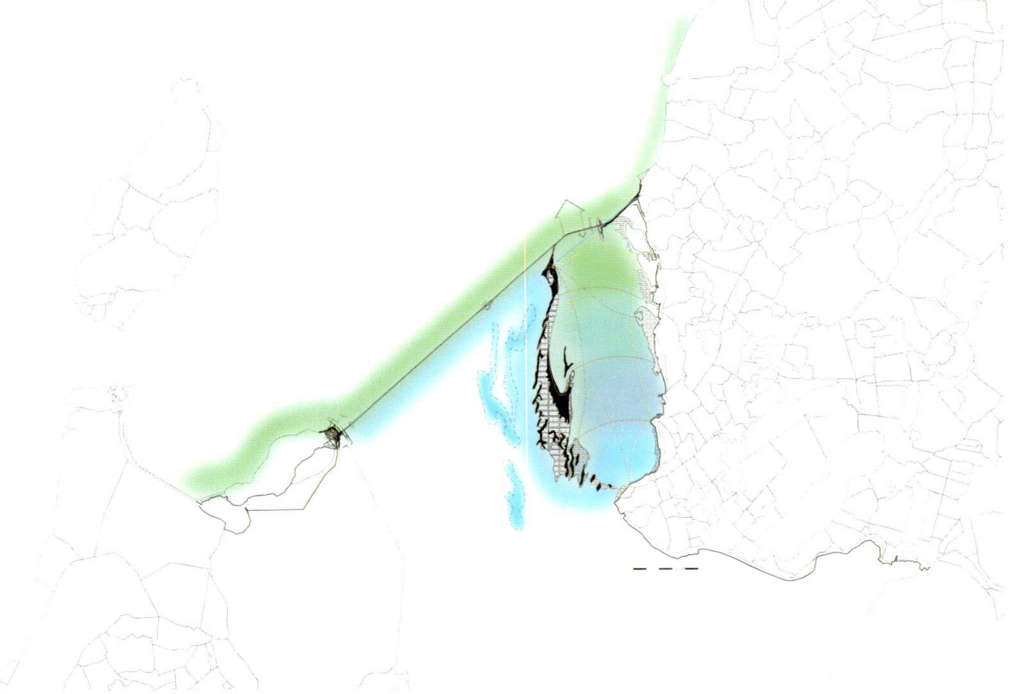

去年，三家富有创造力的荷兰建筑、景观设计以及市政工程事务所共同参与了一个国际性的竞赛，即世界可持续中心Afsluitdijk的项目设计。

他们未获胜的方案在综合可持续发展的角度来说是最为值得注意的。

这是当下最为生态的城市规划。

Afsluitdijk是一个有意义的具有艺术性的堤防，是一个长30千米的大坝。

沙土来自20米深的沟，在平均5米深的艾瑟尔湖的螯鱼，就像有迁移习惯的鸟类在湿地堤防的西部冬眠。WSC Afsluitdijk被期望成为低调的，只有一个东西吸引眼球的科学园区，就是一个大鸡蛋。流动的信息中心，白天面向太阳，形成彩色的田凫蛋，是弗里斯兰省最典型的春天鸟的颜色。不断变化的小规模的实验室沿着Visiteurs小巷过多地建造。餐厅位于Visiteurs中心，在金

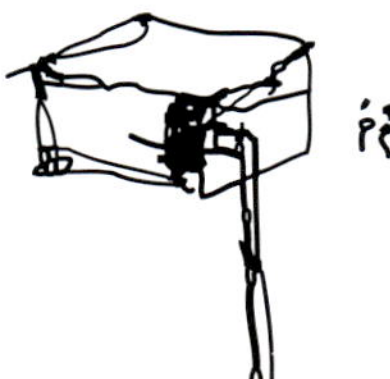

pv

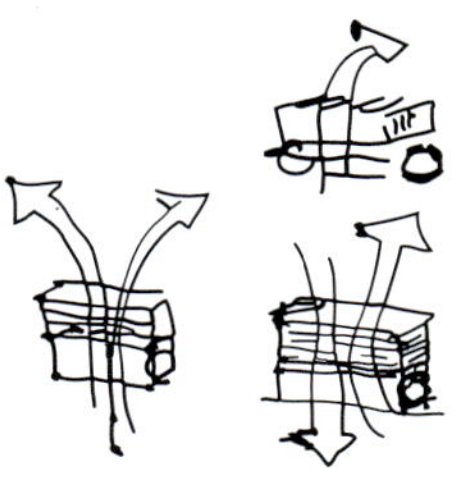

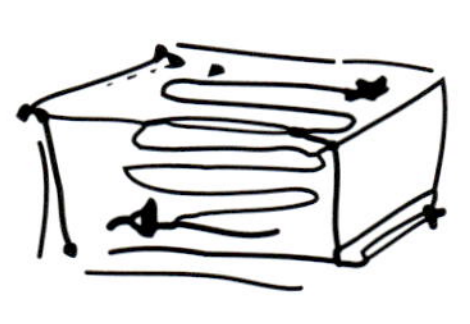

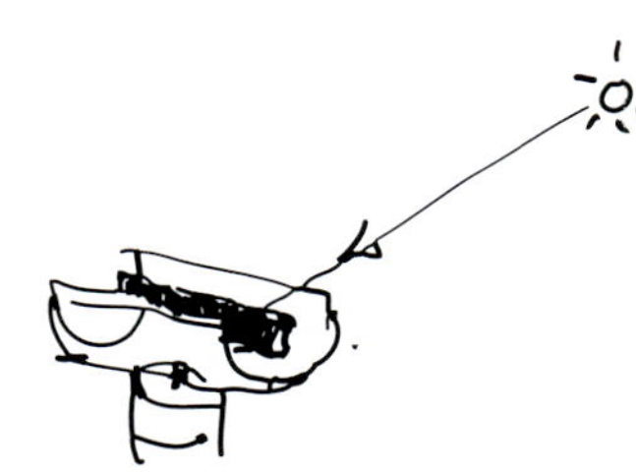

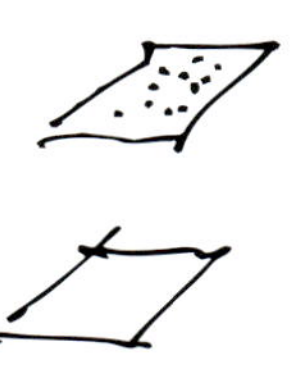

融危机后四星级酒店成为豪华旅馆、酒店船或在马克姆当地的酒店，马克姆村是的一座历史悠久的村庄。海中有盐，水坝内部新鲜的海水及靠盐水制成的蓝能源电力发电站，用反渗透是合乎逻辑的。加热/冷却和复加热平衡ventilatin是一项全新的高科技细线技术，它可以运用25~28摄氏度的低水温，且只在荷兰适用。WSC Afsluitdijk自行发电，在加热、冷却和新鲜的食品加工（除了酒类）方面完全自给自足，节省了至少100吨的幼鱼并得到额外的收获。

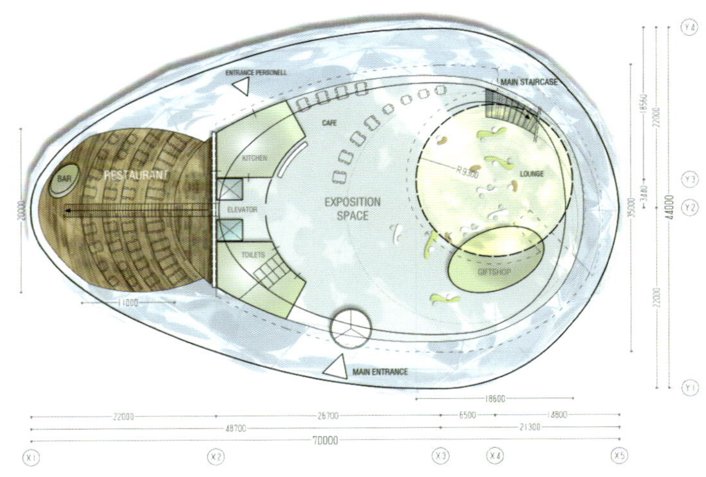

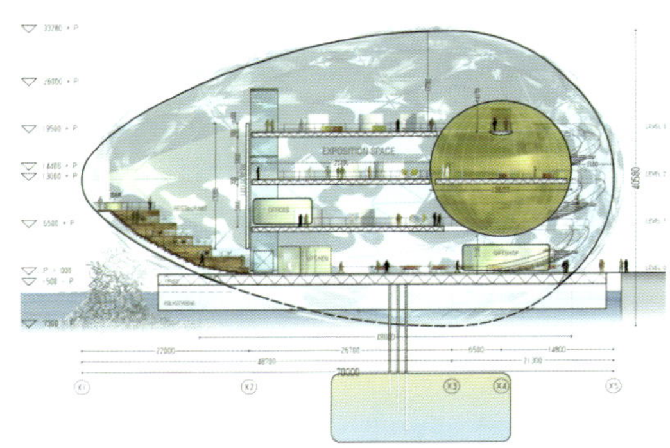

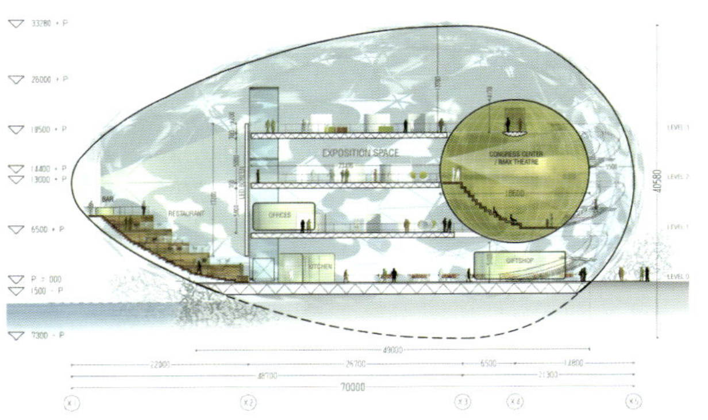

For the first time last year, three creative Dutch architectural , landscape design and civic engineering firms participated in an international competition on World Sustainability Centre Afsluitdijk.

Theirs, not winning proposal is most remarkable from the view of integrated design on all aspects of sustainability.

It is most likely the world state of the art in sustainable landscaping at the moment.

The Afsluitdijk is a magnificent, artificial dike, as a dam 30 km long .

The sand and clay comes from 20 m deep ditches (kuilen) in the average 5 m deep water IJsselmeer for hibernating fish on the west side of the wetland dike also functioning as a furage for migratory birds. The WSC Afsluitdijk is supposed to be a low-profile scientific campus with only one eye-catcher a big egg. The floating information centre, turning sun-oriented during the day, has the form and colours of a lapwings-egg, the most typical springtime bird of the county Fries and. The small scale everchanging laboratories are along the overbuild visiteurs lane. The restaurant is in the visiteurs centre but the financial risky four- star hotel is luxury lodges , hotelboats or a local hotel in Makkum, a historical village on the other site. Having saltwater from the sea, fresh riverwater on the inside of the dike and need for brackish water makes a blue-energy- electrical powerstation , and using reverse osmose is as logical as it can be. The heating / cooling and balanced ventilatin with heatregain is a new high-tech fine-wire-technology using very low temperature water of 25 till 28 degrees that is only applied in the Netherlands. The WSC Afsluitdijk is producing much electricty, and completely self-supporting in heating and cooling and fresh food production (except wine) is saving at least 100 ton of young fish and extra harvesting.

Seoul Commune 2026

首尔公社2026

Mass Studies建筑事务所

由韩国Mass Studies建筑事务所设计的"首尔公社2026（Seoul Commune 2026）"探讨了建立何种可持续性的社区来应对将要面临的城市人口密度过大问题。方案中的塔楼与公园绿化达到了一个平衡状态，形成了具备私密、半公共及完全公共的综合体网络。

据悉，占地39.34万平方米的首尔公社2026将建在全球人口密度最大的地区——首尔市江南区中心的狎鸥亭洞。

在首尔公社2026的15座塔楼里，2 590个巢房状的独立空间将成为人们的私家公寓。相比起传统住宅，它们出格的造型不仅能大大改善自然采光条件，还能就人们的需求组合出不同的户型。

天景酒廊和公共生活设施位于塔楼顶部，宽敞的屋顶则用作空中花园或露天舞台。塔楼中段的球形空间作为公共服务区聚集着各大公司、医疗机构、福利机构和商业服务部门。各楼层中心均设有六台升降梯和六间配电室。两部室外逃生梯不仅是紧急出口，还是多层居民楼中人们平日里休闲娱乐的场所。

为了给这座摩天蜂巢城里的居民提供更多便利，Mass Studies利用三种不同的路径——三条直达电梯中心的地面人行道、连接着各塔楼停车场的地下专用行车道、以各塔楼二层为站点的空中单循环公交车道实现了周围塔楼之间的互通。

独特的空间结构使得塔楼外墙由无数六边形的框

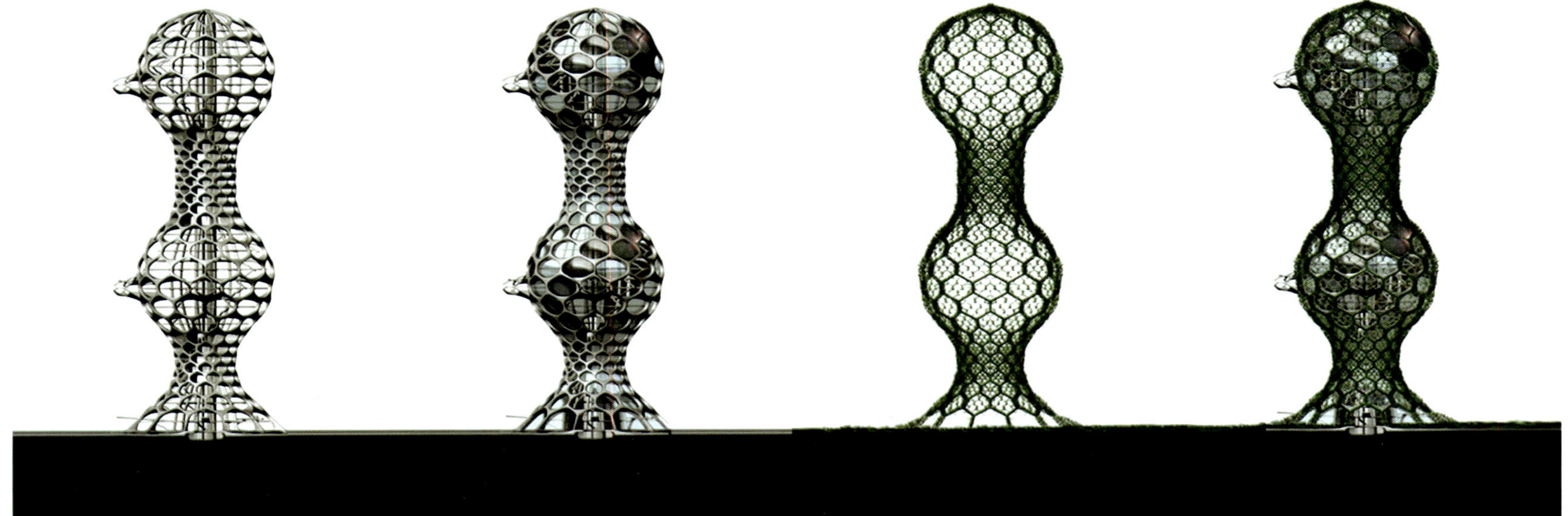

结构组成，从而创造出了塔楼的独特外观，这些六边形的开窗内嵌各种不同种类的玻璃，出于节能的考虑，在朝阳光方向安装了光电玻璃。部分外墙开窗作了内退处理，从而创造了遮阴阳台，外墙结构同时适应于藤本植物的生长，从而在夏季起到了遮阴的效果。循环水系统将在夏季承担30%的制冷量，并可清洁暴露在首尔这座严重污染的城市中的塔楼玻璃开窗。

塔楼表面由无数六边形的框格组成。这些运用了光电玻璃、墙体玻璃面板等高科技的窗户，不仅让塔身看起来像水晶一样通透，还使阳光等天然能源得以充分利用。每到夏天，那些种植在窗户骨架上的攀缘植物将持续生长数月，从而保持楼内部的清凉，并在外观上形成一个绿意盎然的"垂直公园"。

考虑到首尔市正面临严重的环境污染，整个首尔公社2026的设计方案严格遵守了各项环保节能的规定，并体现出可持续、高效率的特点。塔楼表面的玻璃光电板将通过吸收太阳能为公社供电；循环水系统将对生活废水进行二次利用，并帮助降低塔内温度，清洁窗户玻璃，令整座摩天蜂巢城长期保持洁净。

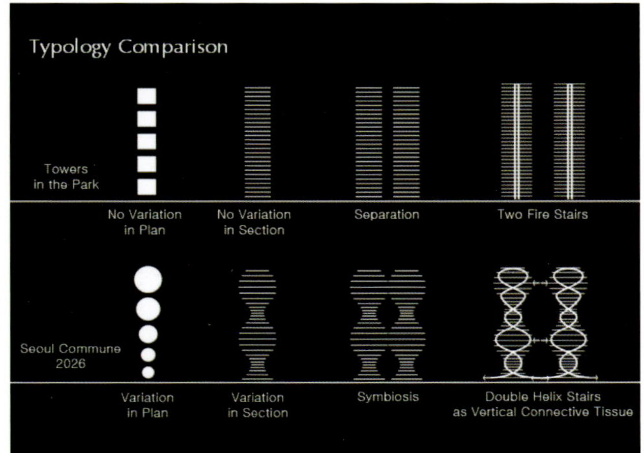

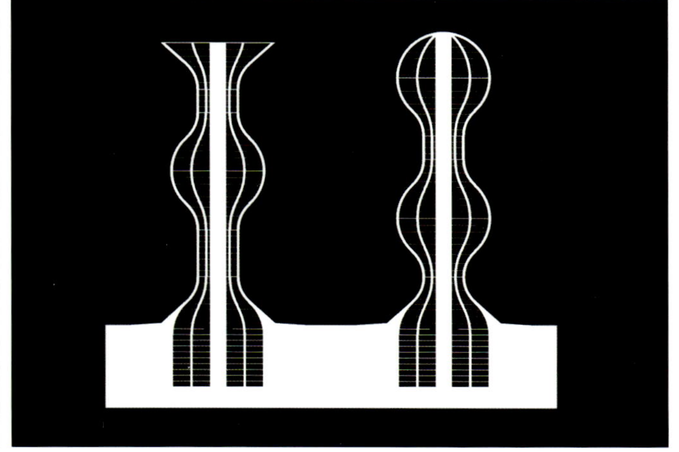

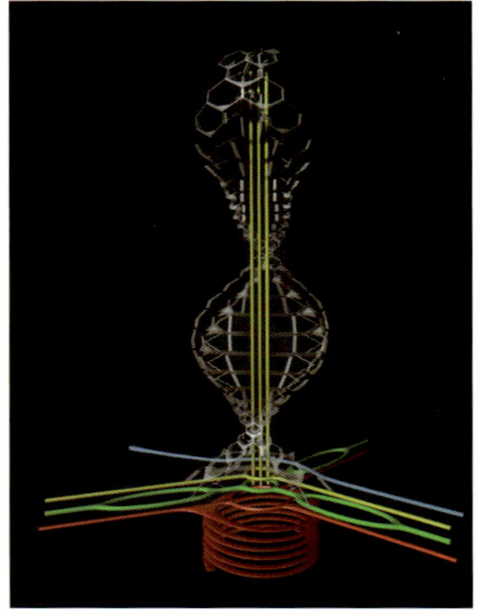

Seoul Commune 2026 (designed by Mass Studies) investigates the viability of an alternative and sustainable community structure in the overpopulated metropolises of the near future. It unites towers and the park in a balanced way and forms a complex network of private, semi-public, and public spaces.

The 393,000m^2 Seoul Commune 2026 is located in Apgujeongdong, a central area in the southern part of Seoul. It is located in a large-scale urban redevelopment zone that is possibly one of the most densely populated places on Earth.

The private spaces in all towers are composed of individually unique beehive-like cells. A total number of 2,590 cells are spread throughout the 15 towers. A single household can consist of a few independent cells with additional functions. The unique honeycomb structure also improves natural light conditions.

The top floors of the towers have a sky lounge for the commune and include shared living and dining facilities. The large public space on the roof is used as a roof garden or as an outdoor arena. Between the top and the base of the tower, the bulb areas serve as offices, medical facilities, public services, welfare facilities, and other supporting commercial spaces. Six elevators and six shaft spaces are located at the center of each floor. Two exterior emergency stairs wind around the building. More than merely an emergency exit, they also serve as a resting area and a garden in the multi-storyed residential buildings.

Three walkways converge the ground floor space which is reserved for pedestrians, and circulate around each tower's elevator core. All vehicular circulation moves below the ground and is connected to underground parking spaces at each of the towers. A monorail loop on the second floor offers public transportation and people movers to connect the neighboring towers.

The exterior skin of the towers consists of hexagonal lattice structures that derive from the unique spatial structure and create the unique appearance of the towers. The hexagonal openings are filled with various types of glass. Photovoltaic glass panels are placed in sunny areas for energy efficiency. Some exterior glass windows are recessed to create shaded balconies. The outer surface covering the lattice structure is made of a geotextile that creates an environment where vines can grow during the summer months to shade the openings. The water distribution system also carries up to 30 percent of the cooling load during the summer and cleans the glass windows of the building in the heavily polluted city of Seoul.

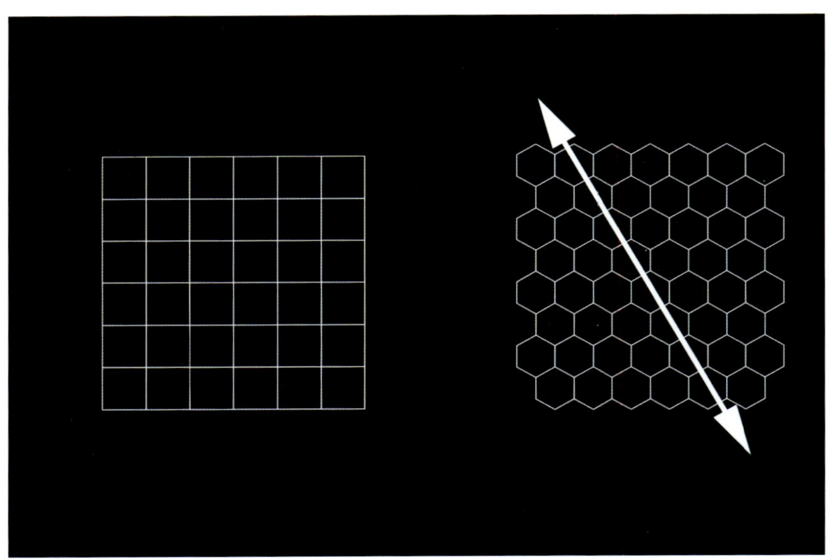

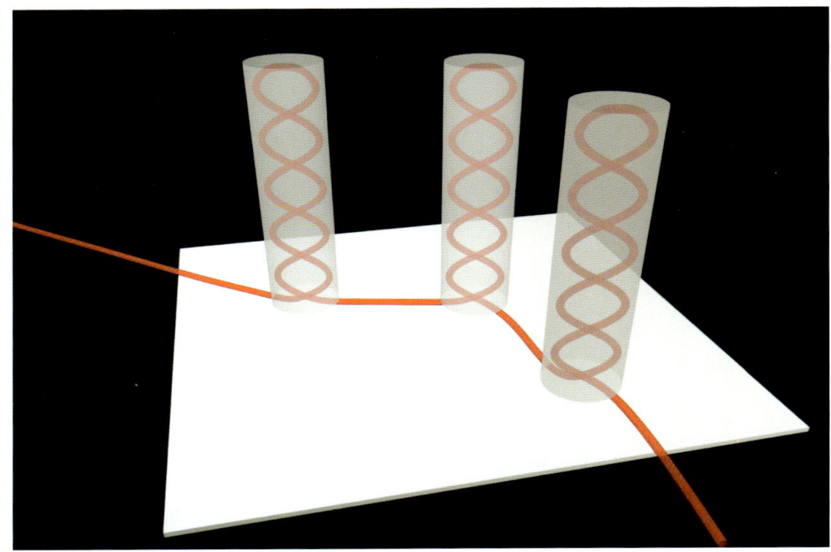

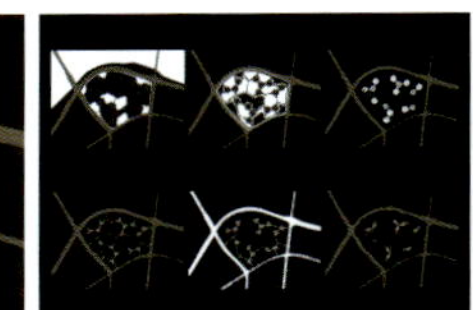

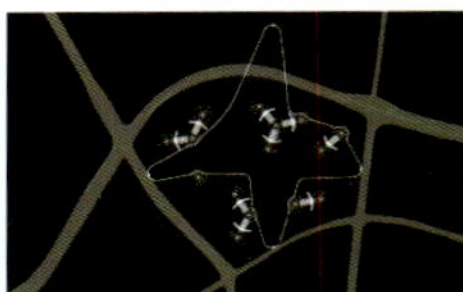

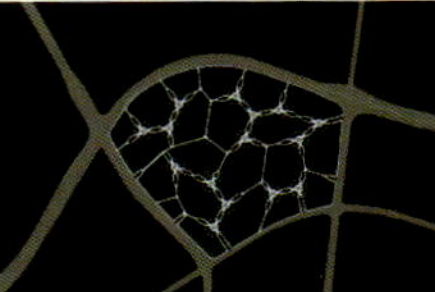

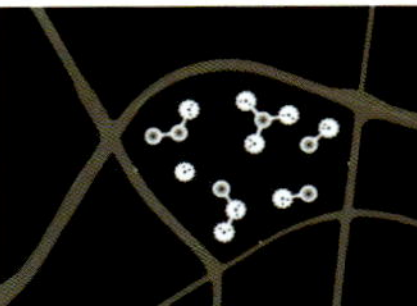

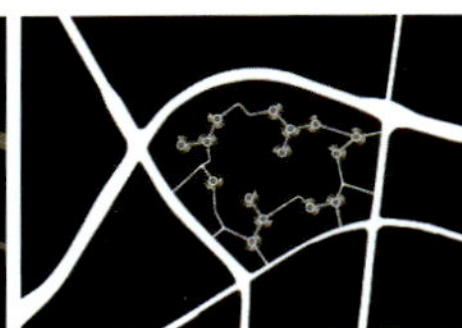

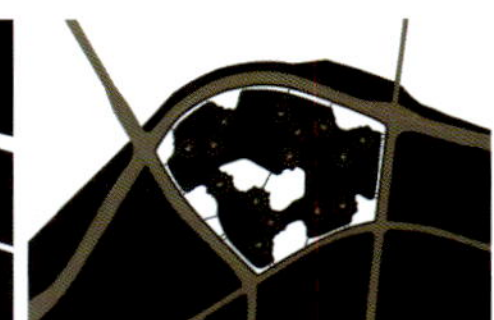

Super Sustainable City
超级可持续发展城市

Kjellgern Kaminsky Architecture AB

项目地点	瑞典，哥德堡
场地大小	城市全貌
主要材料	当地的可持续使用的环保材料，如砖块和木材
设计单位	Kjellgern Kaminsky Architecture AB
主设计师	Joakim Kaminsky& Fredrik Kjellgen

Project Location	Gothenburg, Sweden
Size of Site	Vision for the complete city
Main Materials Used	Localy produced sustainable materials such as brick and wood
Design Company	Kjellgern Kaminsky Architecture AB
Principal Architect	Joakim Kaminsky& Fredrik Kjellgen

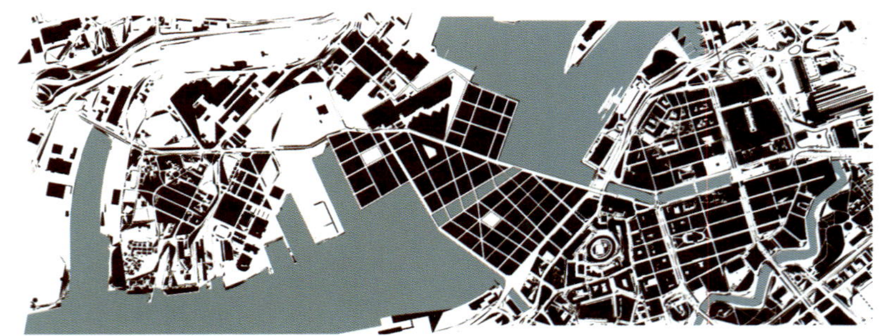

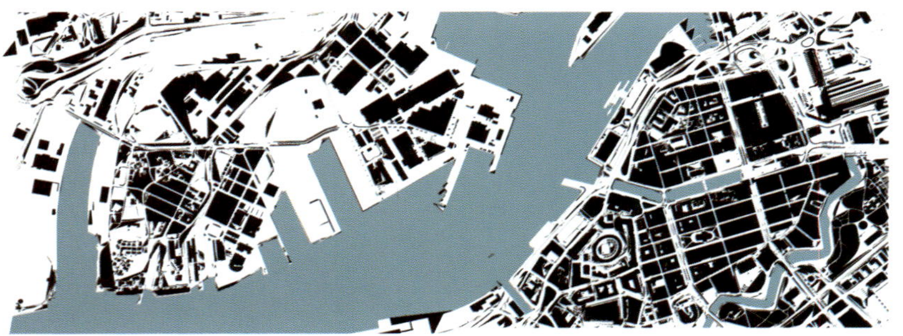

哥德堡正面临着一个可以和工业革命相比的严峻而巨大的挑战，那就是将哥德堡变成一座可持续发展的城市。如今，Super Sustainable智囊团已经设计出了哥德堡未来50年的发展远景。

Super Sustainable City的这个设计始于哥德堡市的一项调查——"2050年的哥德堡"。这个调查显示，人们希望哥德堡成为一座可持续发展的城市。Super Sustainable c ty 吸纳了这一点，并将其作为出发点为哥德堡设计了可持续发展方案。设计师运用了大量的模型、图像、文字和影像来展示和表达他们的想法。

Super Sustainable是一个由建筑设计师Joakim Kaminsky和Fredrik Kjellgren创办的智囊团。它由一个来自整个欧洲的不断成长、积极上进的设计团队组成。

Super Sustainable致力于研究、推进和发展不同规模的可持续设计方案，小到工具，大到城市规划。他们想要领导和探讨可持续发展。

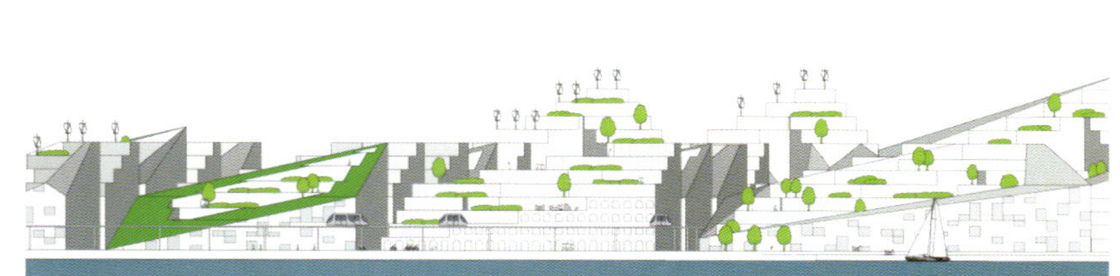

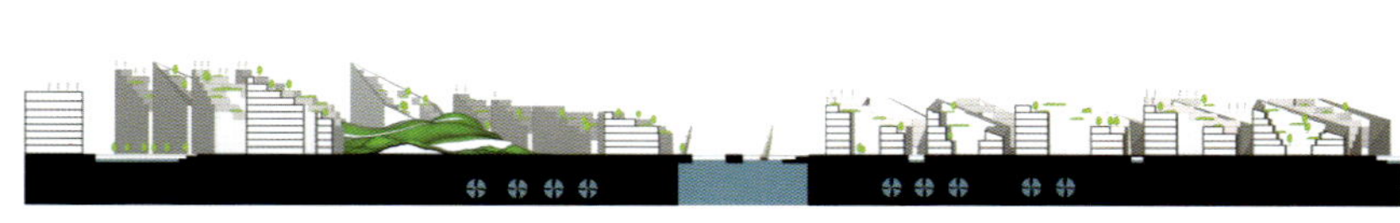

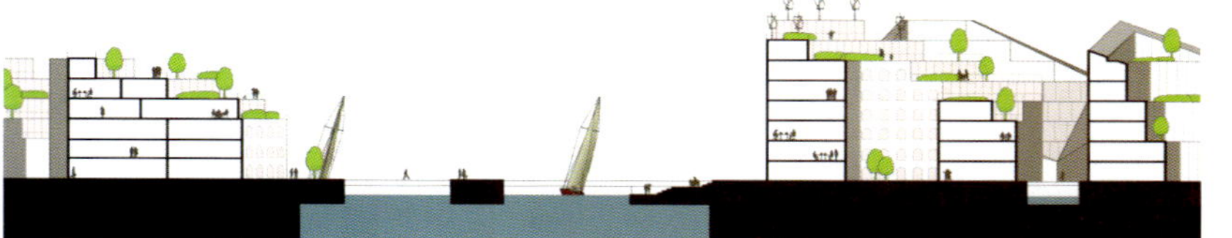

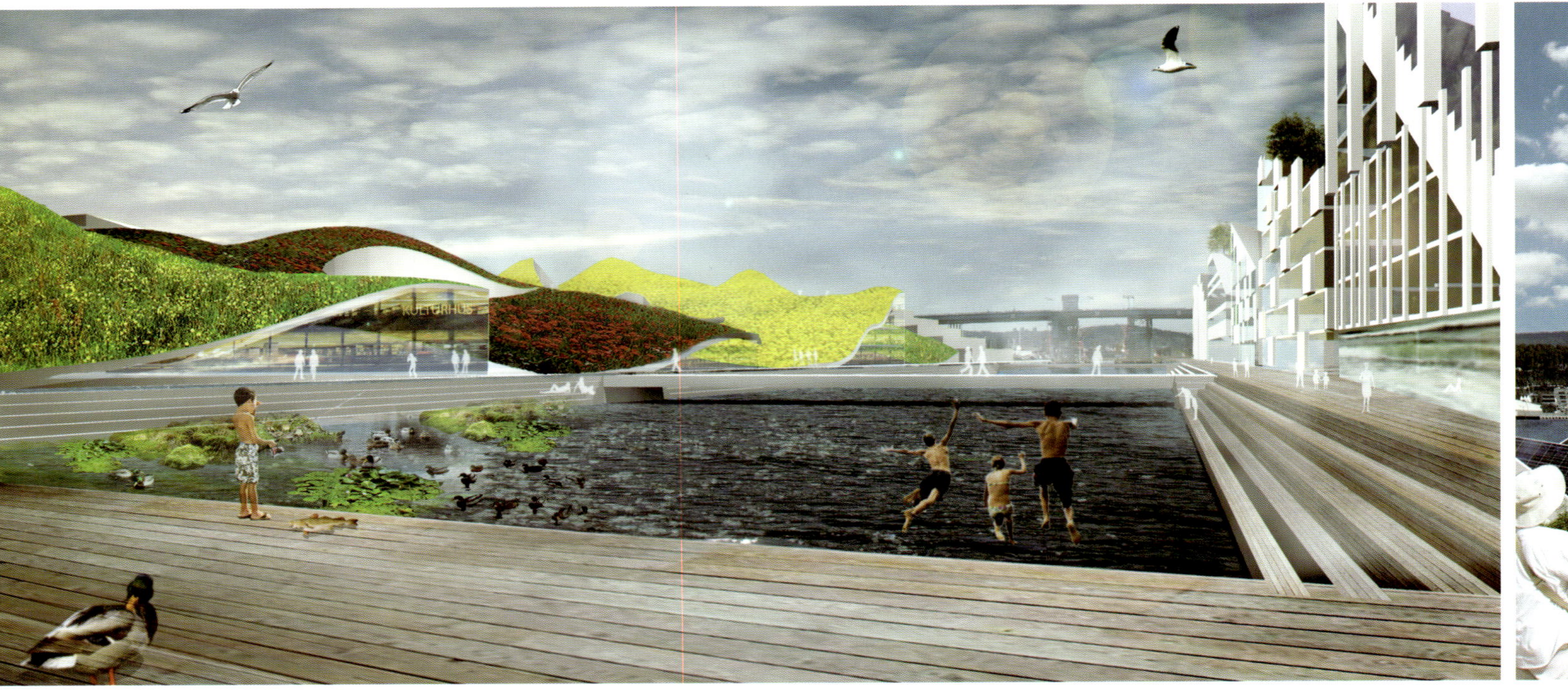

Gothenburg faces a challenge comparable in size with the industrial revolution to become a sustainable city. The Super Sustainable think tank now presents a vision for a sustainable Gothenburg in 50 years.

Super Sustainable City takes its starting point in an investigation conducted by the Göteborg municipality: "Göteborg 2050". In this investigation the goals the city must reach to become a sustainable city are set out. Super Sustainable City takes these facts as a starting point to propose a physical image of a sustainable Gothenburg. They show ideas and visions in the form of models, images, texts and movies.

Super Sustainable is a think tank initiated by the architects Joakim Kaminsky and Fredrik Kjellgren. It is composed by a growing team of engaged persons from all over Europe.

Super Sustainable investigates, promotes and develops sustainable design solutions of all scales; from small gadgets to city planning. They want to lead and discuss the sustainable evolution.

Sun off

Post-Quake Village Rebuild Demonstration Project in Ma'anqiao Village

马鞍桥村庄震后重建示范项目

吴恩融，穆钧

项目团队　香港中文大学建筑学院
建筑师　吴恩融，穆钧

Project Team　Architecture college, The Chinese University of Hong Kong
Architect　Enrong Wu, Jun Mu

家园重建

高科学与低技术战略

出于对当地文化和生计的考虑，以及对于照明、通风和抗震的技术考虑。我们拿了12栋用素土夯实的房屋，选择性地挑选了部分村民根据他们自己的情况重建房屋。环境设计两个关键字是式量热质量和自然通风。我们保持用土砌成的墙，把它改良成适当尺寸的窗，为空气流通提供条件。

传统的用石头砌成的房屋经不起地震的摧残。为了保留这个传统的建筑风格，我们引进了竹子和木材来加强它结构的稳定性和抗震能力。加强过的石砌墙在很大程度上增强了使用功能，在撞击之前对泥土的湿度进行适当的维护，改良了工具和石砌的方法。

房屋雏形修建和培训

每一户家庭的雏形是由村民自己修建的，这样当地居民就可以拥有技能，为自己创造价值，传承下去，独立地重建和发展他们的家园。

房屋雏形在临时的村屋设计和技术的帮助下，33户家庭根据他们原来的住址大小重建了他们的家园。

根据我们的调查，当地村民建造的土砌房屋与传统的实体建筑相比，其舒适性、稳固性、建筑结构和成本都有更好的表现。在冬天不需要用暖气设备，在夏天不需要用空调。沼气被运用于照明和烹饪的能源供应。

90%的建筑原料都是震后材料的再利用或是当地可以见到的自然原料。岩石从附近的河流运来，二手的木材从当地的村民买来，不再使用的二手泥土瓦片一家一

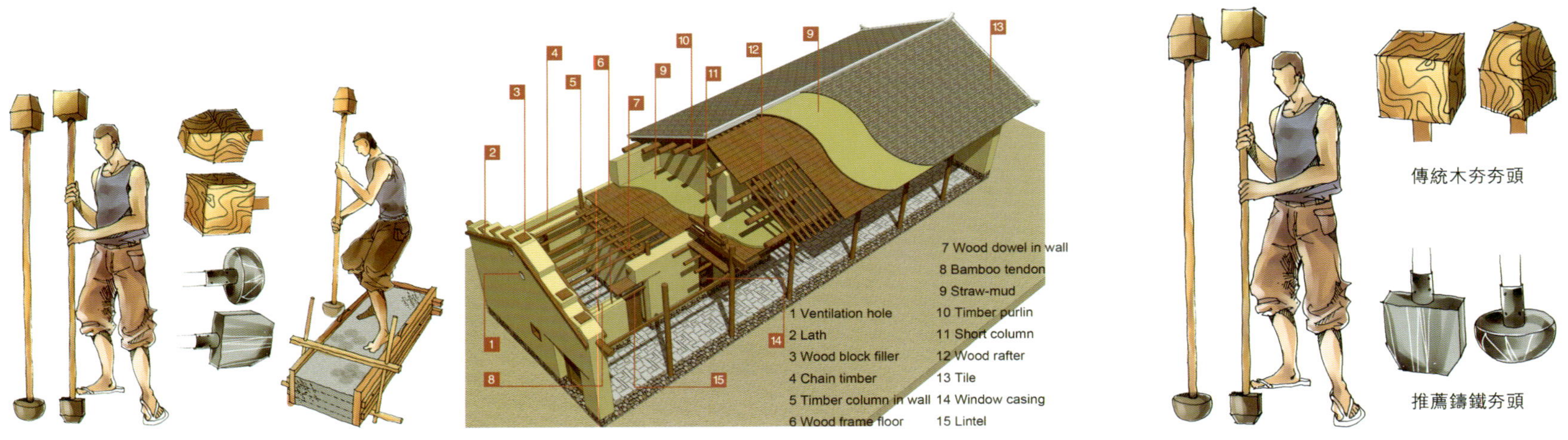
1 Ventilation hole
2 Lath
3 Wood block filler
4 Chain timber
5 Timber column in wall
6 Wood frame floor
7 Wood dowel in wall
8 Bamboo tendon
9 Straw-mud
10 Timber purlin
11 Short column
12 Wood rafter
13 Tile
14 Window casing
15 Lintel
傳統木夯夯頭
推薦鑄鐵夯頭

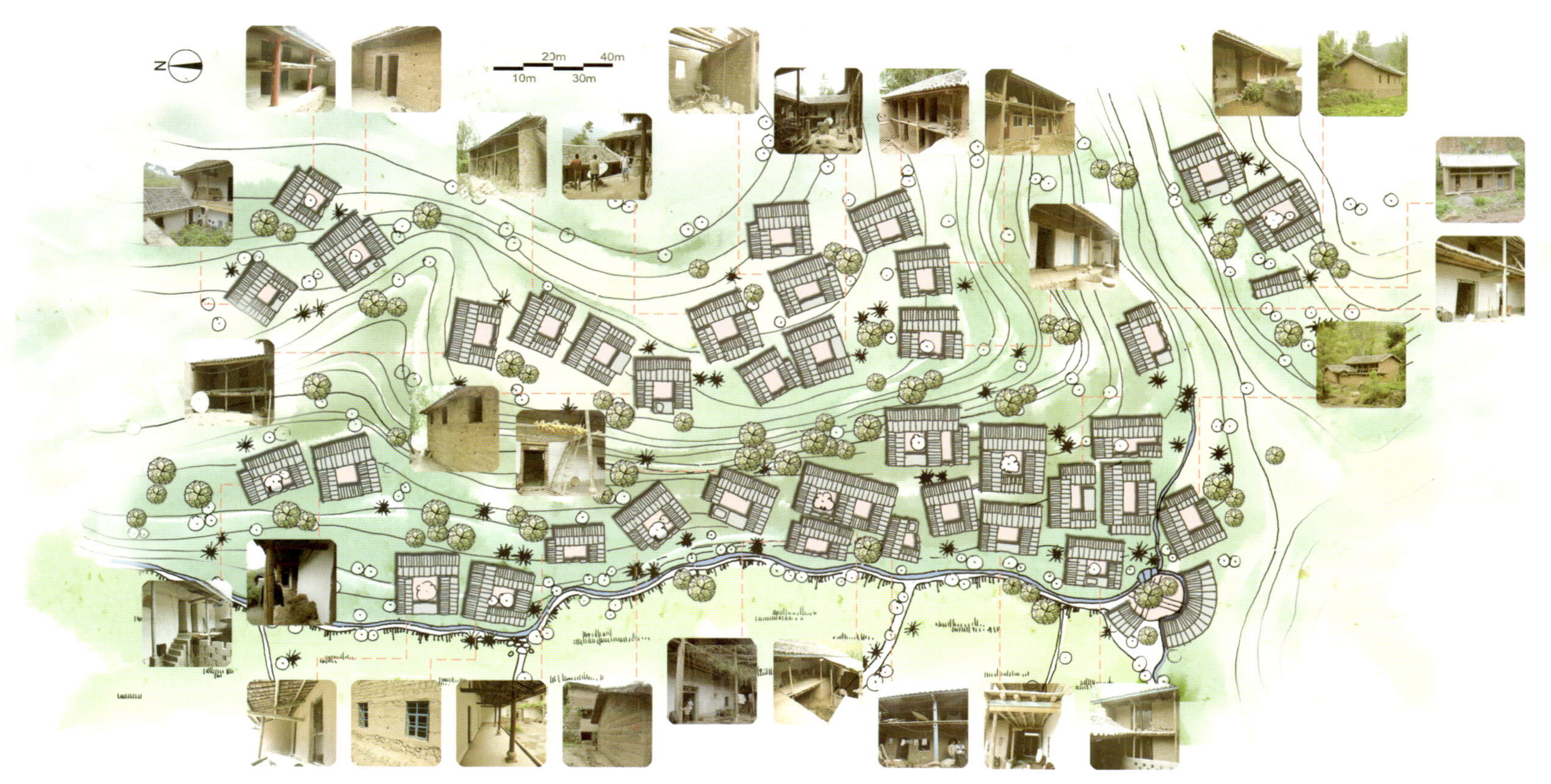

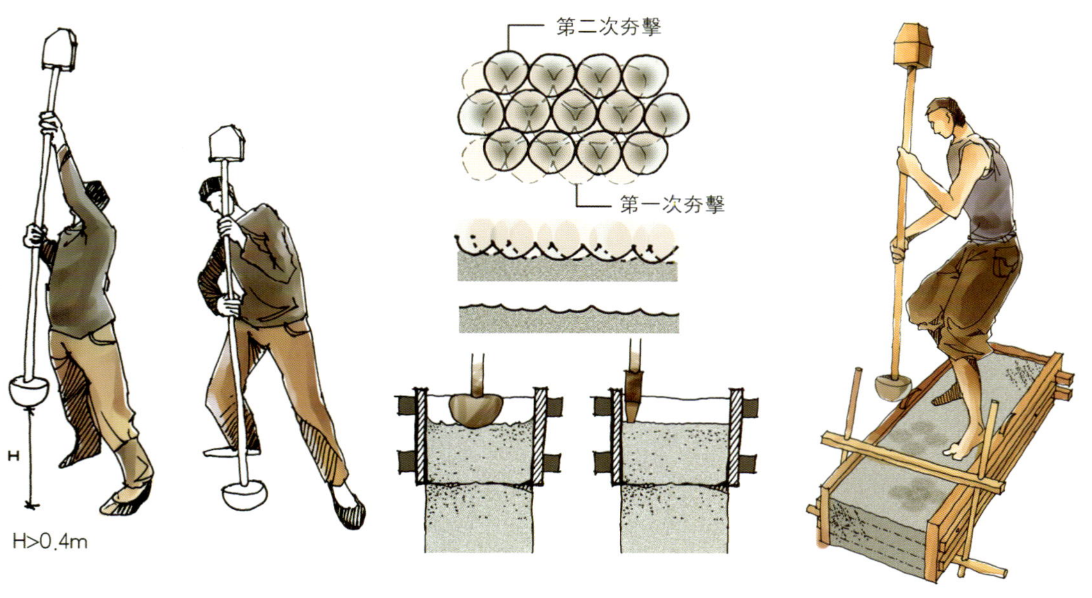

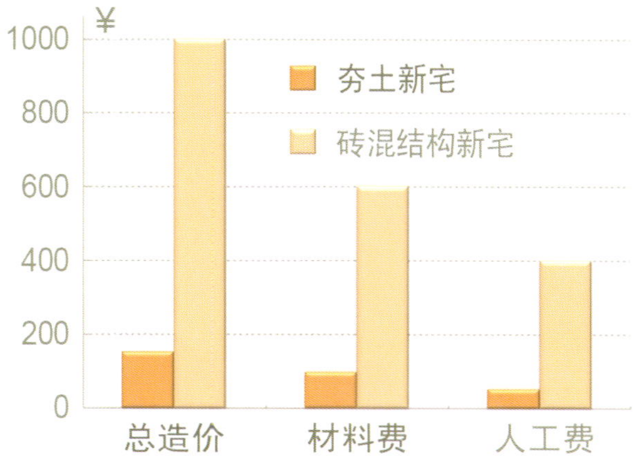

家地收集而来，竹竿和稻草在那样的农村随处可用，只有石灰和水泥是需要从外面引进。马鞍桥石砌住所的重建是中国南部贫困农村最具有代表性的挑战项目。这个项目展示了贫困农村家庭也可以拥有生态居室的可行和有效的方法。

村民说："这里依旧是我们的房子，我们的生活，但是它们变得坚固、安全了。"

社区推广

大桥的建造及改进

大桥是由村民和香港及中国大陆的志愿者建造的，方便这里的村民和孩子可以安全地穿过河流。村庄里主要的小路也进行了重建和加宽。

为了普及基础的公共健康知识，几场公共健康教育，例如保健环境及家畜饲养等讨论会在马鞍桥村庄举行，旨在帮助村民提升他们的居住环境。

村庄中心的建造

为了改善公共服务设施，村庄的中心建有诊所、幼儿园、图书馆和商店。我们渴望为当地村民创造一个公共空间，它为社区提供一个集中的场所，尤其是在节假日期间。

该项目成为土砌抗震建筑技术的展示中心。

2010年夏天，村庄中心的块石面路和绘画艺术已经完成，35个从香港和大陆过来的志愿者自愿效劳，他们与村民一起工作，像一个大家庭般相互协作。

总结与推广

为了在中国西部其他地震灾区广泛宣传此方法，一

个自己制作的抗震石砌村屋指南出版了，大量的图片和图释讲解这些技术与技能。这些技术也被应用到当地实验以做进一步结构稳定和抗震的优化。

此外，在这个示范区内，工匠培训及信息讨论会定期举办，旨在将当地可持续性建筑的技术和理念传播出去。

Homeland rebuilding

High-science and low-tech strategy

With respect for local culture and livelihood, and blend in technical intervention on lighting, ventilation as well as anti-seismic performance, we carry out 12 house types of rammed-earth courtyard, which can be selectively employed by villagers in their stepwise home rebuild according to their own conditions. The two key words of environmental design are thermal mass and natural ventilation. We keep the rammed earth wall and improve it appropriated with properly sized windows and provision for cross ventilation.

Traditional rammed earth building has been challenged by its poor seismic performance. In order to reserve this traditional architectural wisdom, we have introduced bamboo strips and timber to enhance its structural stability and anti-seismic performance. The strength of rammed-earth wall is largely enhanced by applying additives, maintaining proper mud humidity before ramming, as well as improved tools and ramming methods.

Prototype construction and training

A prototype is constructed by the team of villagers from each family, so that locals can own the skill, add value themselves, pass it on, rebuild and develop their homeland independently.

With the help of village house design templates and the techniques learnt in the prototype building, 33 families have rebuilt their home by themselves at their original sites without occupying extra land.

According to our research result, the rammed earth houses built by local villagers demonstrate a better performance in terms of comfort level, sustainability, construction ease and cost when compared with conventional concrete buildings. There is no need to use heating equipment in winter and air -conditioner in summer. Bio-gas has been used to turn wastes into fuel for lighting and cooking.

Most (about 90%) of the construction materials are post-quake recycled materials or local available natural materials. The rocks are carried from the nearby river; second-hand timber is bought from local villagers; un-used second

hand clay tiles are collected from household to household. Bamboo and straw are easily available in that kind of rural area. The only intervention we need from outside is lime and cement. Rammed-earth dwellings and the rebuilding challenges in Ma'anqiao Village are typical in poor rural southern China. This project demonstrates a feasible and effective way for most poor village families towards affordable ecological living.

Villagers said," It is still our houses and our life. But it is solid and safe now."

Community promoting

Bridge building and improve access

A bridge is built by villagers and volunteers from both Hong Kong and mainland China so these villagers and children can across the river conveniently and safely. The main paths in the village are also rebuilt and widened.

Several public health education workshops are held in Ma'anqiao Village to popularize basic public health knowledge with the aim to help villagers to improve their residential environment such as hygiene conditions and conditions of livestock rising.

Village center construction

A village center with clinic, kindergarten, library and shop has been built to improve public service facilities. We aspire to create a common space for local villagers. It provides a focus for the community, especially useful during festive periods.

This building is also a demonstration center of anti-seismic rammed-earth building technology.

In the summer of 2010, the paving and painting work of the village center were accomplished by more than 35 professional and student volunteers mainly from Hong Kong and Mainland. They worked with villagers, helping each other like a big family.

Summarization and promotion

To further promote this method in other earthquake-stricken area in west China, an Anti-seismic Rammed-earth Village House DIY Construction Manual is published. A large number of pictures and illustrations are used to explain the techniques and skills. This is also supported by on-site experiments to further optimize its structural stability and anti-seismic performance.

Furthermore, craftsmen training and communication workshops will be organized regularly in this demonstration center to spread the technology and conception of sustainable vernacular building.

Office Commercial

商业 | 办公

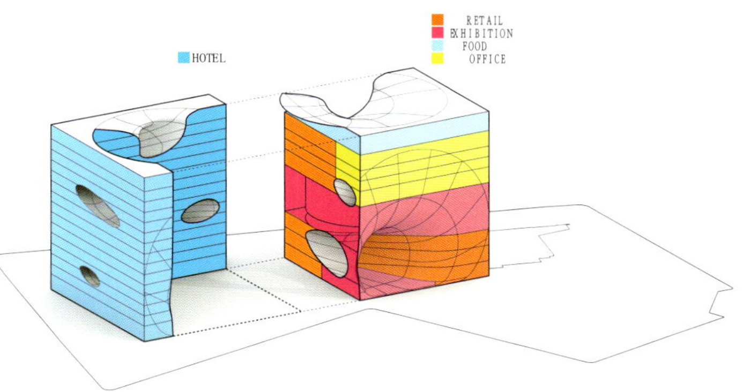
RETAIL
EXHIBITION
FOOD
OFFICE
HOTEL

Masdar Headquaters

Masdar总部

Adrian Smith+ Gordon Gill Architect LLP

项目地点	阿拉伯联合酋长国，马斯达尔市
设计单位	Adrian Smith+ Gordon Gill Architect LLP
客户	马斯达尔市
顾问	Thornton Tomasetti;Environmental System Design,Inc.

Project Location	Masdar City, United Arab Emirates
Design Company	Adrian Smith+ Gordon Gill Architect LLP
Client	The Masdar Initiative
Consultants	Thornton Tomasetti; Environmental System Design, Inc.

在未来，所有的城市都将把Masdar的发展模式作为一个可持续建筑发展的典范。作为城市开发的中心位置，Masdar总部将为整个城市的其他项目奠定基准。

Masdar总部将会成为世界上第一个正能量混合用途建筑，其运用可持续设计和系统产生的能量比所消耗的更多。该项目是Masdar City的核心，是在阿联酋迪拜之外的一个零废品、零排放的发展项目。AS+GG设计事务所与Foster+Patners设计事务所一起制定了Masdar City项目的总体规划。

这栋高7层、134 662平方米的（包括风景美化区在

内）大楼将适应商业、零售、文化等各种用途。大楼的造型符合环境需求，适应古代科技和对阿拉伯风塔、大屏以及其他强调自然通风、遮阳、高热式质量、庭院设计和植被的本土建筑的审美要求。大楼将利用从未被使用过的先进技术来创造极具美感的、功能性强的、高度发展的建筑物，它将代表整座城市。

大楼受到了历史悠久的本土建筑的暗示，将历史上成功的气候策略与最新的技术及创新型的建筑体系相结合。而尤其对于Masdar总部项目，也同时运用了一些成熟的建筑体系。

本次设计，包括无数系统在内将产生大楼的剩余能量，消除碳排放量，减少液体以及固体垃圾。并将利用可持续材料，并综合风电机组、户外空气质量监测器以及世界上最大的楼内置太阳能电池。与其他相同规模的典型混合型用途大楼相比，该大楼将减少70%的用水量。

现代化的风塔是大楼对于伊斯兰建筑的一个参照，是整个复杂的设计的一个基础。它们看起来像风塔，可以抽空室内的热空气以保持大楼的通风，并将城市地下的冷空气抽送到室内。

锥形设计最大化了穿透大楼的日照，通风口处可开启的活动窗使得室内的自然通风成为可能。结构上来说，锥形支撑了大楼的屋顶并且在大楼顶部创造了遮阴的平面。在空间上，在拥有光影和水的池塘的公共领域创造了花园亭台。每个庭院都各不相同，并且为人们提供了设施以及公共空间。

设计使得大楼从建设伊始就做到可持续性及高效性。施工时序上，可先完成建筑的圆锥体和屋顶部分，从而为其他部分提供一个遮盖下的微型气候。由光板覆盖的屋顶将提供足够的能源建造大楼的剩余部分。

大楼包括办公和居住空间，零售和公共花园以及祈祷室和通往城市交通系统的直接通道。

Masdar HQ的符号性的建筑特征是同一系列的十一个风口，提供自然通风以及冷却（将热空气循环到屋顶，这样风可以将热空气吹走）并在一楼形成绿洲形状的室内庭院和/或灵活的空间，每个都拥有自己的主

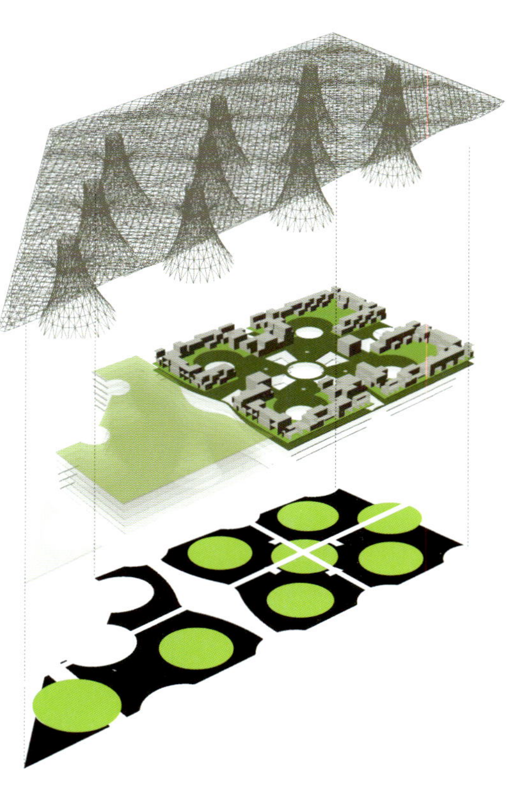

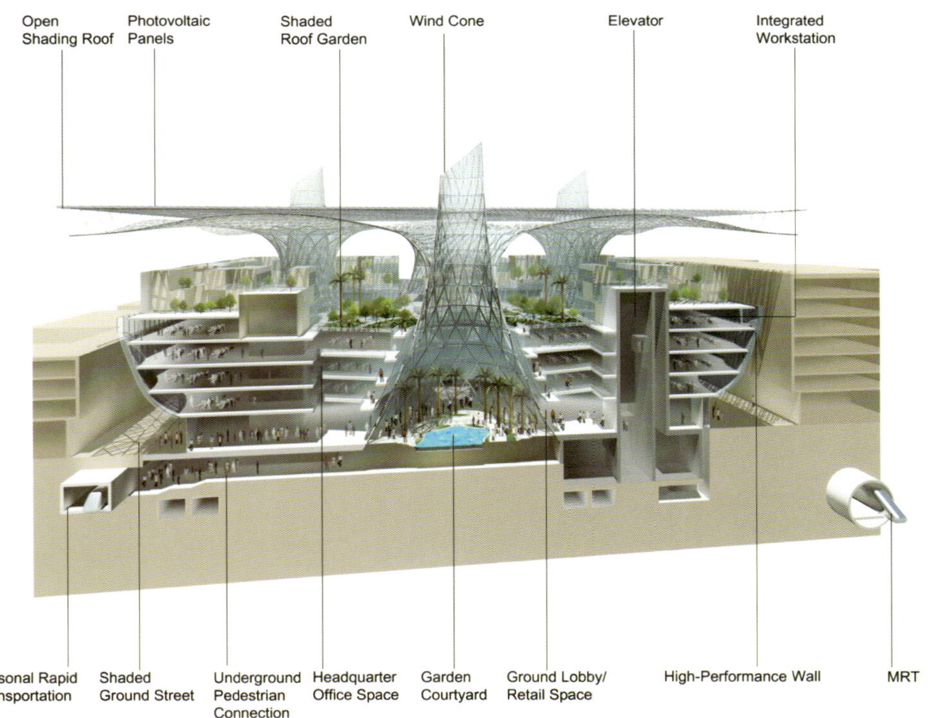

题。通风口同时也为大楼内部提供了柔和的日照。

其他关键的可持续设计特征、系统和策略包括宽敞的屋顶顶盖，它可以提供自然遮阴以及一体化的全球最大的光电以及太阳能面板组，该面板组可以在太阳能集热器提供太阳能制冷的同时搜集太阳能。起伏状的基层设施构成了屋顶墩状的结构性能。

高热式质量外部玻璃幕墙提供太阳能f的同时也能保持透明度以便看到外面的景色。本项目所采用的热工艺还包括了可以减少室外气温并且提供与公共花园相连的具有交通系统的地下人行通道的地下管道。屋顶上茂盛的空中花园创造了一个拥有水域和装点有各种原生植被的安静的社区空间。

Masdar HQ大楼将成为世界上最绿色环保的混合型建筑，零碳排放和零废物（包括液体和固体）以及超越LEED标准的可持续方案。它将比同体积典型的混合型用途建筑少消耗70%的水量，同时也是温热潮湿气候下的现代A级办公大楼中每平方米耗能最低的建筑。

Masdar City将分为七个阶段施工，预计在2016年完工。Masdar的总部大楼是第一期的一部分，已在2010年底完工。

In the future, all cities will look to the Masdar Development as an example of sustainable excellence. As the center of the development, the Masdar Headquarters will be the benchmark of performance for the entire city.

Masdar Headquarters will be the world's first mixed-use, positive-energy building, using sustainable design strategies and systems to produce more energy than it consumes. The project is the centerpiece of Masdar City, a zero-waste, zero-carbon-emission development outside Abu Dhabi in the United Arab Emirates. The AS+GG design will also anchor the Masdar City master plan by Foster + Partners.

The seven-storeyed, 134,662-square-meter structure (which includes landscap areas) will accommodate commercial, retail and cultural uses. The building's form, sculpted in response to extensive environmental analysis, adapts the ancient science and aesthetics of Arabic wind towers, screens and other vernacular architecture, which emphasizes natural ventilation, sun shading, high thermal mass, courtyards and vegetation. The building will utilize pioneering, never-before-seen technology in the creation of the aesthetically astounding, functionally proficient and experientially superior development that will represent the city.

The building takes its cue from the centuries of indigenous architecture, marrying historically successful building strategies for the climate with the latest technology and innovative building systems, including some developed especially for the Masdar Headquarters.

The design, which includes numerous systems that will

平面规划图

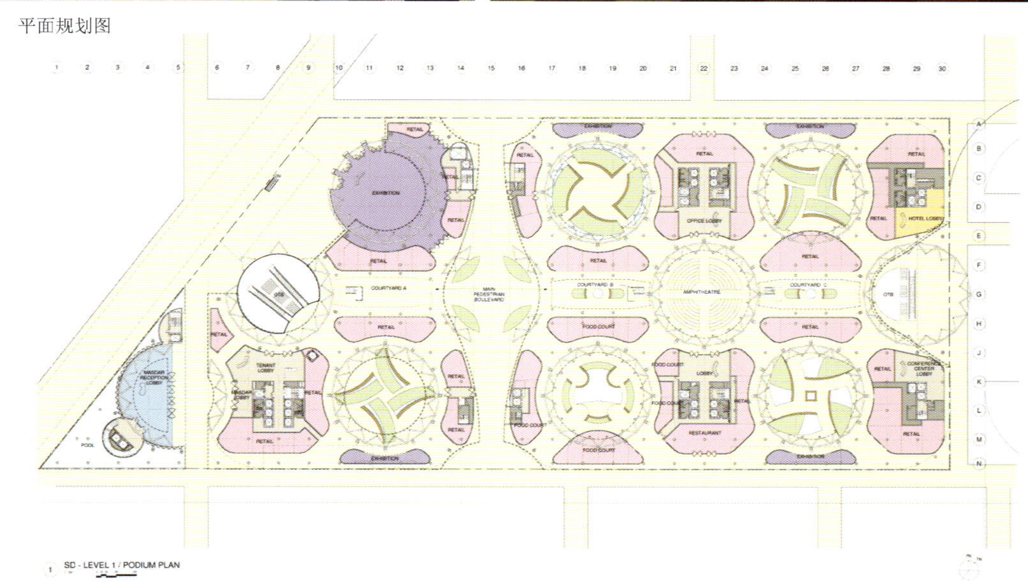

光照系统图

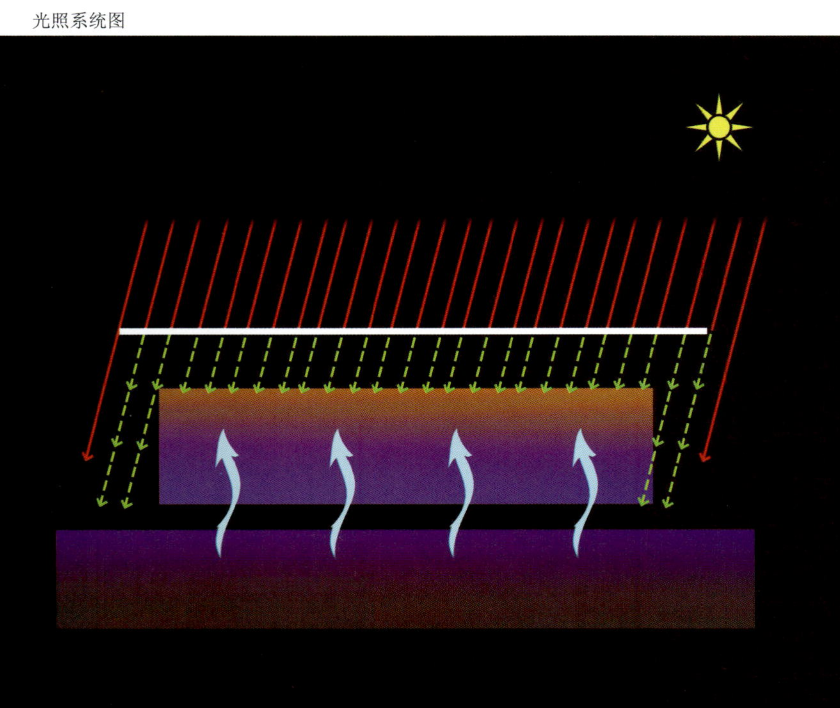

通风系统图

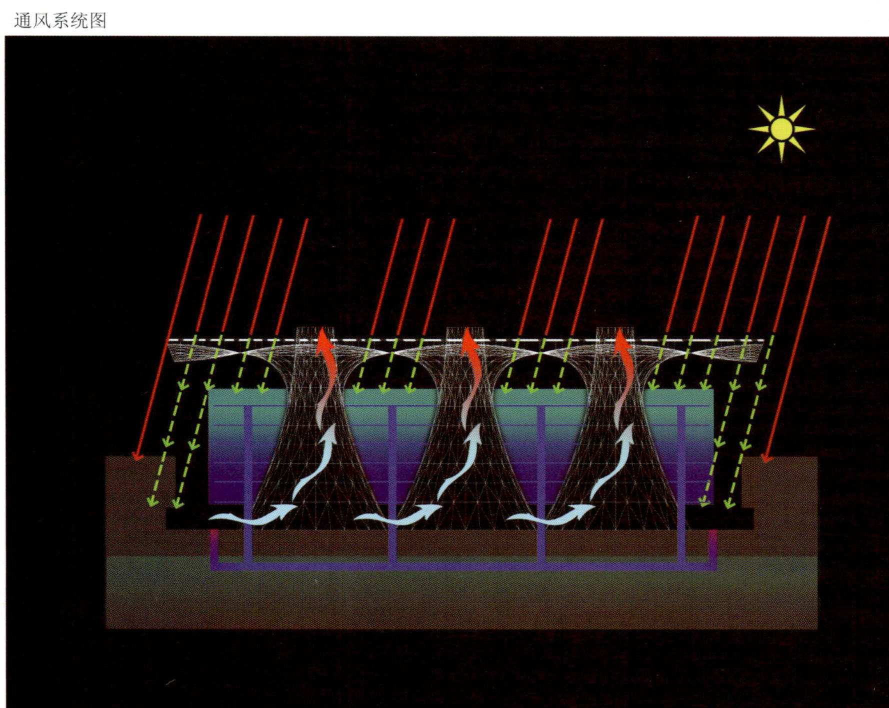

generate a surplus of the building's energy, eliminate carbon emissions and reduce liquid and solid waste. The complex will utilize sustainable materials and feature integrated wind turbines, outdoor air quality monitors and one of the world's largest buildin-integrated solar energy arrays. Compared with typical mixed-use buildings of the same size, the Headquarters will consume 70 percent less water.

Modern wind towers, one of the building's references to traditional Islamic architecture, are the basis for a number of features in the complex design. They act as wind towers, exhausting warm air and naturally ventilating the building, as well as bringing cool air up through the subterranean levels of the city below.

The cones maximize natural daylight throghout the building;the operable windows on the cones allow occupants the option of naturally ventilating interior spaces. Structurally, cones support the building's roof and allow for the creation of a shaded ground plane on the top of the building. Spatially, they create garden countyards at the public realm which has pools of light and water. Each courtyard is programmed differently, providing amenities and public space for occupants.

The building is designed to be sustainable and efficient from the beginning of the construction. The building cones and roof can be built first, creating a shaded micro climated for the remainder of the construction. The roof, covered with photovoltaic panels, will provide enough power to build the rest of the building.

The building includes office and residential spaces, retail and public gardens, as well as prayer hall and direct access to the city's transportation systems.

Masdar HQ's signature architectural feature is a collection of eleven wind cones which provide natural ventilation and cooling (drawing warm air up to roof level, where wind

moves it away) and form oasis-like interior courtyards and/or flexible spaces, each with its own theme, at ground level. The cones also provide soft daylighting for the building's interiors.

Other key sustainable design features, systems and strategies include a vast roof canopy, which provides natural shading and incorporates one of the world's largest photovoltaic and solar-panel arrays, which will simultaneously harvest solar energy while solar thermal collectors provide solar cooling. The roof's undulating understructure facilitates the roof pier's structural performance.

High-thermal-mass exterior glass cladding provides solar heat blocking while remaining transparent for views. Thermal technology in the project also includes earth ducts which reduce temperature of outside air and provide underground pedestrian passages that connect public garden space with the proposed mass transit system. And a lush sky garden on roof level creates a microclimate that includes water features and restful community spaces landscaped with indigenous vegetation.

Masdar HQ will be the world's greenest mixed-use building, yielding zero carbon emissions and zero waste (both liquid and solid) and a sustainable measure beyond LEED platinum. It will consume 70 percent less water than typical mixed-use buildings of the same size, and be the lowest energy consumer per square meter for a modern Class A office building in a hot/humid climate.

Masdar City will be constructed over seven phases and is due to be completed by 2016. Masdar Headquarters is part of phase one and has been completed by the end of 2010.

DIGI Technology Operation Center, Malaysia

马来西亚旋缘大楼，DIGI科技总部大楼

杨经文

项目地点	马来西亚，雪州
楼高	4层（高23.5米）
项目时间	2009年4月（开始时间）
	2010年中旬（计划竣工时间）
项目面积	总面积（GFA） 12 468平方米
场地面积	Lot 42 (8 517平方米) & Lot 43 (8 561平方米)
客户	DIGI 电信中心

Project Location	Selangor,Malaysia
Storeys	4 Storeys (23.5m high)
Project Milestones	April 2009 (Construction start date)
	Mid 2010 (Scheduled completion date)
Project Area	Total gross area (GFA):12,468 m^2
Site Area	Lot 42 (8,517 m^2) & Lot 43 (8,561 m^2)
Client	DIGI Telecommunications Sdn. Bhd

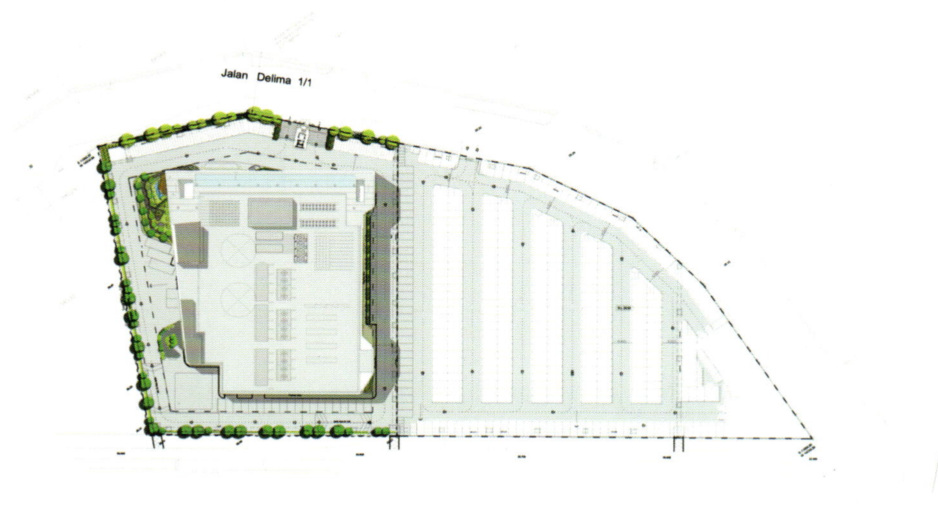

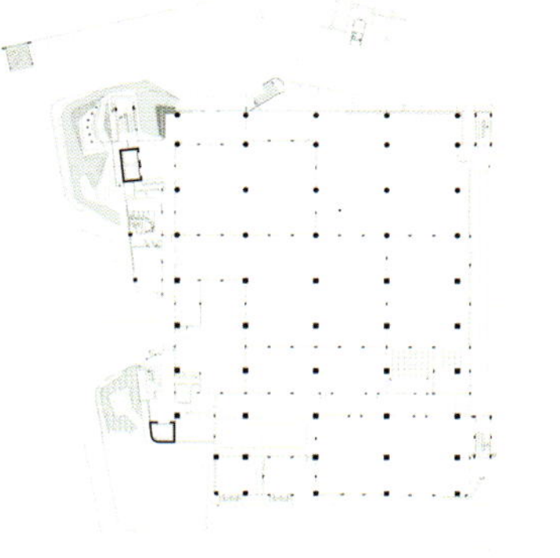
Level 01 Floor Plan

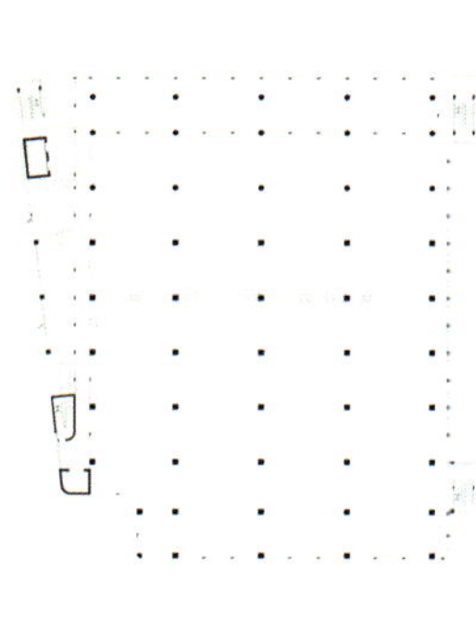
Level 02 Floor Plan

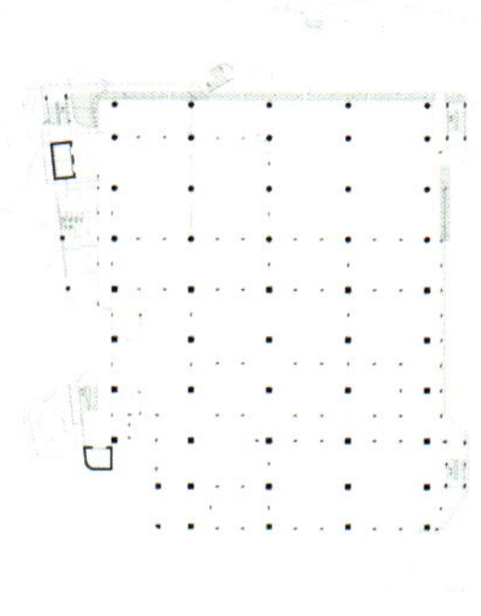
Level 03 Floor Plan

Level 01 Floor Plan

Level 02 Floor Plan

Level 04 Floor Plan

Roof Plan

项目概况

本项目是一个拥有辅助设备的信息中心，例如行政办公楼、接待大厅、电信大楼以及用于电子通信的管理服务中心（指挥中心）。

设计一个数据中心，该中心具备升级为第四安全等级的可能性，并在设计中考虑到生态特征。

设计师遵循客户的指示设计一个可以用于安置电脑、设置有效排水系统和防水系统的优化空间，来保护电脑设备、减少信息中心的热吸收并且拥有严格的安全措施。

信息中心的表面被设计成垂直绿墙，同时作为过滤的屏障提高大楼的室内空气质量以及的居住环境。

绿墙是一个有生命的、能呼吸的、可再生的外壳，它既可以简单如一个居室的艺术设置，同时又像生物空气过滤器般复杂。室内与室外的绿墙，减少二氧化碳排放，增加当地湿度，抵御粉尘，减少噪声，创造一个野外的都市环境。外部绿墙装置减少太阳能（建筑物表面的被动太阳能增益的热量滞留），通过扩建，又减少大楼能量消耗；可防止紫外线辐射以及酸雨的侵蚀，并帮助减少大楼的热岛效应（用种满植被的土地来代替混凝土和沥青，使得这样一个城市中心变成了一片绿色自然的区域）。

生态设计特征

人工植被绿墙：大楼的生态设计原理是通过环绕整栋大楼的“垂直的植被景观系统”体现出来的。植被用作抵挡太阳辐射的有效缓冲带，并通过隔离建筑，对减少热积聚、节约建筑的制冷能耗有重大意义。植物还是一个生态的空气清洁物，并通过一连串的天然生化反应对空气污染物进行降解与清洁，使得现存的温室气体及可挥发的有机合成物得到及时处理，在美化环境的同时还发挥着遮阳的作用。作为24小时使用的办公区域需要遮阳系统来实现直接散热，同时降低制冷负荷。“垂直的植被景观系统”使得建筑可以使用采光良好的透明玻璃，从而减少了人工照明负荷。

日光：主要的办公室和流通空间装有全高玻璃幕墙，提供最大程度的日光渗透，使照明系统发挥更大的能源效果。次要的房间配备能开启的窗户，以保证空气自然流通和充足的日光。

可更新的能量：一系列光伏控制板安装在大楼最高处的屋顶区域。其在屋顶上方用钢铁固定，234平方米光伏阵列生成35.28千瓦时电量，发电直接回输入市电网。这可更新的能源产量意味每年二氧化碳排放量减少12 516千克，帮助抵消资料处理中心设备集中的能源需求。

生态调节沟：自然过滤和排水系统用于减轻公共排水系统的表面压力。它们收集表面流失的污水并再利用，用于中央灌溉系统。项目规划中的不破坏生态平衡的排水战略也包括地面停车场区域草坪道路的使用。这些系统让雨水补充当地的地下蓄水层，远远强于雨水转移到外部蓄水池后而流干。

雨水收集：水管中的雨水从建筑物屋顶沿管道高速下流，雨水途经地面层的蓄水箱而被集中，蓄水箱嵌在对景观影响最小的地面中。配置有新发明的Versi-Tank™系统，这个水箱有100立方米的容量，100%由可回收可循环原材料制成。集中后的雨水经过过滤后再利用于建筑物周围的植物灌溉，为经过整个屋顶灌溉系统的绿墙提供供水需要。该项目的雨水收集系统设计完成后，没有饮用水再被用于灌溉目的。

空调系统：该项目使用一个高效率的分体式空气和水冷却系统（估计0.6千瓦/吨），它有着多种分区风机盘管和精确的空调数据中心。数据中心的冷却需求经过新发明的冷却通道控制系统进行进一步优化。这个系统建立的目的是通过分区冷却来迎合特定设备的需求，而不是通过耗费能量以达到使整个房间均匀冷却的需求。

抽水机：由于超负荷变化，高效能的变速抽水机被采用。变速抽水机预计能够为抽水系统节省30%的能源。

照明：照明配件使用高效率的T5低损耗电子镇流器。这些电子镇流器提供更高的亮度强度，它们的分区制通过特有的使用需求能够降低能源消耗。这个系统由传感器和中央楼宇自动化控制线路支持和控制。

电梯和自动扶梯：电梯将AC VVVF控制和无齿轮电动驱动器与同步电动机和固定磁铁合并在一起。电梯同时配有监控传感器，睡眠模式和节能照明装置。

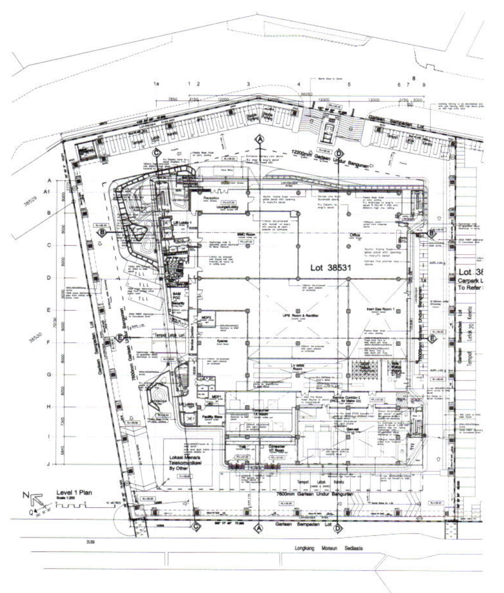

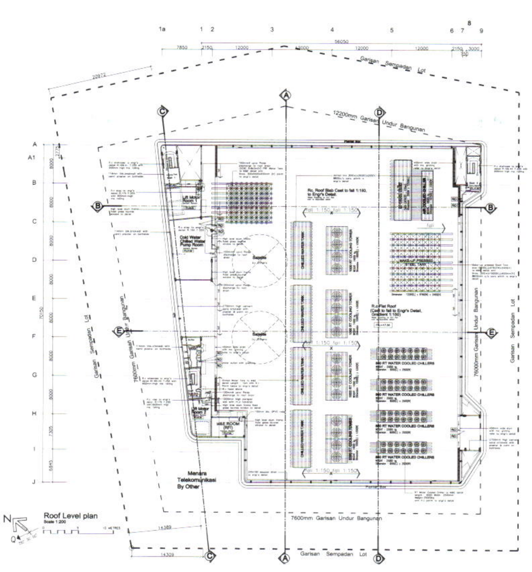

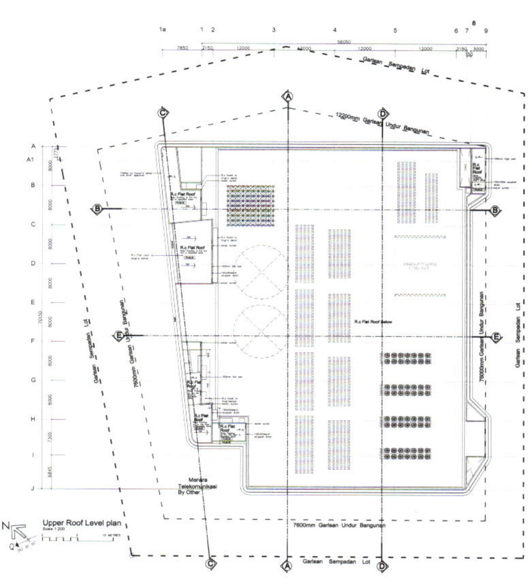

室内空气质量：为了增强居住者的舒适感与健康，空气流通率的最低限度应与ASHRARE 62.1：2007相一致。低挥发涂料、油漆和黏合剂的严格使用能进一步降低内部空气污染对居住者健康的有害影响。被大量的绿色植物包围着的建筑物进一步担当起额外的生物过滤作用以提高周围空气质量。

清洁配件：Dual Flushing WC's在不同使用需求的基础上配备水容量控制系统。所有盆地龙头和其他卫生洁具都是3-tics等级（最高效率评价）的节水装置。除此以外，在男洗手间内规定使用的无水小便池将进一步降低水消耗。建筑物内部全部用水量可减少50%以上。

不破坏生态平衡的运输工具：为了鼓励员工搭乘公共交通上班，从大厦到巴图蒂加KTM站提供定时的班车服务（大约1.6公里）。大厦的主要出入口为合伙使用的汽车优先提供停车位。

绿色原材料

外部的木质甲板是由CT-Wood混合纯聚合树脂和废弃的大米外壳制作而成。这些废弃的原材料再利用会相应地减少空气污染，还可以磨碎成粉状物质回收再利用或再加工。

办公区域使用的是一种日本无臭无味额外的涂料。这种产品被证明没有导致敏感或过敏性病态楼宇综合症的挥发性有机化合物。这样会产生更大的生产力和更健康的工作环境。

主要的屋顶防水系统是Hitchins的"Trafficgard"，是一个应用于物体深处的液体丙烯酸塑料聚合而成的凝胶。这是一个对环境无害的涂层系统，这是一种被鉴定的有绿色标志的产品。屋顶及屋顶下面的空间低RTTV是主要的数据中心和电话区域。

Design Brief

The project is a data center with ancillary facilities i.e. Administration Offices, Reception Lobby, Telco Tower and a Service Management Center (Command Center) for Digi Telecommunications Sdn. Bhd.

The Client's brief is to design a data center based on IT Data Center's Uptime Institute Tier III platform, with the possibility to scale-up to Tier IV security and to have ecological features.

The response to the Client's brief is a building designed essentially to optimize space use for the computers and effective drainage and waterproofing, to protect the computer equipment and to reduce heat absorption into the data center and to have tight security measures.

The façades of the Data Center are designed to have with vertical green walls that act as living habitats and as means of filtering and improving a building's ambient indoor air quality.

Green walls are a living, breathing, regenerative claddings that can be as simple as a living art installation or as complex as a biological air filter. Green walls, both indoors and outdoors, decrease local CO_2, increase local humidity, trap dust, reduce noise and create a habitat for urban wildlife. Exterior green wall installations reduce solar gain (the entrapment of heat by passive solar gain on the building surface) and, by extension, building energy costs; provide protection from the effects of UV radiation and acid rain; and help lessen the building's contribution to the heat island effect (resulting from vegetated land replaced with concrete and asphalt, causing such urban centers to become further than the non-urban natural areas).

Ecological Design Features

Vegetated Green Wall: The ecological design of the building is expressed in a "Vertical Plantscape System" wrapping around the building's envelope. The vegetation acts as an effective barrier against solar radiation and also acts to insulate the building significantly reducing heat build-up and reducing the cooling energy costs of the building. The plants act as bio-purifiers and through a number of natural biochemical processes can break down and remove airborne contaminants which offer immediate environmental advances in reducing existing greenhouse and other volatile organic compounds,and provide solar sun-shading. The office areas are recuped 24 hours and require sun-shading to reduce direct heat gain and to reduce cooling loads. It enables the use of clear glass for better daylight penetration,

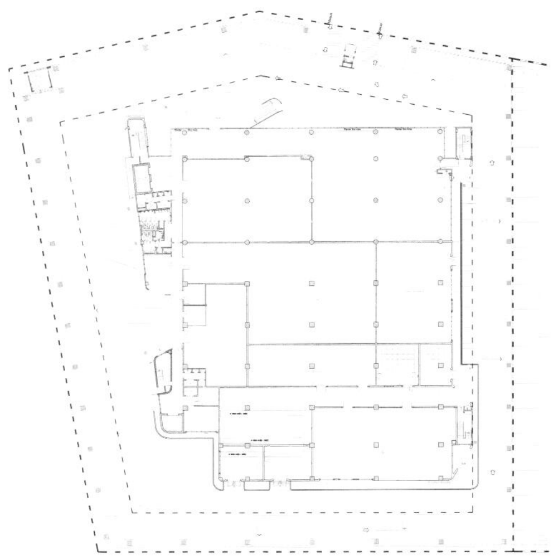

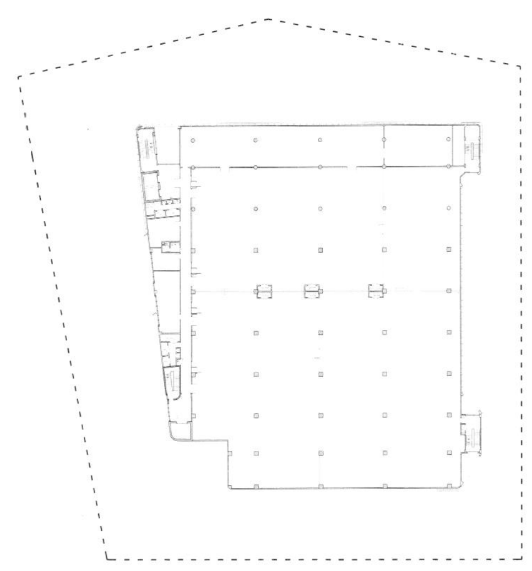

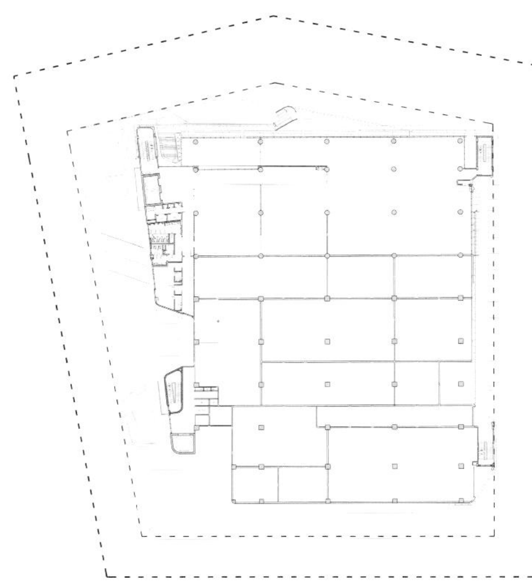

which in turn reduces artificial lighting loads.

Daylighting: The main office and circulation spaces are glazed using full-height curtain wall to provide maximum daylight penetration and enable energy efficient lighting systems within the spaces. Secondary rooms are also fitted with openable windows for natural ventilation and daylight.

Renewable Energy: An array of photovoltaic panels is installed on the building's uppermost roof area. Mounted on a steel trellis above the rooftop, the 234 m^2 PV array generates 35.28kwh of electricity on site with all power generation feeding directly back into the municipal power grid. This renewable energy production represents an overall reduction in CO_2 emissions of 12,516 kg per year and helps offset the intensive energy demands of the datacenter facility.

Bioswale: Natural filtration and drainage systems are used to reduce the burden of surface run off to the public drainage system. They also act to collect surface run off for reuse in the central irrigation system. The project's sustainable drainage strategy also includes the use of permeable grass pavers for all ground level car park areas. These systems allow rainwater to replenish the site's local aquifer rather than being diverted into an external storm-water drain.

Rainwater Harvesting: Siphonic rainwater down pipes are used for high velocity rain-water run-off from the building's roof. This rainwater is channeled into a collection tank situated at ground level, buried in the landscape for minimal impact. Comprised of an innovative Versi-Tank™ system, the tank has a capacity of 100 cubic meters and is made of 100% recycled content materials. Collected rainwater is filtered and reused to irrigate plantscapes around the building and caters for the water requirements of the vegetated green wall via an integrated gravity-fed rooftop irrigation system. The project's rainwater harvesting system is designed so that no potable water is used for irrigation purposes.

Air-conditioning: The project utilizes a highly efficient dual-mode air and water cooled chiller system (0.6KW/tone estimated) with multiple zoning FCU and precision air-conditioning in the data center areas. Zoning of FCU creates savings in energy requirement due to variations on zoning usage. The cooling requirements of the data center are further optimized via an innovative cold aisle containment system. This system establishes targeted cooling zones that meet the requirements of specific equipment rather than expending the energy otherwise required to cool the entire room uniformly.

Pumps: Variable speed pumps for energy efficient usage are used due to variation in load. This estimates to provide 30% energy savings for the pump systems.

Lighting: Light fittings are energy efficient Florescent T5 using low loss electronic ballasts. These offer higher luminance levels and their zoning enables energy conservation through specific usage requirements. This system is bolstered through the use of occupancy sensors and controlled via the central Building Automation control scheme.

Lifts and Escalators: Lifts incorporate AC VVVF control and gearless motor drives with synchronous motor and

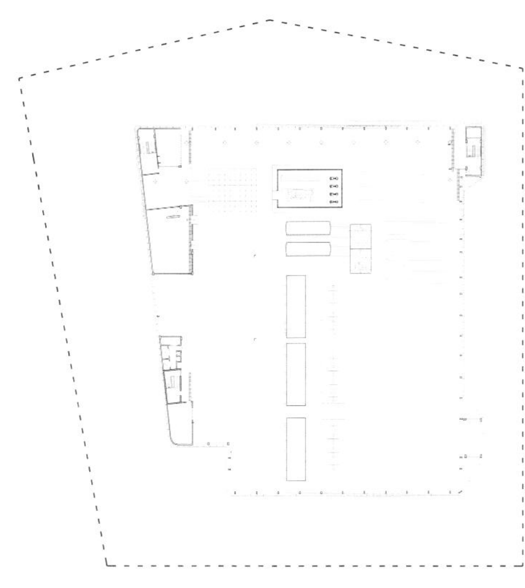

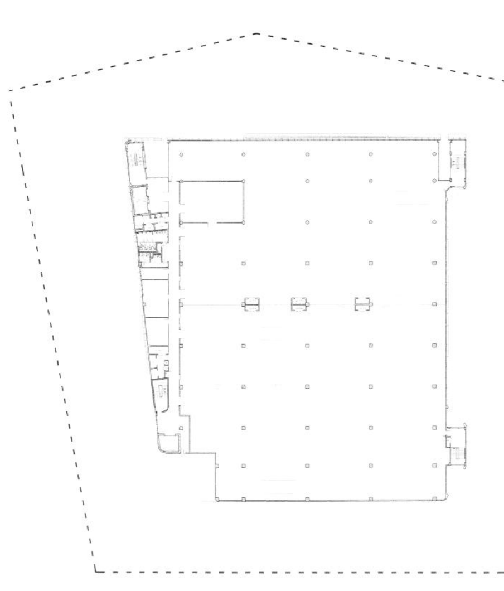

permanent magnets. Lifts are also fitted with detection sensors, sleep modes and energy saving light fittings.

Indoor Air Quality: There are minimum requirements for ventilation rates in accordance with ASHRARE 62.1:2007 which are designed to enhance the comfort and well-being of the occupants. Strict use of low-VOC coatings, paints and adhesives further reduces detrimental impact on occupant's health from finishes that emit internal air pollutants. Extensive greenery at the building envelope further acts as an additional biological filter to improve ambient air quality.

Sanitary Fittings: Dual Flushing WC's are fitted allowing control of water volume based on different usage needs. All basin taps and other sanitary wares are water efficient fittings with certified WELs ratings of 3-tics (maximum efficiency rating). In addition, waterless urinals are specified in the male toilets, further reducing water consumption. Overall use of potable water within the building is reduced by more than 50%.

Sustainable Transportation:Regular shuttle service is provided between the building and the Batu Tiga KTM station (1.6 km away) in order to encourage employees to commute via public transportation. Preferred parking for carpools is provided adjacent to the building's main entrance.

Green Materials

External timber deck is made from "CT- Wood", a composite of pure polymer resin and bio-waste (rice husk). This material reuses bio-waste which in turn reduces air pollution and can be re-grinded into powder to be recycled or reprocessed.

Office areas are to receive Nippon Odour- less premium all in 1 paint. This product has almost zero Volatile Organic Compounds which have been indicated to lead to sick building syndrome while ideal for those sensitive or allergic to pain odour. This engenders greater user productivity and a healthier working environment.

The primary roof water-proofing system is Hitchins "Trafficgard" , a liquid applied membrane of heavy bodied acrylic polymer gel. This is an environmentally-friendly coating system and is a Green Label certified product. Low RTTV from the roof as the space below the roof is mainly data center and telco areas.

The interior walls and Ceiling lining in the office areas "Boral EnviroBoard" are recycled, environmentally-friendly plasterboards, composed of High Purity Synthetic Gypsum (HPSG) plaster core encase in Heavy Duty face a backing liner and is non-hazardous to work with. It also contains low OTTV value and complies with MS 1525.

Office areas receive Carpets with 30% of more recycled products .

Solaris

Solaris生态项目

杨经文

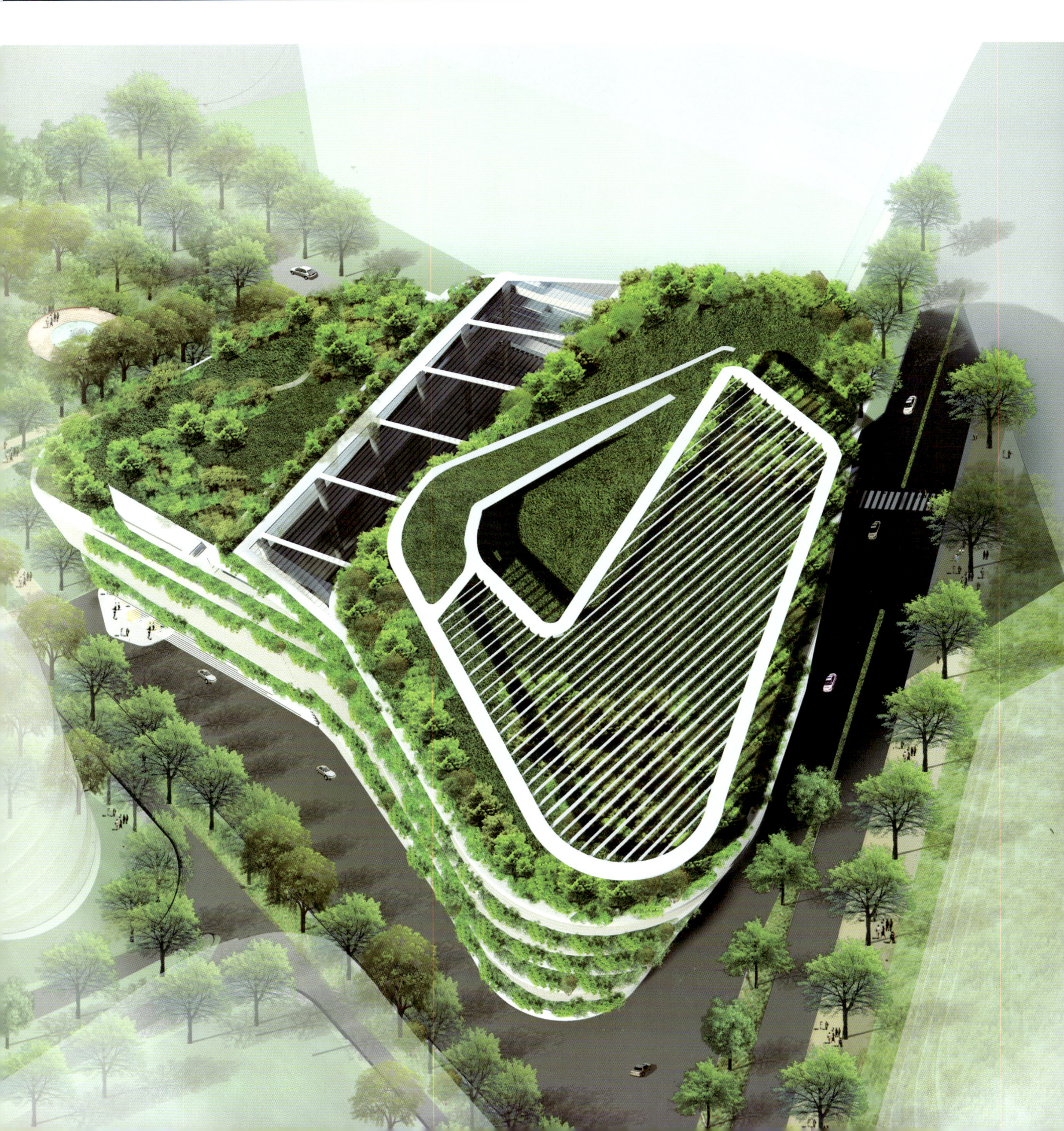

项目地点	新加坡	Project Location	Singapore
项目时间	2008年9月（开始建造时间）	Project Milestones	September 2008 (Construction start date)
	2010年9月（计划竣工时间）		September 2010 (Scheduled completion date)
建筑高度	80米	Building Height	80 m
A座	15层+楼顶花园	Tower A	15 Storeys + roof garden
B座	9层+屋顶花园	Tower B	9 Storeys + roof garden
总建筑面积	51 282 平方米	Total GFA	51,282 m²
基地面积	7 734 平方米	Site Size	7,734 m²
园区总面积	8 363平方米	Total Landscape Area	8,363 m²
项目总设计	T·R·哈姆扎、杨经文建筑师事务所	Lead Designer	T. R. Hamzah & Yeang Design Team

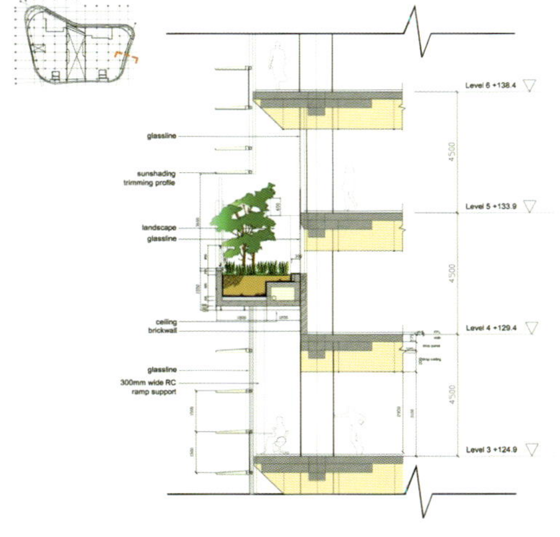

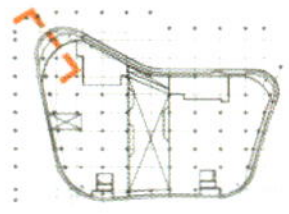

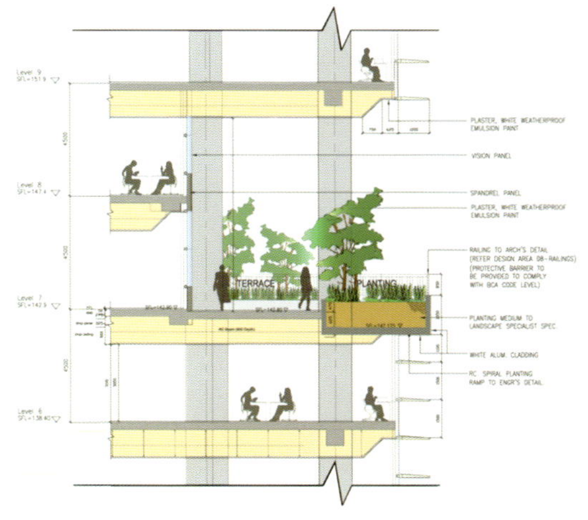

项目概况

项目位于新加坡市中心一北社区中的研究和商业公园，该项目二期在当地是一个旗舰项目。纬壹启汇资讯传媒园是一个为促进通信技术、媒体、自然科学和工业领域的企业家进行创新发展而研究与开发的项目。项目总体规划及未来发展由Zaha Hadid建筑公司规划。

建筑总体能量消耗与当地建筑范例相比减少36%之多，其高性能墙面有小于39瓦特/平方米外部热转移系统。该项目有8 000平方米的景观绿化，同时Solaris也引入了建筑物本身规划区之外的植物。

Solaris最具有代表性的规划是一个固有的生态方式建筑设计。项目被一个大型的自然通风中心分割成两栋塔楼。办公楼层与天桥相连，天桥横跨高区的中心。此幢建筑物将会成为一北社区的焦点，天窗和庭院创造出的自然光，空气流通和持续不断的景观道路，穿越大街的一北公园的延伸段，使天台和建筑物的生态结构互相贯通。建筑物大量的生态基础设施，都有着不破坏生态平衡的设计特色和绿色概念覆盖的创新想法，Solaris努力加强现存的生态系统，而不是取代它们。

生态设计特色

连续周边园景的匝道：一条1.5公里长的生态线与邻近的一北公园从地面、地下室和建筑物高层的屋顶花园水平相连。斜坡最小宽度3米。盘旋景观的保养舷梯

通过一个平行路得以使用，园丁进入无须穿过业主的私人空间。小道也被设计成一个长形的公园，所有的道路都从地面向最高处的屋顶区域伸展。景观美化的连续性是项目生态设计概念一个重要的组成部分，连续的景观使得各种生物体和植物能够在建筑内部的绿化区域运动通畅，在加强生物多样性的同时，也为生态系统的整体健康发展作出贡献。坡道利用深远的悬挑与大规模的遮阳植物，成为整个建筑立面冷却系统综合战略的关键要素。这个生态基础设施为建筑高层的居住者提供社交、相互交往和创造性的环境，同时使得建筑结构的内在无机性与有机质之间达到平衡。

太阳能轴：斜角光线穿越A栋大楼的高层，让日光可以穿越建筑物的内部深处。当日光可以充分利用时，内部照明系统会感应并关闭照明灯以自动减少能源浪费。太阳光景给相邻的空间增强特色，使人们从街上可以更好地观赏大楼景观。

生态屋：坐落于建筑物东北角螺旋形的斜路与大街相接处，生态屋使绿化、日光照射和自然通风可以延伸至地下停车场。生态屋最低层有控制储藏箱和雨水收集系统的水泵房。

自然通风与大厅日照点：二栋高楼之间的公共广场为社会活动和表演提供了公共空间。第一层的自然通风运转起来就好像一个复杂模式（非空调），该模式利用中庭上空可开启的玻璃百叶屋顶来强化通风，同时在需要时提供应对自然气候因素的保护。CFD用于分析中庭的热环境和风速。这些研究成果被用于中庭正面建筑的优化，以加强空气流通，提高舒适度。

袖珍公园/广场：地面层的景观美化与街对面的一北公园连接，使广场一层形成十字通风，该景观为社交活动提供了一个聚集场所。

遮阳百叶窗：项目外立面设计起源于当地太阳线分析。新加坡位于赤道，太阳线完全东西走向。分析太阳线的外立面研究决定遮阳天窗外形和深度，就好像双重灯架。这种遮阳方法进一步降低穿越建筑物底层的热能量，实现39瓦每平方米的低外部热量转移量。与螺旋形的风景斜坡、空中花园和深处的悬挂物相协作，屋顶天

窗同样为适于居住的空间增加了舒适度。联合长线使得建筑物遮阳天窗长度延长了10公里。

屋顶花园和转角空中露台：垂直的景观美化担当起热量缓冲器，创造出放松、活动的区域。这些广阔的花园使大厦居住者与自然相拥，欣赏邻近的一北公园带来的舒适环境体验。由于螺旋形的斜路能够通往大厦的任何一个角落，它变成了一个大大的天梯。项目快完成时，植物的总量超过建筑物本身的规划。作为一个充满视觉效果的绿化设计，该项目95%的绿化带在地面以上。

雨水收集/循环利用：建筑物大量景观区被大范围的雨水系统灌溉，雨水经由B栋大厦污水管道再从周边风景斜道的地下排水管道集合而成。雨水由Eco-cell底层地下室和屋顶水箱储存。储藏室的容量有700立方米，经过专门处理的雨水可以使建筑物内的种植区得到灌溉，一套整体的施肥体系将帮助维持灌溉循环中的有机营养水平。

环保标章评级：2009年9月，Solaris系统被授予绿色白金等级证书，是BCA绿色标记授予的最高级别的证书，成为新加坡不破坏生态平衡的建筑标本。

Design Brief

Solaris is located in the research and business park in central Singapore's one-north community. The building is a flagship project in the second phase on this locality. Fusionopolis is a R&D hub for Infocomm Technology, Media, Physical Sciences & Engineering industries which is intended to facilitate innovation and entrepreneurship in these fields. The masterplan for the visionary mixed-use development is prepared by Zaha Hadid Architects.

The building's overall energy consumption represents a reduction of over 36% compared to local precedents and the high performance façade has an External Thermal Transfer Value (ETTV) of less than 39 W/㎡. With over 8,000 ㎡ of landscaping, Solaris also introduces vegetation which exceeds the area of the building's original site.

Solaris stands as a dramatic demonstration of the possibilities inherent in an ecological approach to building design. The project comprises two tower blocks separated by a grand naturally-ventilated central atrium. Office floors are linked by a series of sky bridges which span the atrium at upper floors. The building will become a vibrant focal point for the one-north community through the introduction of open interactive spaces, creative use of skylights and courtyards for natural light and ventilation and a continuous spiral landscaped ramp, an extension of one-north Park across the street, which forms an ecological nexus tying together an escalating sequence roof gardens with sky terraces that interpenetrate the building's façade. With its extensive eco-infrastructure, sustainable design features and innovative vertical green concept, Solaris strives to enhance its site's existing ecosystems, rather than replace them.

Ecological Design Feature

Continuous Perimeter Landscaped Ramp: An uninterrupted 1.5 km long ecological armature connects the adjacent one-north Park at ground level and the basement Eco-cell with the cascading sequence of roof-gardens at the building's highest levels. The ramp has a minimum width of 3m. Maintenance of the spiral landscaped ramp is achieved via a parallel pathway which allows for servicing of the continuous planters without requiring access from internal tenanted spaces. The pathway is also designed to serve as a linear park that stretches all the way from the ground plane to the uppermost roof areas. The continuity of the landscaping is a key component of the project's ecological design concept as it allows for fluid movement of organisms and plant species between all vegetated areas within the building, enhancing biodiversity and contributing to the overall health of these ecosystems. The ramp, with its deep overhangs and large concentrations of shade plants, represents a key element in the comprehensive strategy for the ambient cooling of the building façade. This eco-infrastructure provides social, interactive and creative environments for the occupants of the building's upper floors while balancing the inherent inorganicness of the built-form with a more organic mass.

Solar Shaft: A diagonal shaft that cuts through the upper floors of Tower A allows day-light to penetrate deep into the building's interior. Internal lighting operates on a system of sensors which reduces energy use by automatically turning-off lights when adequate day-lighting is available. Landscaped terraces within the solar shaft bring added quality to adjacent spaces and enhance views up into the building from the street below.

Eco-cell: Located at the building's north-east corner where the spiral ramp meets the ground, the Eco-cell allows vegetation, daylight and natural ventilation to extend into the car-park levels below. The lowest level of the Eco-cell contains the storage tank and pump room of the rainwater harvesting system.

Naturally Ventilated and Day Lit Grand Atrium: A public plaza between the two tower blocks provides a space for communal activities and creative performances. This naturally-ventilated ground floor operates as a mixed-mode (non-air conditioned) zone with an operable glass-louvered roof over the atrium enhancing ventilation while providing protection from the elements when needed. CFD (Computational Fluid Dynamics) simulations are used to analyse thermal conditions and wind-speed within the atrium. The results of these studies are used to optimize the atrium façade design to improve air flow and enhance comfort levels.

Pocket Park / Plaza: Ground level landscaping, linking to one-north Park across the street, allows for cross ventilation of the ground-floor plaza and provides a venue for social and interactive events.

Extensive Sun-Shading Louvers: The project's climate-responsive façade design originates with analysis of the local sun-path. Singapore is at the equator and the sun-path is almost exactly east-west. Façade studies analyzing the

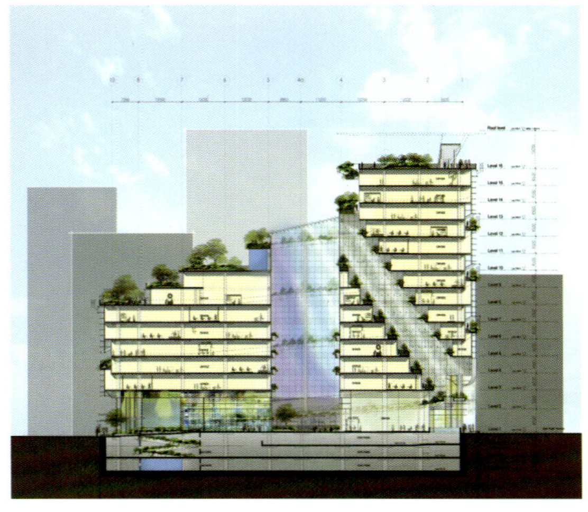

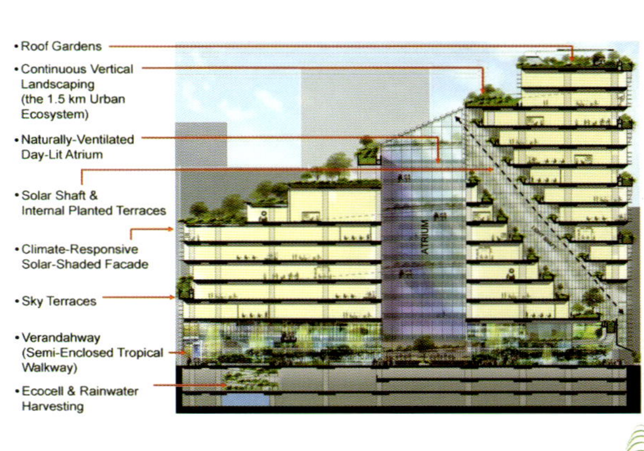

ecological design features

solar-path determine the shape and depth of the sunshade louvers, which also double as light-shelves. This solar shading strategy further reduces heat transfer across the building's low-e double-glazed perimeter façade, contributing to a low External Thermal Transfer Value (ETTV) of 39 W/㎡. In conjunction with the spiral landscaped ramp, sky gardens, and deep overhangs, the sunshade louvers also assist in establishing comfortable micro-climates in habitable spaces along the building's exterior. The combined linear length of the building's sun-shade louvers exceeds 10 km.

Roof Gardens and Corner Sky Terraces: Vertical landscaping acts as a thermal buffer and creates areas for relaxation and event spaces. These extensive gardens allow building occupants to interact with nature and offer opportunities to experience the external environment and enjoy views of the treetops of the adjacent one-north Park. As it reaches each corner of the building the spiral ramp expands into generous double-volume sky terraces. Upon completion, the sum of its vegetated areas will exceed the footprint of the site on which the building sits. A dramatic vision of the possibilities inherent in skyrise greenery design, 95% of the project's total landscaped area is above ground level.

Rainwater Harvesting/Recycling: The building's extensive landscape areas are irrigated via a large-scaled rainwater recycling system. Rainwater is collected from the drainage downpipes of the perimeter landscaped ramp and from the roof of tower B via Siphonic drainage. It is stored in rooftop tanks and at the lowest basement level, beneath the Eco-cell. A combined storage capacity of over 700 m^3 allows the building's vegetated areas to be irrigated almost exclusively via harvested rainwater. An integrated fertigation system helps maintain organic nutrient levels throughout the irrigation cycle.

GreenMark Platinum Rating: In September 2009, Solaris was awarded a Green Mark Platinum rating, the highest level of certification granted by BCA's Green Mark, Singapore's sustainable building benchmark.

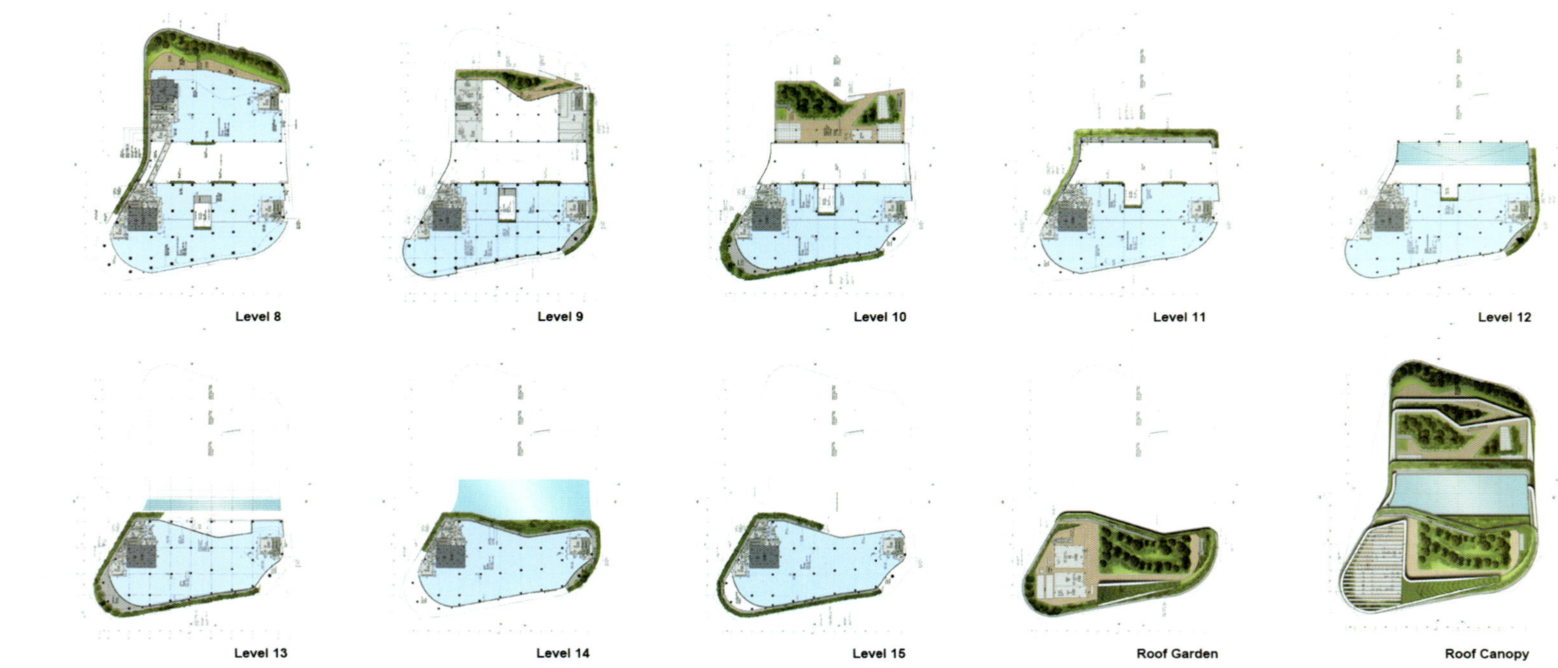

© 2009 T. R. Hamzah & Yeang Sdn. Bh

KANENA Building in Amman Jordan

约旦KANENA建筑

Kois Associated Architects–KAA

项目地点	约旦，安曼
项目面积	1498平方米
总建筑面积	8885平方米
设计单位	Kois Associated Architects–KAA
总设计师	Stelios Kois
主要材料	现浇混凝土和三聚氰胺涂层的CNC切口木材框架；定制型多边形中空玻璃；绿墙

Project Location	West Amman , Jordan
Project Area	1,498㎡
Total GFA	8,885㎡
Design Company	Kois Associated Architects-KAA
Principal Architect	Stelios Kois
Materials	In situ concrete casted with melamine coated CNC cut plywood formwork, custom made polygonal insulated glass panels, green wall

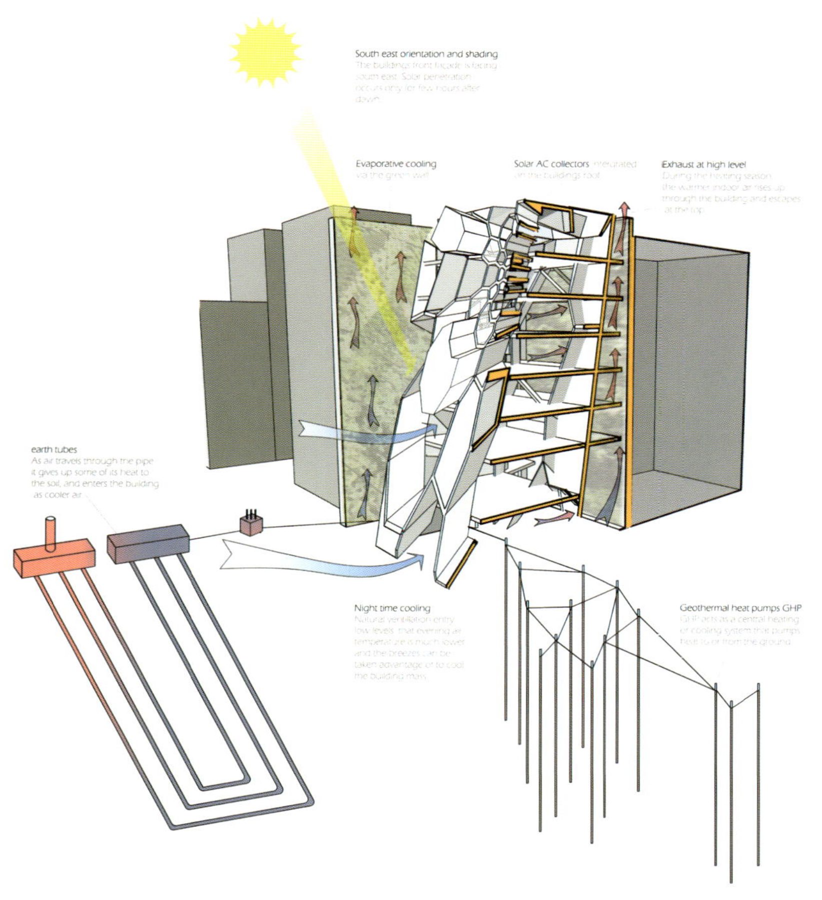

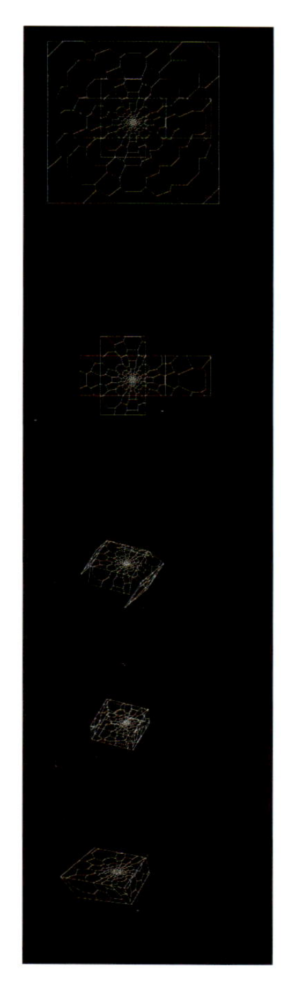

本方案是为一栋拥有办公室以及顶层阁楼的多层大楼做设计，其位于约旦首都中心。设计进程始于对安曼城市环境的研究，对其组建原则以及结构类型学的研究位于邻区。对于约旦建筑的分析和研究，提醒了始于平面六面体的对于体积的调查，平行于样地形状并依附于毗邻建筑物。

建筑下缘连续不断地接近棋盘式街道，然而体积从场地的内边界分开。精心设计的建筑外表，作为城市的最后边界以及能够准确传达现代城市意义的标志，可形成与特定地域之间的联系。出于这个理由，设计师研究了各种约旦传统建筑的形式，以配合不同的设计方案。

穆斯林有种信仰，人们可以在动物、植物细胞、植物形状以及地质构造中窥见设计中所需要的几何形状，从而表达“几何形状是上帝的旨意”。伊斯兰宗教风格的建筑设计是基于夸张的几何图形，由非常简单的圆形或是方形图案组成或生成。这些几何图形被混合、复制、交织并且以各种复杂的方式进行重组。伊斯兰教有种信仰，即对这些自然模式、形状、关系以及相互之间关联的研究会增加其神秘感——宇宙间的法则和定律。

设计师最后选用的设计方案是基于最基本的体积表面，很好地展现了作用于之上的力。适应大楼各种功能的原始构造被围栏圈起，这样一来也在一定程度上使大楼可以很好地融入整个城市中，在室内和室外之间逐步产生了一种交互作用，楼层与楼层之间也各不相同，以符合相当层次的需求。内边界，地板与邻近建筑物的小区相邻，由一层植被覆盖，以响应相邻街区之间环绕的绿色边界。用这种方式为大楼创造一种特殊的环境，形成了大楼综合的背景基础，树立了大楼鲜明的形象。

大楼上层（包括办公空间以及公寓）的主要垂直和水平环流位于由场地边界的大楼腾出的空间，以这种方式，大楼外部主要中心的循环就完成了，这样一来游客只能从小门进入大楼围栏，逐步呈现内部空间。

空间的层次感符合所要求的次序并反映了对社会私密性的逐步体现。整个项目最公共的部分，即商店，则位于大楼的地下层与一层。这两层之间通过一条单一的通道相连，从主入口处开始，一直延伸到二楼，代表了整座城市在大楼的延伸。商业单元的组织模拟传统集市的结构。设计师将集市作为一种基础结构，当地商业活动的一种尝试，也作为交流互动的一个流动的枢纽。集市的关键因素，大楼结构的支柱，通往商店的主要路线，都沿着原路线清楚的展现出来，而商业大楼则有组织地围绕在其周围。考虑到办公室的要求，设计过程中的主要目标即为用户提供最大限度的方便。将单一的桌子作为基础元素，而图书馆书架是第二个因素作为区分课桌的边界，设计师研究了各类几何图形以对各种不同单元进行整合。这种有组织的组织方式的好处是根据不同的人数以及需求使每一个工作团队都可以拥有独立定制的工作空间。大楼的最高层则被单纯地用于居住并拥有两个宽敞的家庭屋顶。

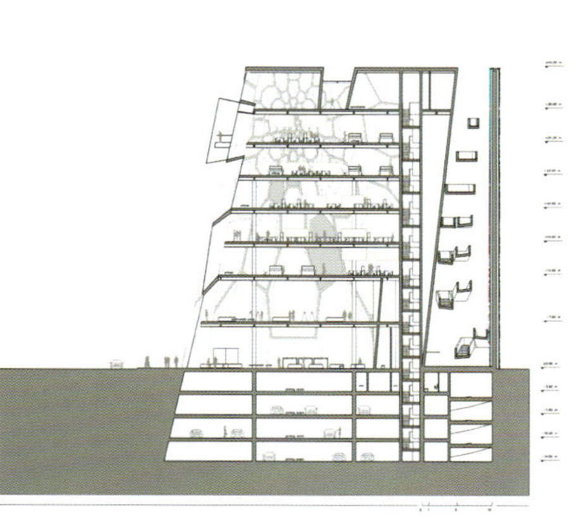

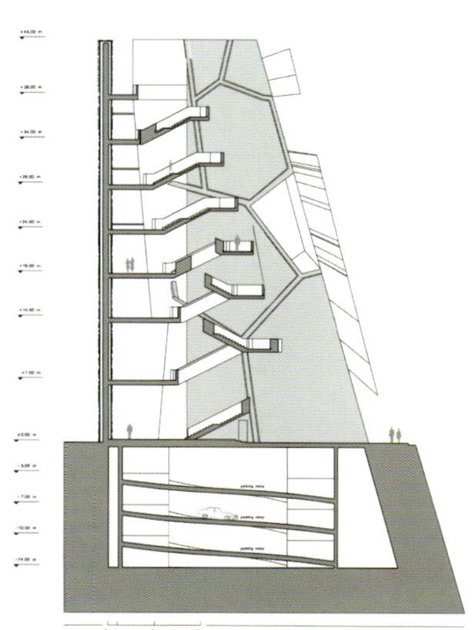

生物气候策略

地形与气候

安曼位于约旦西北部的丘陵地区。整座城市位于七座山上，现在已经发展成十九座山了。由于地处高原位置，受冷空气影响，安曼与同地区的其他地方相比，四季更加分明。该地区夏季的温度持续在二十八摄氏度至三十五摄氏度的范围之内，但却伴随着低湿热并且经常刮风；春天和秋天温度适中宜人；冬天夜间温度接近零摄氏度，而且每年这个时候都会降几次雪。年平均雨天是5~45天，雪天是0~8天。通常在六月至九月初这段期间不会降雨，主要是以晴好天气为主。

环境保护

避免阳光直射：目标是避免阳光直射。在白天日照时间完全避免太阳光直射入大楼。

避免白天空气流通：高温天气下白天的空气十分干燥并且炎热。让热空气直接流通入大楼是不明智的。

促进夜晚空气流通：沙漠气候无疑夜晚要比白天凉快许多。因此加大夜间空气流通有利于将热空气驱逐出建筑物，使建筑物内温度降低。

利用反射以及对浅色非吸热颜色的运用来实现采光：通过对浅色的运用，大楼可以尽可能比使用深颜色吸收更少的热量。浅色可以发散光线，并且不利用直接的电子束辐射就可以使大楼有良好的光线照明。

实施

通过对大楼的定位以及将建筑外墙作为遮盖设施来避免阳光直射。大楼前立面面向东南处。日出后光线也仅能穿透几小时。早上的太阳辐射并不如正午那么强烈。从大楼东北方向照射进的太阳光线在配有电板式有色玻璃的门窗作用下扩散开来。大楼白色的混凝土表面使直射的阳光反射并随之扩散开来。建筑外墙的深度使得它在充当了遮阳设备的同时也具备了照明的作用。这意味着它既阻挡了太阳光的直射，也将光线通过反射发散改变了照射到室内的方向。建筑表面的厚厚的硅藻符合光照路线，这样就可以达到期望的照明和庇荫条件。类似地，下连续墙上的硅藻层之类的其余的外表面也是浅色的，可以避免太阳的过度曝晒。大楼内部也采用了浅色以提高内部亮度。

白天的通风则是通过利用地管、长的地下金属以及空气可以流通的塑料管来实现的。当空气通过管道的时候空气中的热量则被传播到了地下，这样进入大楼的空气就是冷却过的了。会发生这样的情况，主要是因为在干燥天气状况下，空气的自然流通会均衡室内室外的温度。这会导致热舒适度降低以及不必要的大楼内部温度的升高。

约旦的高温、湿热以及夹杂着灰尘的热风常常让人感到极度的不舒适。为了缓解气候因素给居民带来的不舒适感，设计师创造了一个小型气候区。微型气候指的是定位于一小块区域与其他附近的区域在某种程度上都不一样的天气模式。其变化可以是温度、湿度、降雨、风或是这几种任意的组合。改善建筑物周边地区的局部气候可以提高人们生活的质量，给使用者也提供了舒适。设计师想要实现的改变包括：相对湿度、空气温度、流动风以及光照。为了达到这一目标，设计师在大楼与场地边界处开了一个缺口。在边界的边缘处设计师设立了一个垂直结构来促进垂直公园的建设和嵌入式的

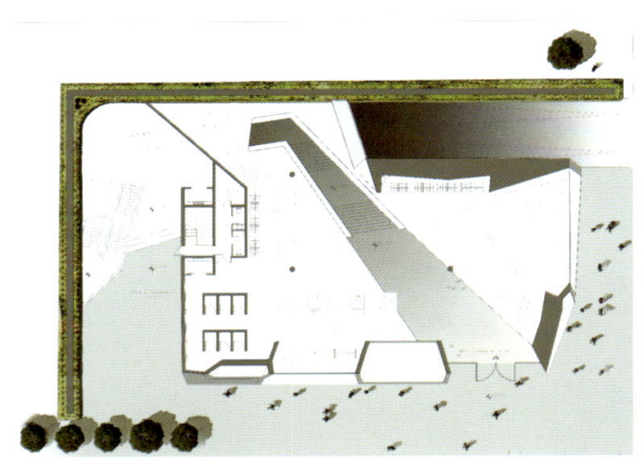
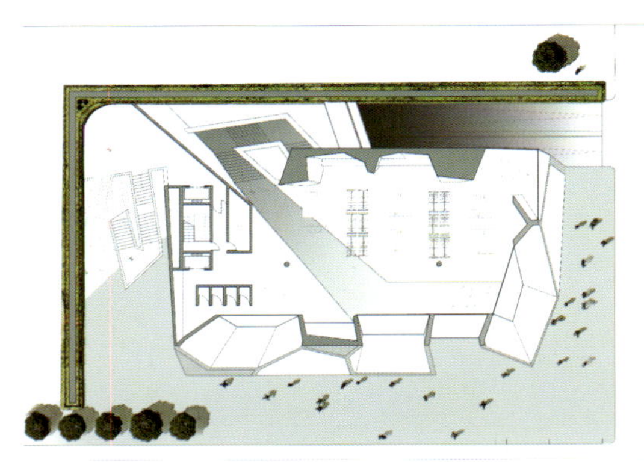
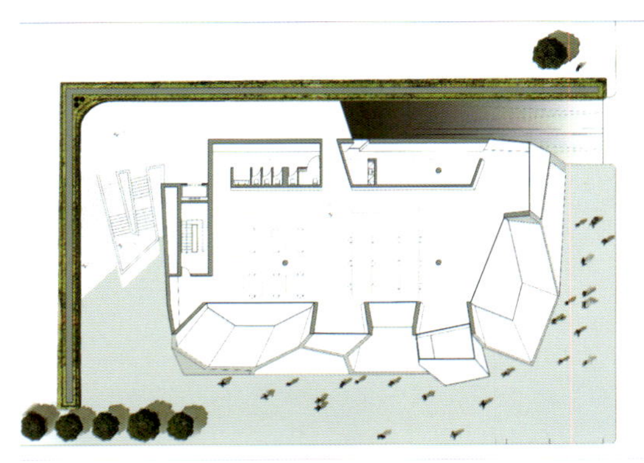

注水系统。墙上的植被不但可以吸收二氧化碳和释放氧气，而且还可以通过改变温度、相对湿度、过滤空气和发散气流来调节微气候的因素。建筑与绿墙之间的缺口则使得气流可以流通并且驱除饱和的风量以使冷却率保持在高水平，为居民创造一个舒适的环境。

大楼的主动加热系统由底层地板下的加热系统控制，这一系统由地热热泵以及后备锅炉供给。地热热泵作为一个中心的加热或冷却系统将热量输入或输出地面。它把地面作为一个热能源（冬天）或者是一个吸热设备（夏天）进行利用。本设计利用了地表适度平衡的温度来提高效率，减少加热和冷却系统的操作成本。超出地面三米地方的顶部气温相对比较稳定，保持在十摄氏度和十六摄氏度之间。地热热泵系统在寒冷的冬夜可以达到相当高的效率（300%~600%），相对于在凉爽天气中的空气源热泵的175%~250%的效率。地源热泵作为最高效的技术可以提供暖通空调和热水供暖。对于大多数安曼的大楼都选择垂直系统主要是由于有限的空间以及不断增长的需求。被挖的地洞（直径大约10厘米）每隔六米有一个，深20米。在这些洞里有两根连接到底部的管道并弯曲成U形形成一个环。这些垂直的环与水平管道相连，被放置在管沟里，与大楼的热泵相连。

被动式降温

在夏天，利用喷雾可以更好地达到自然冷却。水沿着大楼顶部被压到喷洒器里，然后沿着斜立面流下来。由于夜间空气接触到更大的表面面积，在这样一个系统里蒸发率也大大提高了。屋顶的喷雾器依靠外部的力量将水传送到屋顶，因此它并不是一个完全的被动系统。用来抽送的总能量相对于为提高冷却指数所节约的能源是非常少的。多余的水可以采集并再利用。同样的技术被用来喷洒面对南立面的?

废水再循环

大楼配备了污水收集以及净化系统。污水主要来自于洗衣服、洗碗和洗澡所用的水，这些污水可以用于灌溉等在其他方面的循环利用。这些污水50%~80%来自由建筑物卫生系统设备（不包括卫生间）产生的居民生活污水。该污水处理系统主要是处理污水中的有毒物质——漂白剂、人工染料和清洁剂，避免土壤和灌溉污染。

总结

在气候干旱的地区进行设计是非常具有挑战性的，尤其是在设计过程中还要考虑恶劣的环境条件和能源标准。该项目的宗旨是不仅要创造一个舒适的建筑物还要为这个城市打造一个具有视觉享受、文化内涵和环保型的建筑大楼。

The proposal is for a multi-storeyed building which comprises of offices and lofts and is located in the centre of Jordan's capital. The design process began with the study of Amman's urban environment, its organizational principles and the study of the typology of the structures located in the neighboring area .

This analysis and the study of Jordanian architecture, informed the volumetric investigation which started with a

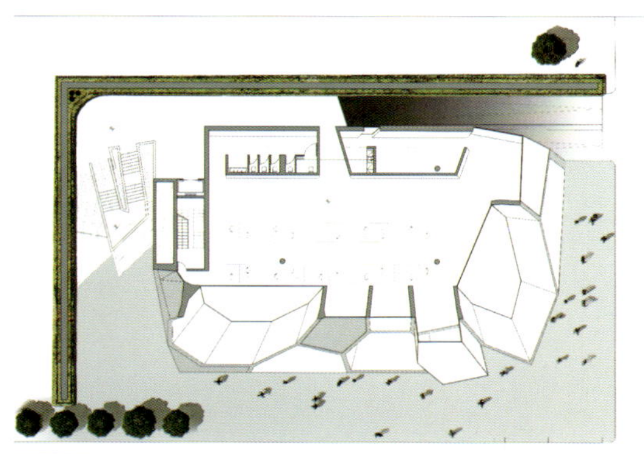

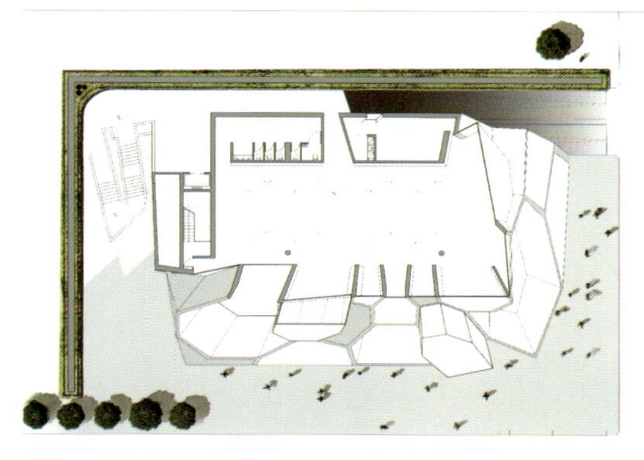

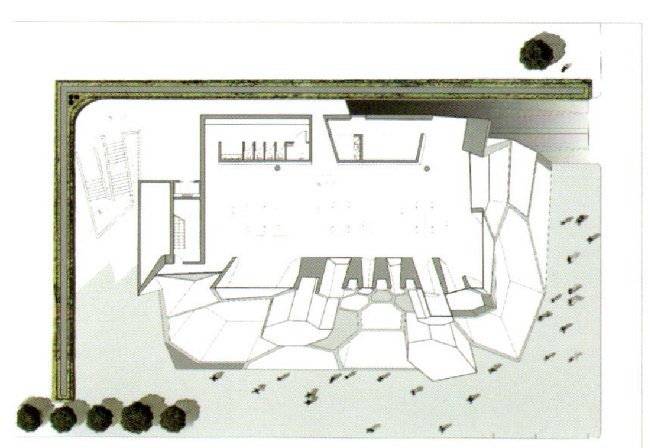

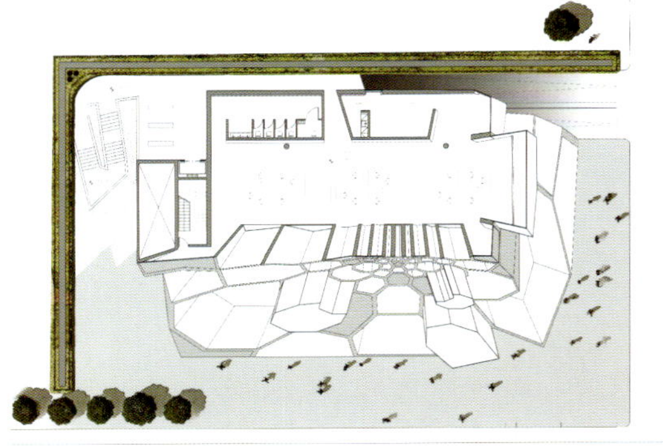

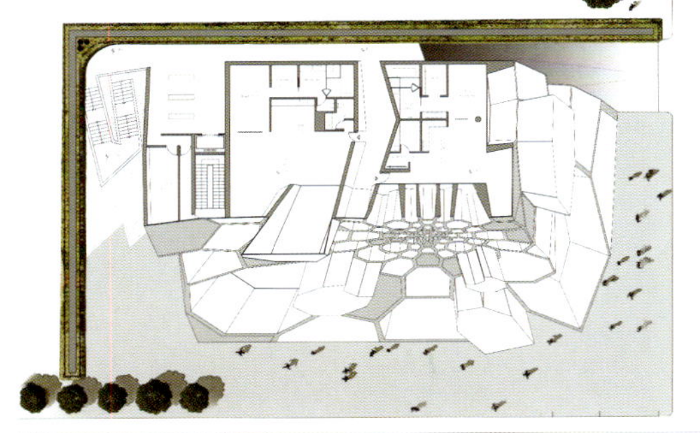

plain parallelepiped, parallel to the building plot's shape and attached to the adjoining buildings.

Consecutively, the lower edges are drawn towards the street grid, while the volume is detached from the inner boundaries of the site. The elaboration on the volume's skin, as the final border with the city and the one that determinately conveys meanings and symbolisms to the contemporary city, is what will form the connection with the specific location. For this reason, various patterns observed in traditional Jordanian architecture were studied, along with the various methods of their scheme reproduction.

There is a Muslim belief that one can find geometry in the design of all life forms from animal and plant cells, plant forms, and geological structures hence the expression "geometry is God manifest". Islamic religious architectural design is based on 'sacred' geometry, consisting of, or generated from, such simple forms as the circle and the square. These geometric patterns were combined, duplicated, interlaced, and arranged in intricate combinations. There is an Islamic belief that studying the nature of these patterns, forms, relationships and their connections, insight may be gained into the mysteries — the laws and lore of the Universe.

The pattern that was finally selected to be processed was eventually projected onto the primary volume's faces, expressing the forces that act upon it. The primitive box, the one that will accommodate the building's functions, is enclosed in this veil, which also defines the extent of the city's flow into the building, allowing a gradual interplay between indoors and outdoors, differing from floor to floor, in accordance to the needs of the equivalent level. The inner border, the back planes neighboring with the adjacent building plots, is covered with a layer of low planting, responding to the green border encountered between the neighboring building blocks. It is in this way that this relieving green breath creates a special environment for the building, forming the background of the architectural synthesis and cumulates towards its intended clearly recognizable identity. The main vertical and horizontal circulation to the upper floors (offices, apartments) is situated in the space created by the detachment of the building from its site's boundaries. In this way the circulation is accomplished outside the main core of the building and allows the visitor to entry the enclosure only through small openings, gradually revealing the interior spaces.

The layering of the spaces follows the required sequence and reflects the gradual accession of the social privacy. The stores, as the most public part of the program, are situated on the ground and the first floor. The connection between the two floors is established through a single route beginning from the main entrance and resulting at the second floor, symbolizing the incoming of the city into the building. The organization of the commercial units mimics the structure of the traditional bazaar. We consider

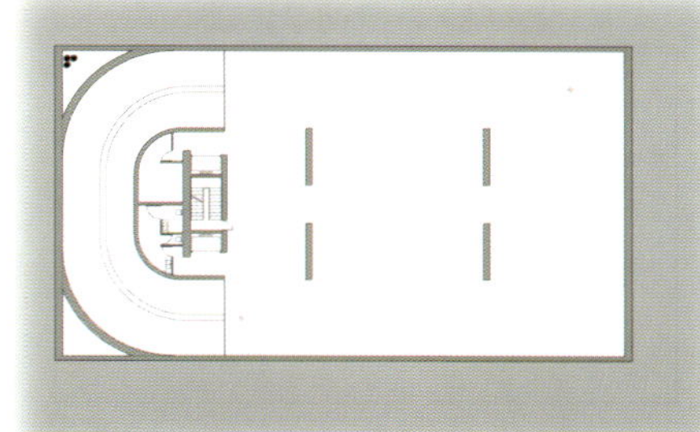

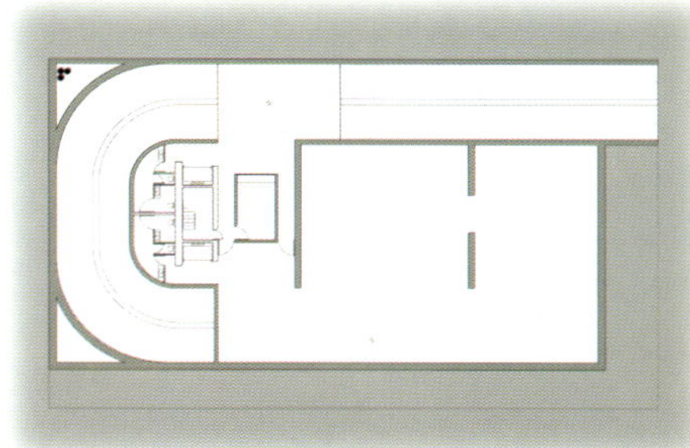

the bazaar as a paradigm structure of the merchandising local practice, which comprises a vibrant hub for social interaction. The core element of the bazaar, the spinal of its structure, the main route that fuels the stores, is interpreted here with this very path, with the commercial units being organized around it. As far as the offices are concerned, the main goal during the design is to achieve maximum organizational flexibility for the user. Considering a single desk as the primary unit, and a library stack as the secondary element that acts as a dividing boundary between two desks, various geometries are studied, resulting to various combinations of the congregation of the units. The benefit of this organizational method is that each working team will be free to customize the working space according to its number of members and their requirements. The top level of the building is purely residential in use and houses two spacious family lofts.

Bioclimatic strategy

Topography- Climate

Amman is located in a hilly area of north-western Jordan. The city was founded on seven hills, but it now spreads over an area of nineteen hills. Because of the cooling effects of its location on a plateau, Amman enjoys four seasons of temperate weather compared to other places in the region. Summer temperatures range from 28 °C -35 °C, but with very low humidity and frequent breezes. Spring and fall temperatures are extremely pleasant and mild. The winter sees night time temperatures frequently near 0 °C, and snow usually falls a few times each year. The yearly average number of days with rain is 5-45 and with snow it is 0-8. It typically will not rain from June to the beginning of September, with cloudless blue skies prevailing.

Environmental control

Solar avoidance. The goal is to keep direct solar gain out of the building. During the daylight hours completely avoid letting direct sun enter the building.

Avoid daytime ventilation. The daytime air in hot climates is dry and very hot. It is not wise to encourage it to enter the building.

Promote nighttime flushing with cool evening air. The evening air in a desert climate is substantially cooler than in the daytime. The promotion of ventilation in the cool evening hours will draw heat out of the thermal mass of the building and cool it down.

Achieve day lighting by reflectance and use of light non-heat absorbing colours. By the use of light colors the building will not attract or hold as much solar heat as dark colors absorb heat. Light colors will also help light to bounce around and achieve a good level of brightness without the use of direct beam radiation from the sun.

Implementation

The solar avoidance is achieved by the orientation of the building and the use of the building's skin as a shading device. The building's front façade is facing southeast. Solar penetration occurs only for few hours after dawn. The early morning solar radiation is not that intense as mid day. The solar rays that enter the building from northeast facade are diffused by the fenestration equipped with electro chromatic glass. The building's white concrete skin deflects and diffuses the direct solar radiation. It is the depth of the building skin that allows it to act both as a shading device and a light self. This means that it blocks the direct sun rays and at the same time it diffuses and redirects them to the interior. The diatoms of the skin and its thickness were designed in accordance with the solar path in order to achieve the desired lighting and shading conditions. The rest of the exterior surfaces like the caps of some diatoms of the diaphragm wall are also light colored to avoid the overheating via solar radiation .The interior finishes are light colored too to promote the light enhance to the interior.

The ventilation during the day is achieved by the use of earth tubes, long, underground metal or plastic pipes through which air is drawn. As air travels through the pipe it gives up some of its heat to the soil, and enters the building as cooler air. This happens because the natural ventilation in hot arid climates during the day would equalize the interior exterior aerial mass temperature. This would lead to thermal discomfort and unnecessary rise to the building mass temperature.

In Jordan, high temperatures, humidity and dusty hot winds lead to excessive discomfort for people. In order to relieve the residents from the harshness of the climatic elements we decide to create a microclimatic zone. A microclimate is a weather pattern that's localized in a small area and different in some significant way from the weather of nearby areas. The variation can be one of temperature, humidity, rainfall, wind, or any combination of these. An improved microclimate in

the vicinity of the building can improve the quality of life and comfort for its users. The changes we want to implement regard relative humidity, air temperature, wind flow and light exposure. To achieve that, we create a gap between the building and the site boundary. At the edge of the boundary a vertical structure is erected in order to facilitate a vertical garden and its embedded watering system. The plants on the wall not only absorb carbon dioxide and exhale oxygen but play an active role in the regulation of the microclimatic factors as they alter the temperature, the relative humidity, filter the air and diffuse the air flow. The gap between the building and the green wall lets the air drafts move within it and drives away the saturated air volume keeping the cooling rates at a high level and maintaining the comfortable conditions for the residents.

The active heating of the building is conducted by under floor heating system fed by geothermal heat pump (GHP) and a supplement boiler. GHP acts as a central heating or cooling system that pumps heat to or from the ground. It uses the earth as a heat source (in the winter) or a heat sink (in the summer). This design takes advantage of the moderate, stable temperatures in the ground to enhance efficiency and reduce the operational costs of heating and cooling systems. The top 3m of Earth's surface maintain a relatively constant temperature between 10°C and 16°C. The geothermal pump systems reach fairly high efficiencies (300%-600%) on the coldest of winter nights, compared to 175%-250% for air-source heat pumps on cool days. Ground source heat pumps (GSHPs) are among the most energy efficient technologies for providing HVAC and water heating. For the building in Amman a vertical system is selected mainly because of the limited space provided and the increased demand. Holes (approximately 10 cm in diameter) are drilled about 6 meters apart and 20 meters deep. Into these holes fit two pipes that are connected at the bottom with a U-bend to form a loop. The vertical loops are connected with horizontal pipe (i.e., manifold), placed in trenches, and connected to the heat pump in the building.

Passive cooling

In the summer, sprays can be used to achieve optimum natural cooling. The water is pumped to sprinklers along the top of the building and allowed to trickle down the sloping facade. The rate of evaporation is greatly enhanced in such a system because a much larger surface area is exposed to the night air. Roof sprays rely on a little external power to get the water to the roof and hence do not qualify as completely passive systems. But the total amount of energy consumed for pumping is very minimal compared to the energy saved by the added cooling rate attained. Excess water can be captured and reused. The same technique is used for the watering of the green wall opposite the south façade.

Gray water recycling

The building is equipped with gray water collection and purification system. Gray water is considered the wastewater generated from washing of clothes and dishes and bathing which can be recycled on-site for uses such as irrigation. Greywater composes 50–80% of residential wastewater generated from all of the building sanitation equipment (except toilets).The grey-water system is essential to be fed nothing containing toxic substances—no bleaches, artificial dyes, cleansers in order to avoid soil and irrigation contamination.

Conclusion

Designing in a hot and arid climatic zone can be very challenging especially when strict environmental and energy criteria are taken into consideration in the design process. The aim here is not only to create a functional comfortable building but also an icon for the city in terms of visual appearance, cultural connotations and environmental performance.

Vitus Bering Innovation Park,Horsens

维图斯Bering创业园

Arkitektfirmaet C. F. Møller

项目地点	丹麦，霍森斯
项目时间	2008年~2011年
项目面积	8000平方米
设计单位	Arkitektfirmaet C. F. Møller
客户	University College Vitus Bering Denmark
合作方	Pihl & Søn A/S
景观设计师	C. F. Møller Architects

Project Location	Horsens, Denmark
Project Milestores	2008-2011
Project Area	8,000 m^2
Design Company	Arkitektfirmaet C. F. Møller
Client	University College Vitus Bering Denmark
Collaborators	Pihl & Søn A/S
Landscape Architect	C. F. Møller Architects

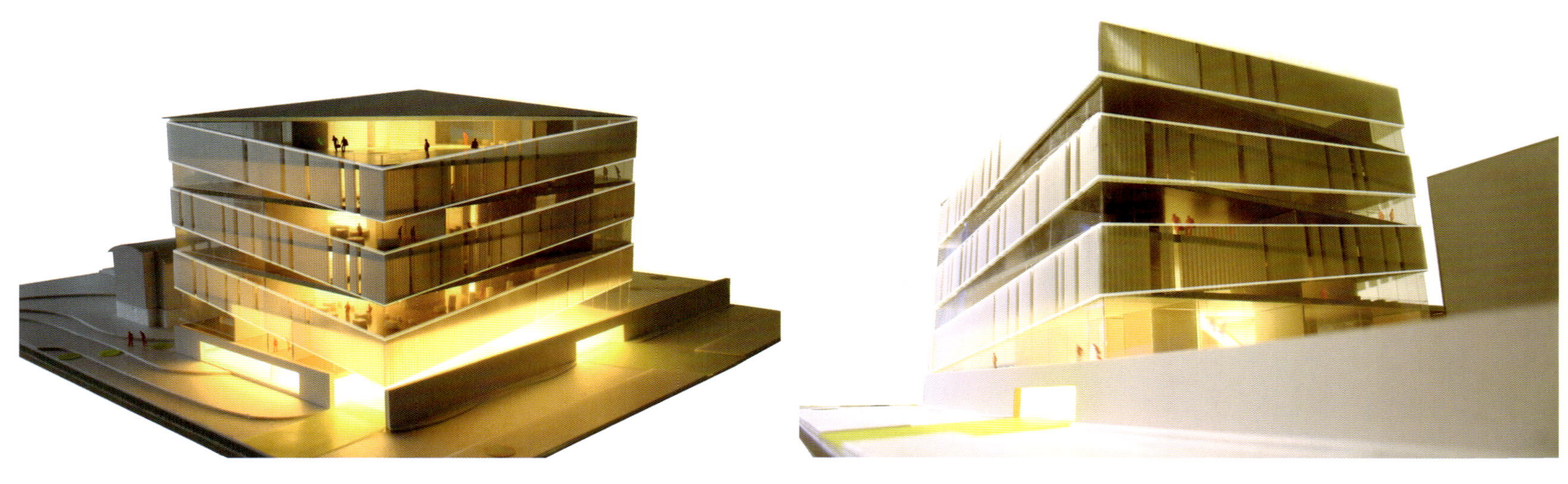

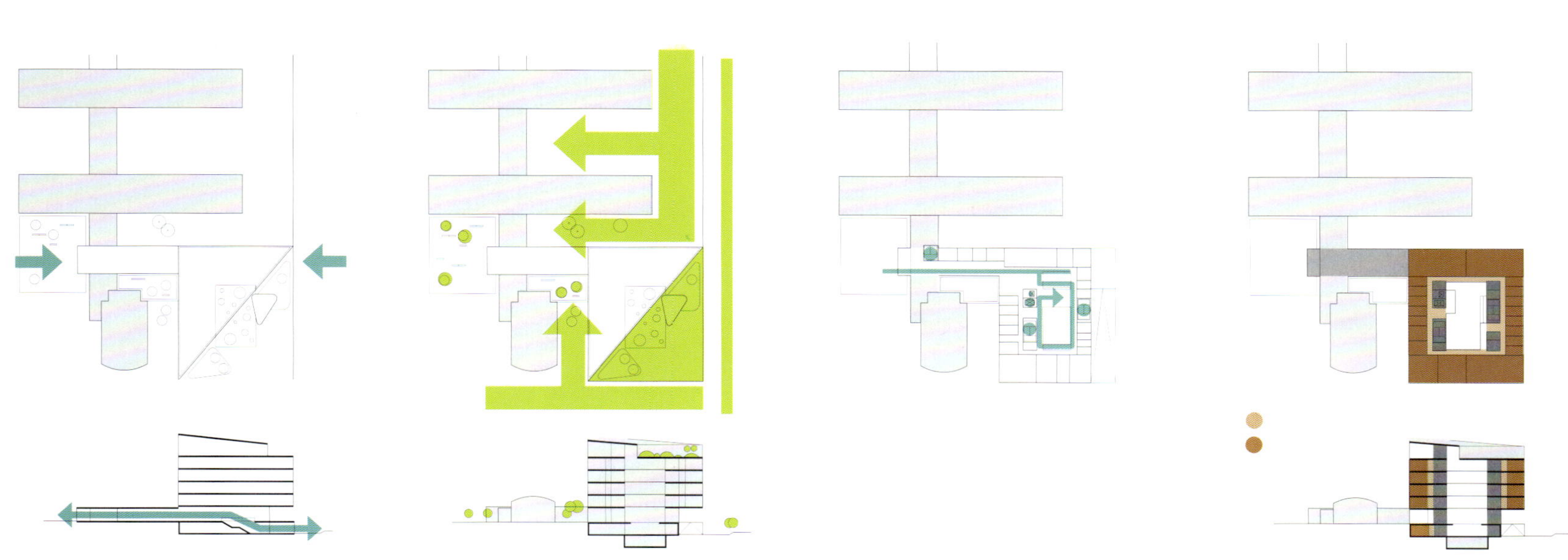

这是一栋教学与企业相长的办公大厦——是丹麦霍森斯维图斯白令学院对现存的建于1970年的一座独特建筑的扩建和翻修。新设计的建筑坐落于一砖体结构上，与现有的建筑直接相连，但与原建筑仍然有着鲜明的区分。

该建筑的多样性和创新性通过其螺旋形外型得到了很好的展现。在表面，玻璃带一直穿过六层大楼延伸向天空，形成螺旋状；在内部，由绿色纤维水泥制成的楼梯呈螺旋状在楼层与内部中庭之间通过。大楼倾斜的形式同时也很实用，可以形成一条贯穿全楼的必要的逃生通道。

该建筑的基本平面简单且灵活，并且实用性与适应性兼具。巨大的可移动绿色阶梯可以通往会议室以及一个有着Horsens Fjord美丽景色的屋顶平台。阶梯可以通往每一层的不同位置，以此作为建筑的中央枢纽激活了整个中庭。中庭由可移动的对角木条板屋顶面覆盖，它有着圆形天窗，并可形成屋顶阳台。

维图斯Bering创业园是丹麦第一个被评定为一级低碳的办公综合楼，低能耗是通过高度隔热窗和建筑表面的绝缘体等因素实现的。它的另一个特点则是大楼的智能空气调节系统可以根据每个房间的人数对温度进行调节。

维图斯Bering创业园是丹麦第一个被评定为一级低碳的办公综合楼。这就意味着它的能源效率是丹麦建筑规则所要求最低限度的2倍（这表示它与不超过50 kWh/平方米/年的能源消耗大致相等，包括照明、加热和通风等）。

比较小体积的建筑通过尽量缩小与内部空间和体积有关的表面来减少建筑物的主要热量损失。通过周围场地的照明来照亮建筑物的内部空间，减少电光源补充的需求量。而且只有该空间的顶楼有一个可以阻止热量损失的外表。

能源的低消耗是通过高度隔热窗和幕墙外表面（利用低能的三层玻璃和涂层玻璃面板）以及建筑其他表面的绝缘体等因素实现的。另一个特点是建筑物智能空调系统能够根据当前每个房间人的数量进行自动调控，而这主要是通过人们释放出的热量和二氧化碳实现调控的。

该建筑的内部照明由阳光和运动感应器控制，避免了不必要的能源消耗。所有照明装置都是经过特别的低能设计，来满足高质量的灯光和每平方米最大的能源消耗的要求。

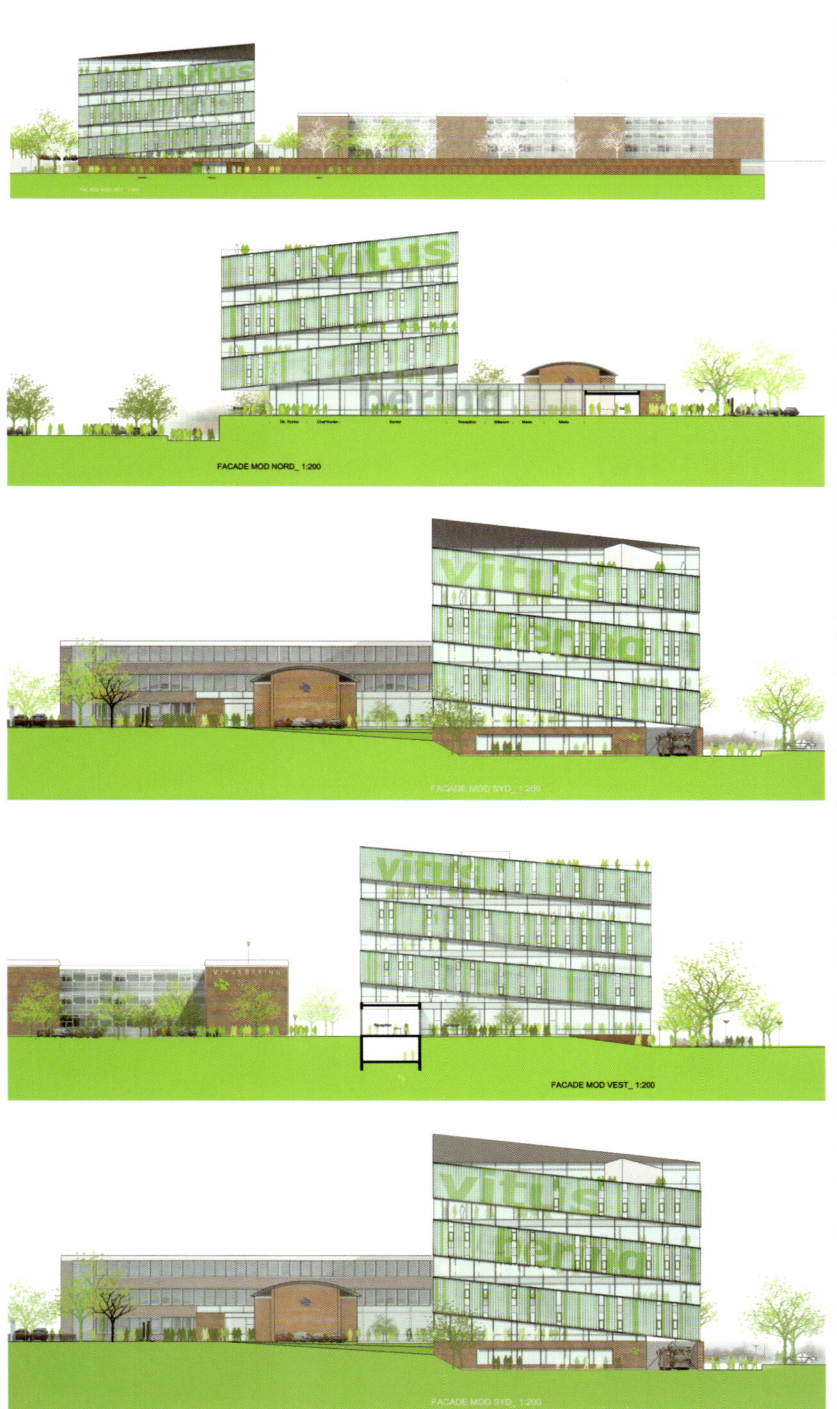
vitus
FACADE MOD NORD_ 1:200
FACADE MOD VEST_ 1:200

HEDE NIELSENS FABR

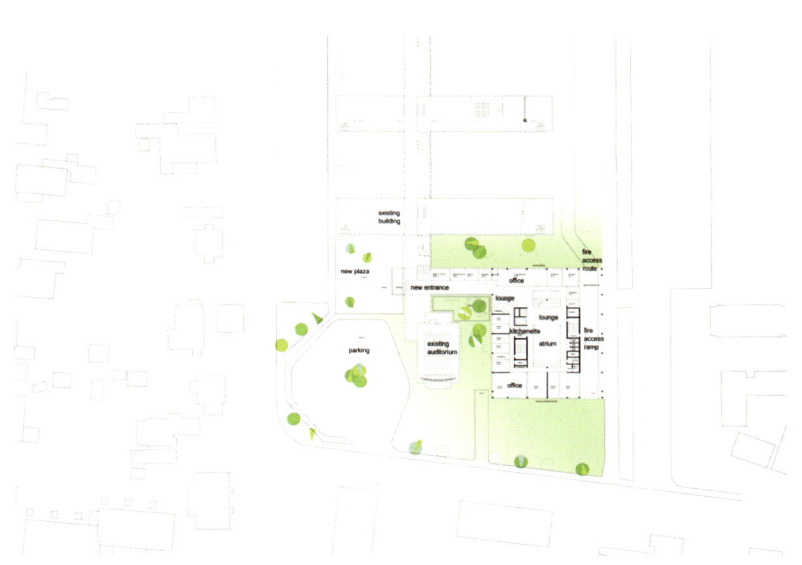
existing building
new plaza
new entrance
fire access route
office
lounge
lounge
kitchenette
atrium
fire access ramp
parking
existing auditorium
office

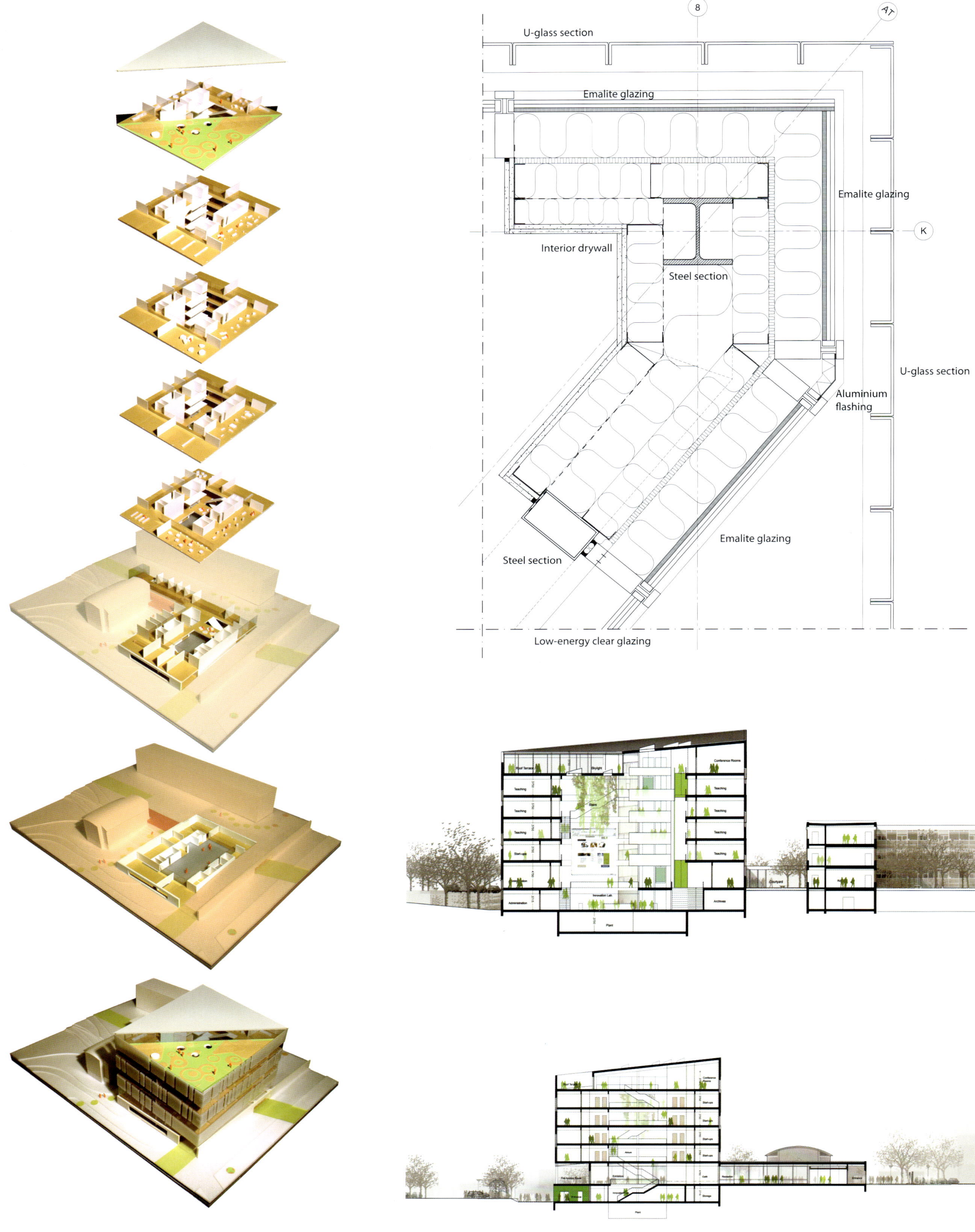
8
AT
U-glass section
Emalite glazing
Emalite glazing
K
Interior drywall
Steel section
U-glass section
Aluminium
flashing
Emalite glazing
Steel section
Low-energy clear glazing

Teaching and entrepreneur start-up office facilities side by side — that's the philosophy behind the distinctive extension to the existing 1970's structure of the University College Vitus Bering Denmark in Horsens. The new innovation building is designed to sit on a brick base, which is a direct continuation of the existing complex architecture, but from there on it is distinctly different and unique.

The building's dynamic and innovative character is expressed via its spiral shape. On the facades, the movement is seen in the glazing strips that stretch towards the sky across the six storeys of the building and create the impression of a spiral sequence, while intern ally it is expressed via the main staircase in green fibre cement, which runs in a spiral form between the storeys in the unifying internal atrium. The inclined forms of the building also have the practical advantage of allowing a necessary fire escape route to be cut through the building.

The basic floor plan of the building is a simple and flexible layout, to allow the integration of numerous uses and adaptations. The large and dynamic green stairway element leads to common meeting facilities and a roof terrace with a beautiful view of the Horsens Fjord. The stairs land in a different position on each level, thus activating the entire atrium as the central hub of the building. The atrium is covered by a dynamic, diagonally split roof-plane with circular skylights, of which one half forms the common roof terrace.

The Vitus Bering Innovation Park is one of the first office complexes in Denmark to be classified as low-energy class. The low level of energy consumption is achieved through such factors as highly insulating windows and extra insulation on all of the building's external surfaces. Another feature is the building's intelligent air conditioning system, which adjusts itself according to the number of people present in each individual room.

The Vitus Bering Innovation Park is one of the first office complexes in Denmark to be classified as low-energy class, which means that its energy efficiency is twice that of the minimum required by the Danish building regulations (which means it is roughly equivalent to an energy consumption of no more than 50 kWh/㎡/year including lighting, heating, ventilation, etc.).

The compact cubical volume of the building helps achieve this by minimizing the exterior surface in relation to the interior areas and volume, thus reducing primary heat loss from the building. The internal atrium is largely lit by means of the transparency of the surrounding spaces, reducing the need for supplementary electrical light, and only the top floor of the atrium has an external facade which keeps heat loss from the atrium minimal.

The low level of energy consumption is furthermore achieved through such factors as highly insulating windows and curtain wall facades (using low-energy triple-glazed

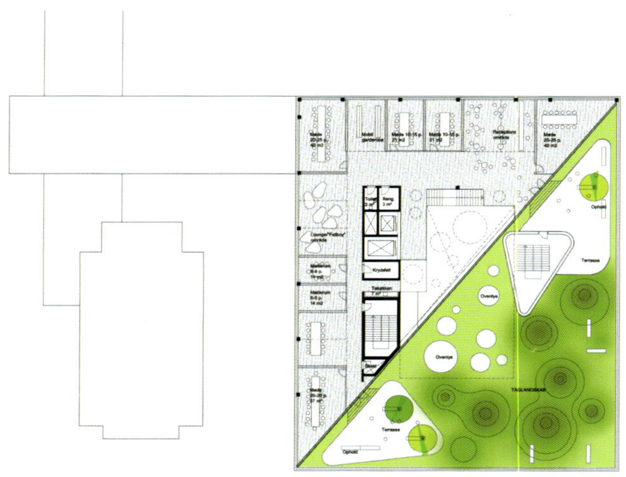

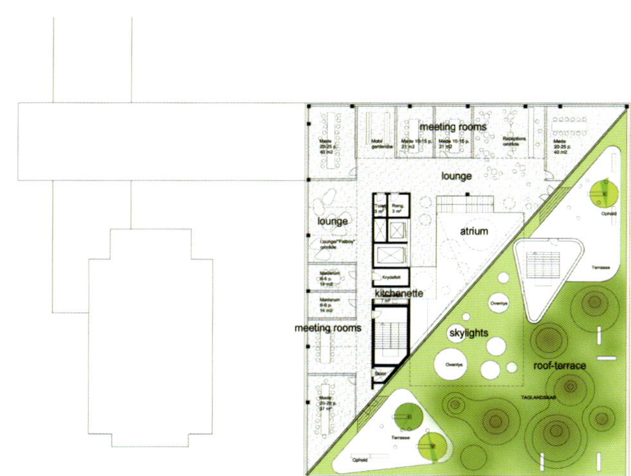

and coated glazing panels) and extra insulation on all of the building's other external surfaces. Another feature is the building's intelligent air conditioning system, which adjusts itself according to the number of people present in each individual room by reacting to heat and CO_2 levels.

Internal lighting is controlled by daylight and motion sensors, avoiding unnecessary energy consumption. All lighting fixtures are specially selected low-energy designs, to fulfill the requirements for internal lighting levels and quality as well as the overall maximum of energy consumption per square meter.

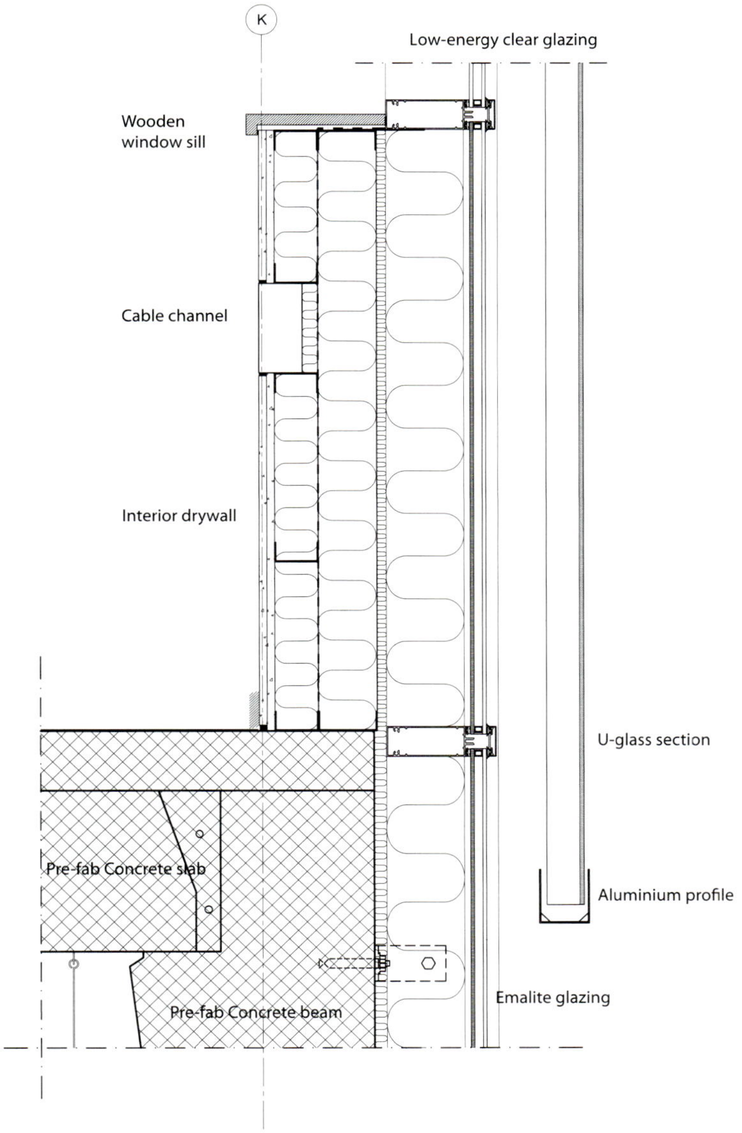

K

Low-energy clear glazing

Wooden window sill

Cable channel

Interior drywall

U-glass section

Pre-fab Concrete slab

Aluminium profile

Emalite glazing

Pre-fab Concrete beam

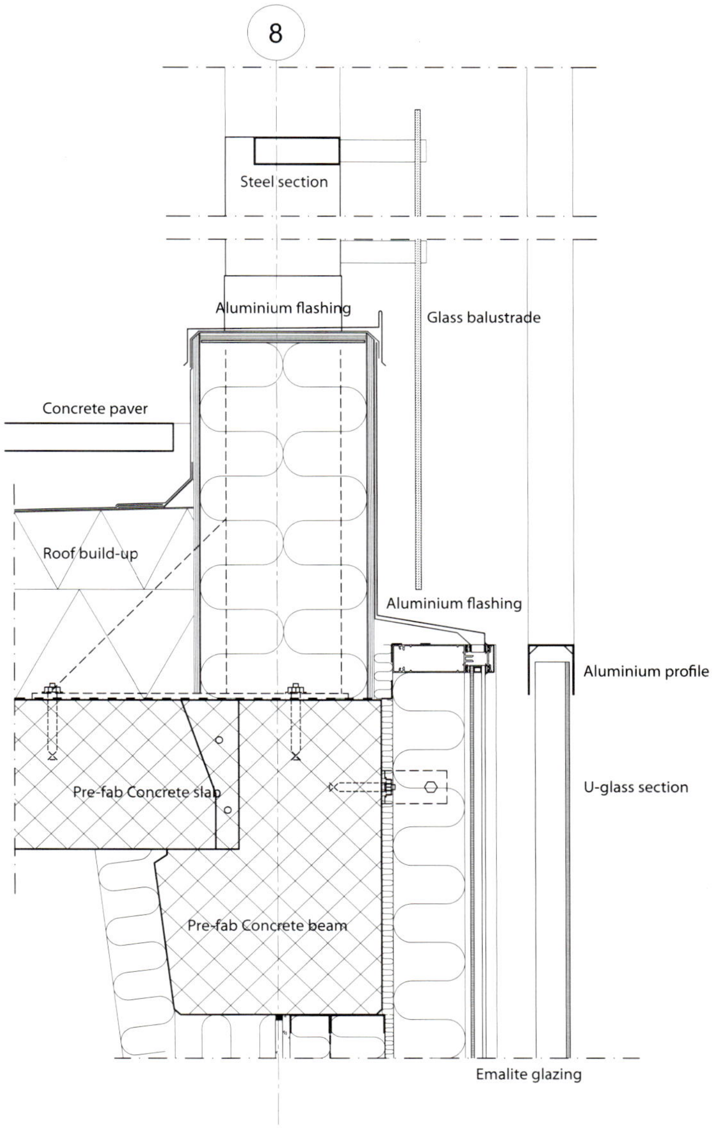

8

Steel section

Aluminium flashing

Glass balustrade

Concrete paver

Roof build-up

Aluminium flashing

Aluminium profile

Pre-fab Concrete slab

U-glass section

Pre-fab Concrete beam

Emalite glazing

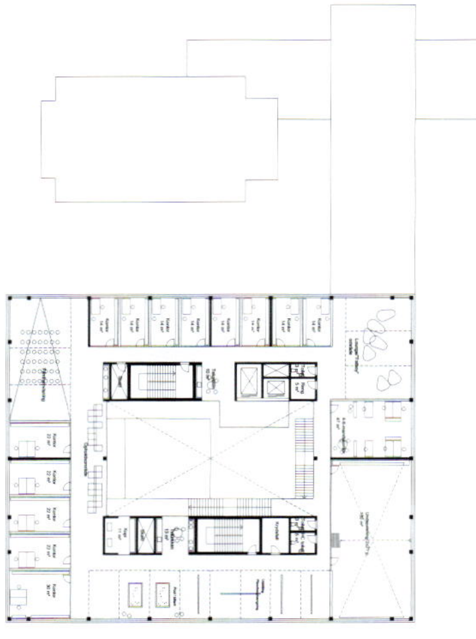

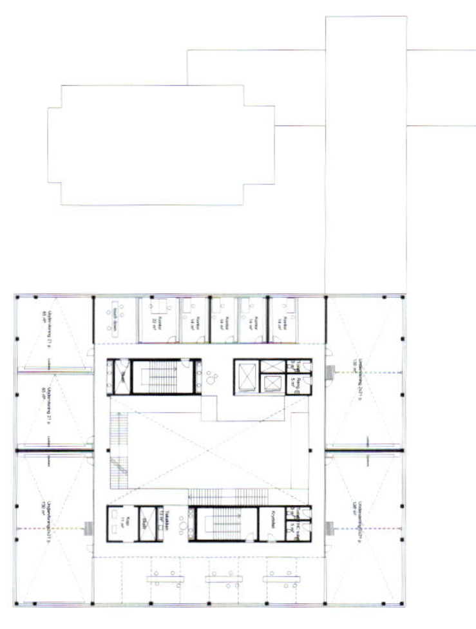

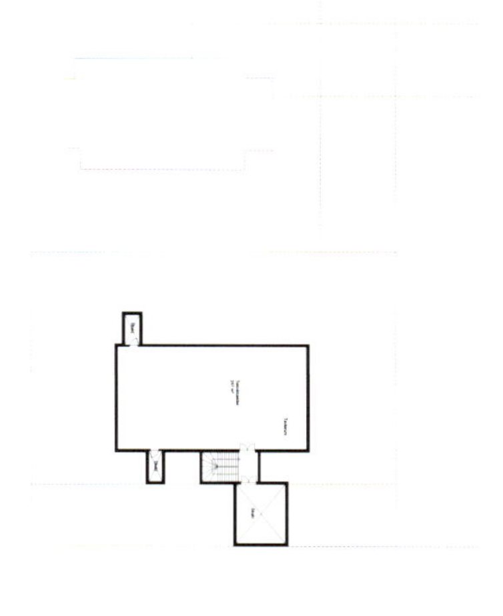

TEK3 TaiPei

TEK3 台北创意产业园

Bjarke Ingels

台北 TEK3：科技、娱乐及知识中心——又称作台北TEK3——是集中了各类当代科技和媒体相关活动的人口稠密的城市街区。

聚焦点：地处市民大道新生高架道路中间，坐落于以台北市站和新机场快线为中心的公共发展轴，台北TEK3坐拥台湾都市图景的地标性位置。

螺旋线：为充分利用场地并最大化建筑量体，建筑内部的公共街道以螺旋的方式盘绕上升，从一楼贯穿至屋顶花园。

立方体 = TEK3：作为市民公共生活的渗透，螺旋上升的媒体功能，被包含进一个57米x57米x57米的立方体中。

115步：立方体由混凝土薄板层构建而成，薄板层既作为遮阳设施也作为公共入口。该芯片层层后退以形成宽敞的公共楼梯供市民从正立面进入，一路走上屋顶。

街道边界：面向新生路与市民大道的交会点以及步行街，我们建议提升薄板层，从而创造开放的店面以及进入旅馆的大堂与旗舰店的开放入口。

四广场：建筑坐落于中间，稍微旋转为四个广场腾挪出空间。

功能：建筑由两个主要功能体组成，一部分是L型的旅馆朝向西面与北面，而另一部分是由方形基本平面叠加的用于商业零售和展示的，并围绕着中间以供信息发布和讲座的会议展演厅。

效率：方形的平面和连续的各层平面使得一个高效灵活的一般性布局成为可能，从而可以应对各种变化。

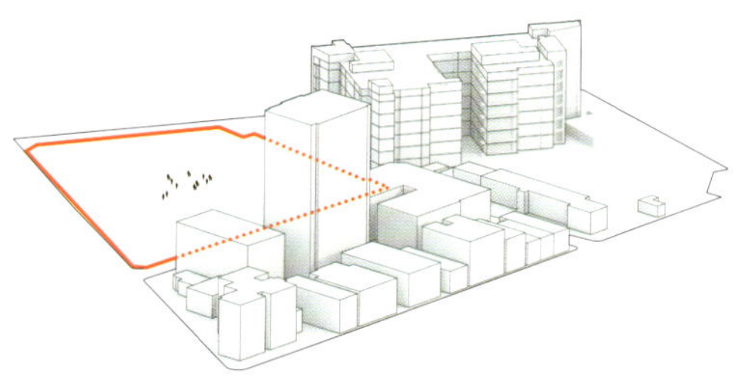

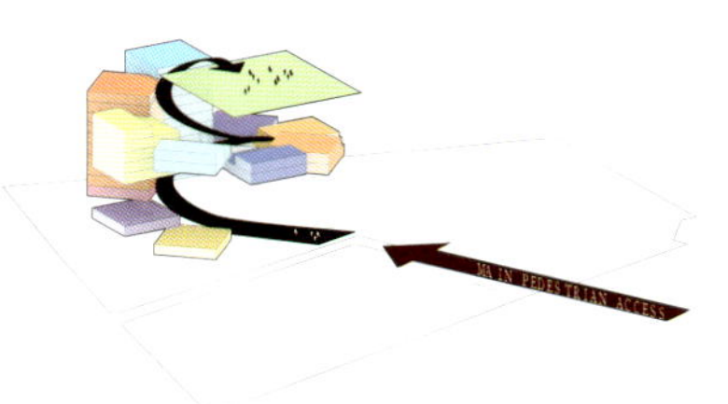
MAIN PEDESTRIAN ACCESS

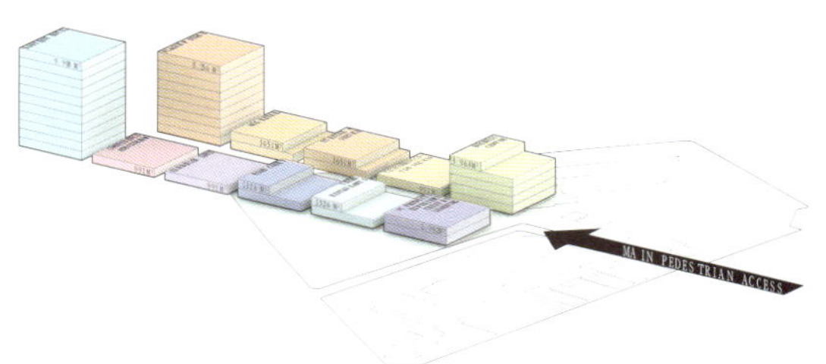
MAIN PEDESTRIAN ACCESS

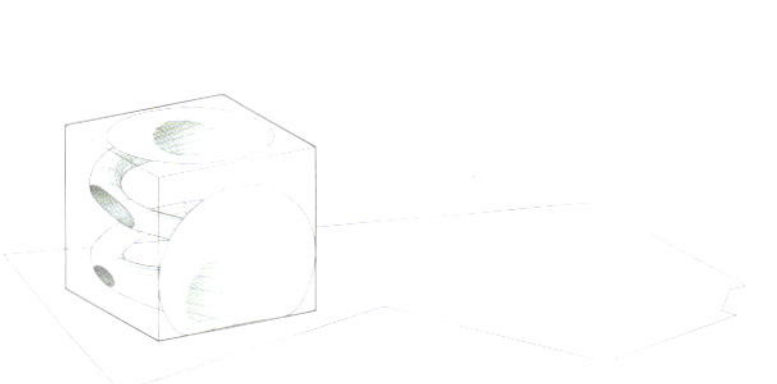

具有日光感应系统的功能如酒店客房及办公室分布在建筑四周，而零售和展览空间则位于核心筒部分。

流线：流线分为三个互补的系统。主要的是从地面至屋顶连续贯穿的核心螺旋走廊。其次为一系列连接各个楼层的于螺旋走廊下部连续上升的自动扶梯。因此两条流线在室内与室外平行进行，外部流线即为螺旋走廊，而作为内部流线的自动扶梯，最大化地利用了螺旋走廊的底部空间，特别适合在酷热的天气使用。最后，四个楼梯和电梯竖井位于建筑的四角，提供了一种严谨而直观的垂直交通方式。

可持续设计

热质：通过增设一个位于建筑底部的热质（水箱或混凝土），百叶的被动式散热可以提高效率。

用于自然冷却的屋顶树林：屋顶树林的树木形成了屋顶上的冷却罩，通过利用自然遮阴和蒸发散热创建出一个低于本地室外温度几摄氏度的环境。相对凉爽的空气将会自然下降并通过螺旋楼梯排出，从而营造出一个有微风的舒适的公共空间。

旅馆流线：建筑体西北角楼电梯竖井将服务于旅馆流线，通过这组楼电梯可以到达所有的旅馆客房。为满足疏散要求，两个相邻的楼电梯竖井以及螺旋走廊将补充主要通道。

TEK3 TaiPei: The Technology Entertainment Knowledge Center—aka TEK3 Taipei—is a dense urban block of all kinds of activities related to contemporary technology and media.

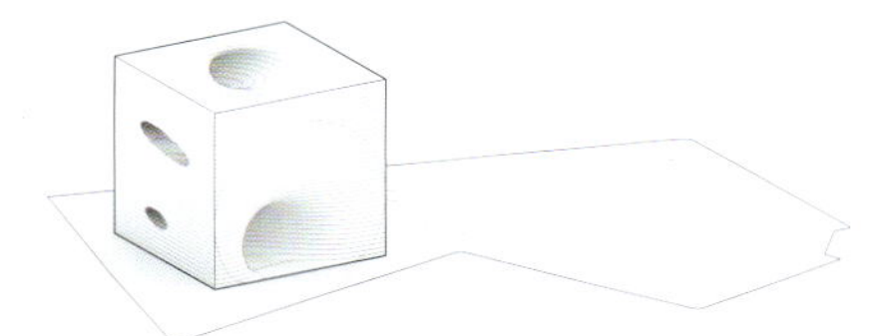

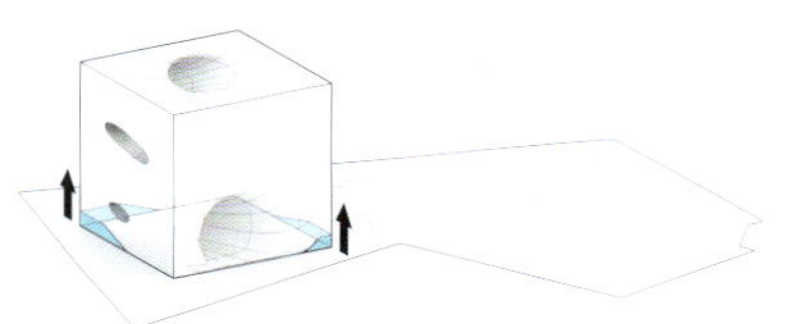

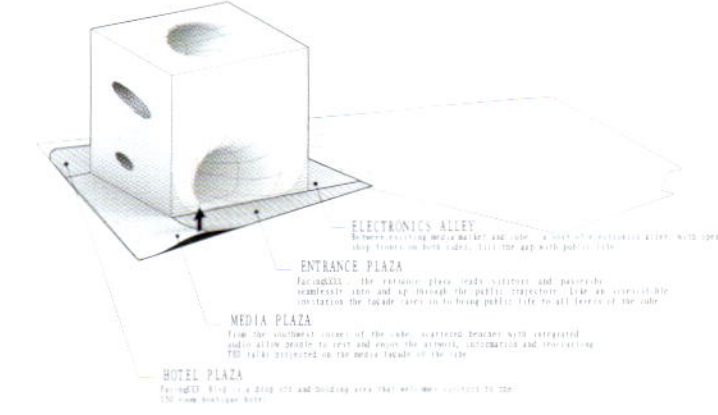

Focal point : Located at the the intersection between Shimin avenue and Xinsheng elevated way and the axis of public developments with the Taipei station and the New Airport Express at its center, TEK3 Taipei holds a landmark position in the urban image.

The coil: To remain within the site and the maximum building volume, the public street is coiled up in an ascending spiral leading from the ground floor to the roof garden.

The cube= TEK3 : The spiraling street of media programs is consolidated into a $57x57x57m^3$ cube of program permeated by a public trajectory of people life.

115 steps : The cube is finished in concrete lamellas serving as solar shading as well as public access. The lamellas recede inwards forming a generous public staircase allowing the public to walk into the façade and all the way to the roof.

Street edge: Towards the Xinsheng Road and Shimin Blvd intersection and the pedestrian street we propose to lift the lamellas to create open storefronts and uninhibited access to the hotel lobby and the ground floor flagship stores.

Four plazas: The cube is positioned in the middle of the site, slightly rotated to make space for 4 dedicated plazas.

Progam : The cube is composed of two main organizational diagrams. An L-shaped hotel with two wings faces west and north, and a square stack of generic floors of retail and showrooms is sandwiched around a central auditorium for launches and lectures.

Efficiency : The square plan and deep continuous floor plans allow for a very efficient, flexible generic layout capable of accommodating countless conversions. The daylight sensitive programs such as hotel rooms and offices reside on the perimeter, while retail and exhibitions occupy the core.

Circulation : The circulation is split in three complementary systems. Primary is the central trajectory that provides seamless continuity from the ground to the roof and back again. Second is a chain link of escalators underneath the trajectory that connects every single floor internally. Circulation happens in parallel inside and outside the building. Exterior circulation is created by the spiral trajectory while an interior path of escalators optimizes the underside of the spiral. Internal circulation is especially provided for times of inclement weather. Finally 4 stairs and elevator cores,

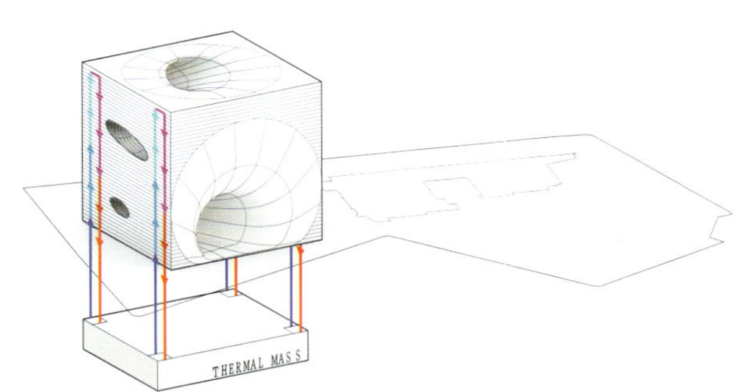

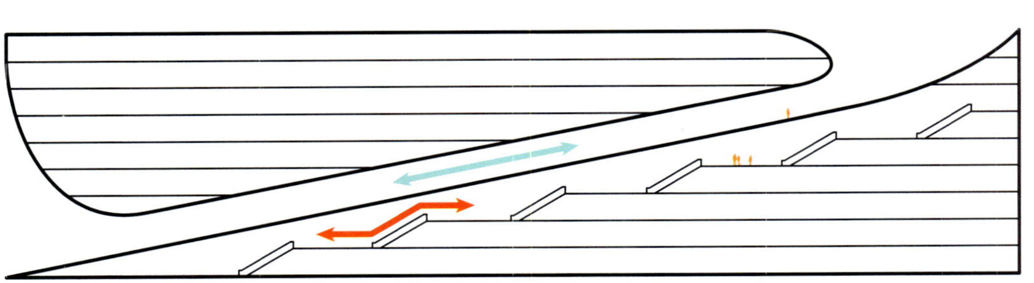

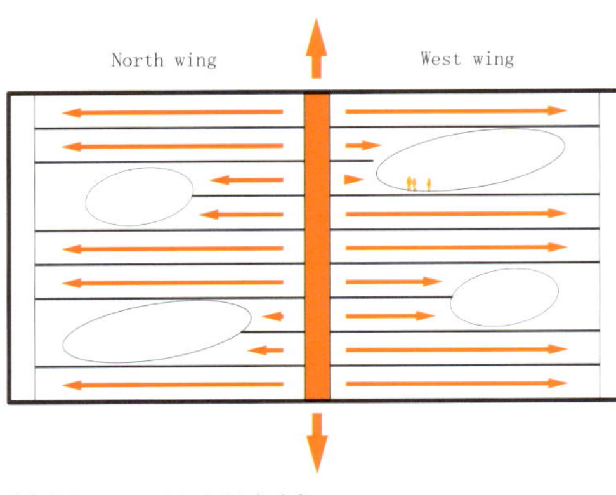

平行流線

因此兩條流線在室內與室外平行進行，外部流線即爲螺旋走廊，而作爲內部流線的自動扶梯，最大化的利用了螺旋走廊的底部空間，特別適合在酷熱的天氣使用。

PARALLEL PATHS : Circulation happens in parallel inside and outside the building. Exterior circulation is created by the spiral trajectory while an interior path of escalators optimizes the underside of the spiral. Internal circulation is especially provided for times of inclement weather.

旅館 -- 旅館流線

建築體西北角樓電梯豎井將服務于旅館流線，通過這組樓電梯可以到達所有的旅館客房。爲滿足疏散要求，兩個相鄰的樓電梯豎井以及螺旋走廊將補充主要通道。

HOTEL CIRCULATION : The northwest core of the cube is dedicated to hotel circulation. From this central elevat core all hotel rooms are reached. For escape purposes the two adjacent cores as well as the trajectory supplements the main access core.

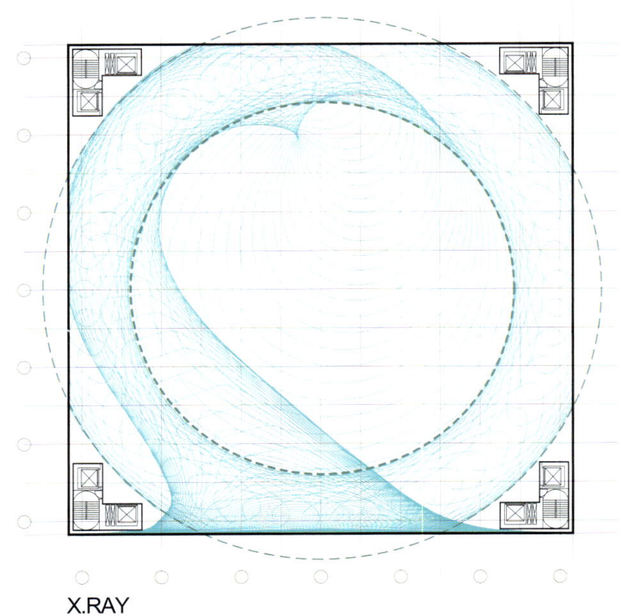

X.RAY

A perfect circular spiral inscribed within a cube

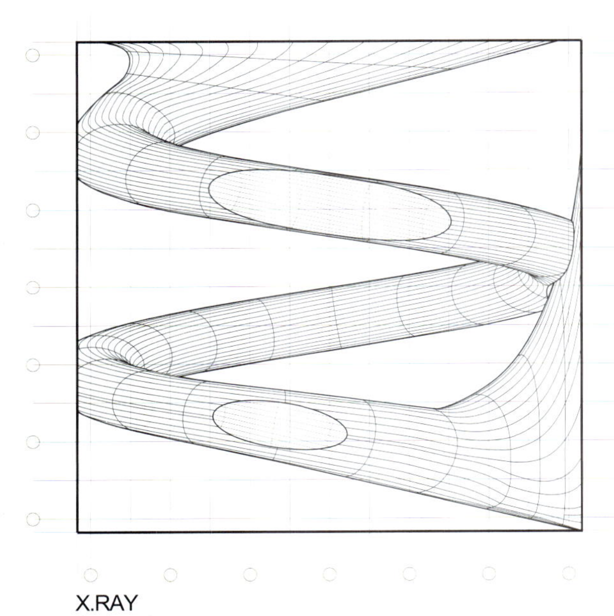

X.RAY

Consistent slope

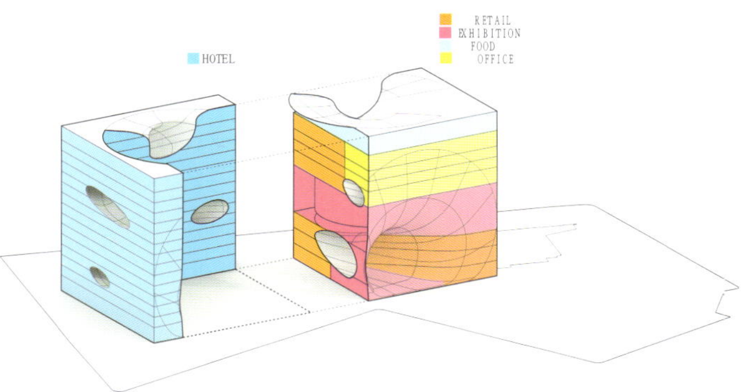

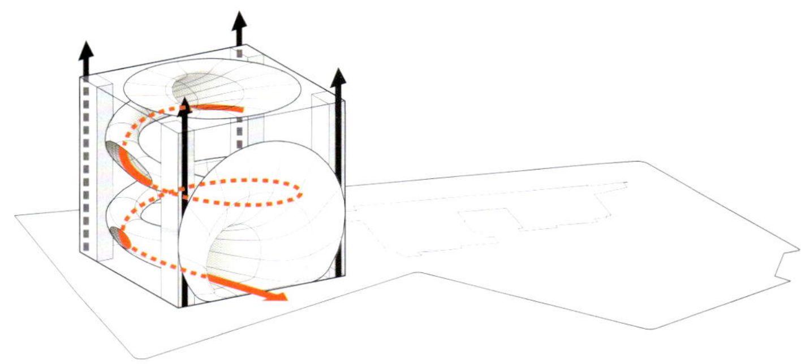

one in each corner, constitute a rigorous and intuitive means of vertical movement.

Sustainability:

Thermal mass : Passive cooling from the louvers can be enhanced by the addition of a thermal mass (water tank or concrete) underneath the building.

Sustainability:

Roof grove for natural cooling : A roof grove of trees form a cooling canopy on the roof exploiting natural shade and evaporative cooling to create a local drop of exterior temperature by a couple of degrees. The relatively cooler air will naturally drop down and drain through the trajectory creating a comfortable breeze throughout the public space.

Hotel circulation: The northwest core of the cube is dedicated to hotel circulation. From this central elevator core all hotel rooms are reached. For escape purposes the two adjacent cores as well as the trajectory supplement the main access core.

TWELVE|WEST

TWELVE|WEST综合大楼

Zimmer Gunsul Frasca (ZGF) Architects

项目地点	美国，俄勒冈州，波特兰市
项目面积	23000平方米
设计单位	Zimmer Gunsul Frasca (ZGF) Architects, LLP
建筑类型	餐厅、商业办公、零售空间、综合住宅楼
竣工时间	2009年7月

Project Location	Portland, Oregon, U.S.A.
Project Area	23,000 m^2
Design Company	Zimmer Gunsul Frasca(ZGF) Architects, LLP
Building Type	Restaurant, Commercial office, Retail, Multi-unit residential
Completion Date	July 2009

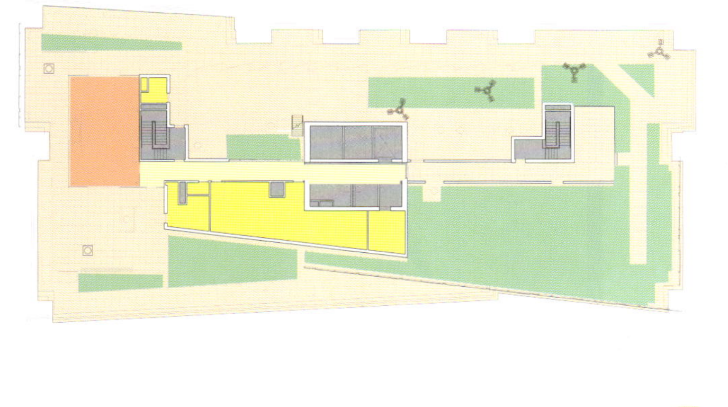

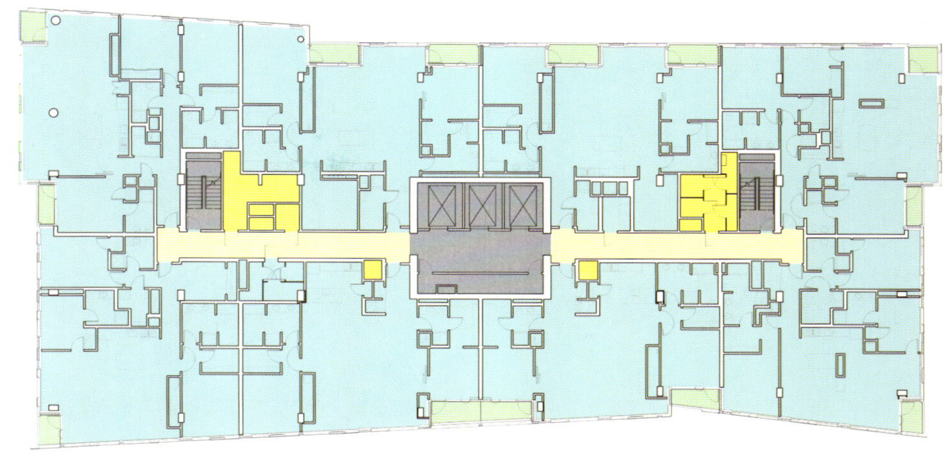

项目

在美国俄勒冈州的波特兰市，升起了一座23层高的建筑，本项目是一座综合性大楼，作为先进的可持续设计策略的实验性建筑，其设计有望达到LEED白金认证。整栋大楼拥有沿街的零售空间，其中4层楼作为ZGF设计事务所的办公空间，17层为公寓，还有5层作为地下停车场。大楼的屋顶是生态屋顶，并有屋顶式花园以及露台空间，大楼内还有健身会所以及一个剧院。四组风力发电机高高地位于大楼顶部，它是美国高层建筑上的第一组风力发电设备。Twelve | West 不仅仅是在城市街区迅速转型时期的前瞻性建筑，更是未来可持续建筑设计的典范。

场地

Twelve | West是首个在美国西部社区扎根的具有历史意义的全新建筑，它位于波特兰市中心，大约占据了SW十二和十三大道以及华盛顿大街二分之一的街区。West End是一个相对成熟的市区，它常常被人描述为“前卫的”。在这里你可以看到融合了不同文化和社会背景、不同体制的高密度的住宅楼、多样化的零售店、餐厅以及夜店。

本项目位于波特兰著名的街区Pearl南部——这是一个融合了居住、工作和娱乐等各种功能的多样化再造街区，Twelve | West将建造地址选在这里，通过贯穿Burnside街道的四车道将Pearl街区与West End街区区分开来，达到加速West End附近街区转型的目的。其目标是在位于城市东南部的中心商业区及位于北部和西部的West End综合性社区间建立一种商业的以及近在咫尺的联系。

大楼设计

设计团队建大楼时设定了一系列目标：

• 创造一种既可以为波特兰的城市天际线增色又可以巧妙地将生活、工作、学习和娱乐融合起来的建筑结构。

• 建一栋透明大楼——引人注目且有活力的——通过充分利用自然光线将城市景观与大楼完美地结合起来。

• 采用接地建筑的方式使大楼可以更好地提升街道生活。

• 建造用于交流互动与休闲的室外花园和露台。

• 利用先进且共生可持续的大楼系统来促进对自然资源的保护。

• 设计使空间、光线、视角以及节能设施最大化的居住空间。

这些目标具体体现在透明精确的幕墙上，它为大楼营造了一种不同的结构式美感。为了增加差异感，大楼南部的外立面被设计成倾斜的。设计师还做了许多精妙的结构上的修饰，包括活动窗、阳台、有着不同颜色变换的脊状凹槽的不锈钢，并利用半透明的可反射熔块隔板营造出动感活跃的封闭空间。当阳光及明亮的天空反射到大楼外立面时，其效果更加突出。在冬天，这种反射效果尤为动人，西北方向深灰色雨濛濛的天空在大楼的幕墙玻璃上神奇地呈现出漂亮的蓝色。

大楼底座也鼓励流动性和社区生活。以大型的木质窗户为特点的双层零售空间与大楼的商业和住宅大厅共享空间。为了更进一步加强与街道的联系，零售店安装了巨大的透明玻璃。大楼的南边是欧式的咖啡馆和餐厅，玻璃顶棚则可以为户外用餐的客人以及街边的行人遮风挡雨。

街边树木点缀着大楼底层，创造出另一种样式与空间，并与街边停车场附近的铺设方式相呼应。大楼内分别有两个入口与公共大厅供办公人员与居民出入，开辟出额外的交流与会面空间。

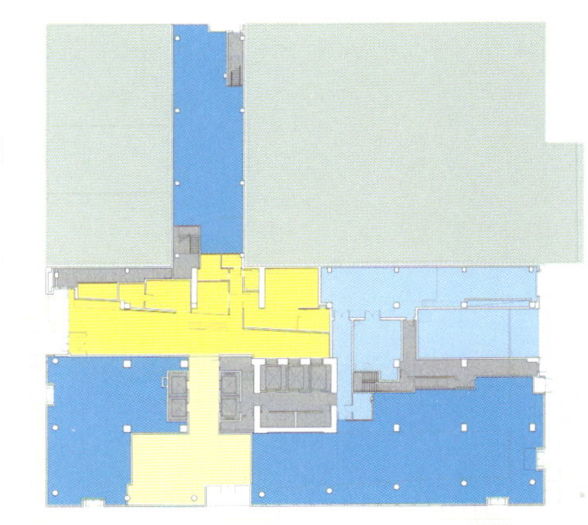

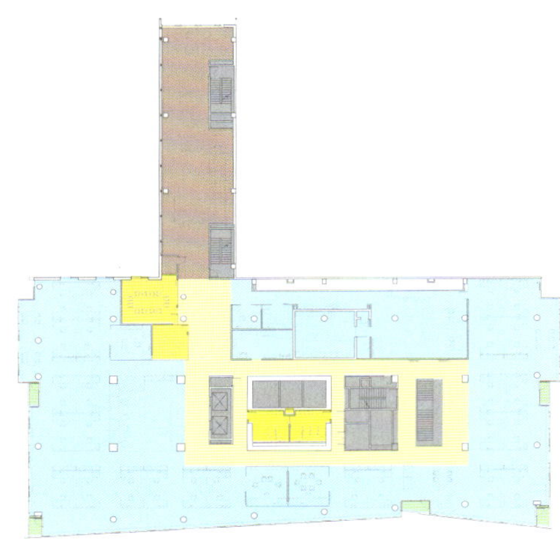

Program

Rising 23 stories above Portland, Oregon's evolving West End neighborhood, Twelve | West is a mixed-use building designed to meet two LEED Platinum Certifications and serve as a laboratory for cutting-edge sustainable design strategies. It features street-level retail space, 4 floors of office space for ZGF Architects LLP, 17 floors of apartments and 5 levels of below-grade parking. The building has an eco-roof, rooftop garden and terrace space, complete fitness studio and a theatre. Four wind turbines sit prominently atop the building representing the first U.S. installation of a wind turbine array on an urban high-rise. Twelve | West serves as not only an anchor in a rapidly transforming urban neighborhood, but also as a demonstration project to inform future sustainable building design.

Site

Occupying roughly a one-half block site between SW 12th and 13th Avenues and Washington Street in downtown Portland, Twelve | West is one of the first significant new buildings to take root in the West End neighborhood. The West End is a mature urban district that is often described as "edgy." It is home to a mix of cultural and social, institutional, high density residential buildings, diverse retail, restaurants, and nightclubs.

Located immediately south of Portland's well-known Pearl District—an acclaimed urban renewal district that is considered a model of live, work and play mixed-use development—Twelve | West's site is selected in part to serve as a catalyst for additional transformation in the West End neighborhood by drawing new development across the four-lane Burnside Street dividing the Pearl District from the West End. The goal is to create a significant retail and pedestrian connection to the Central Business District in the southeast and the mixed-use neighborhoods to the north and west of the West End.

Building design

The design team set a number of goals to inform the architecture of the building including:

- Creating a structure that enriches the Portland skyline and carefully unites the live/work/learn/play components of the building.
- Constructing a transparent building—inviting and active—that connects the building's inhabitants to the urban landscape, while taking advantage of natural light.
- Grounding the building in a manner that promotes active street life.
- Accommodating exterior gardens and terraces for areas of

ZGF

interaction as well as respite.

- Integrating advanced and symbiotic sustainable building systems to promote natural resource conservation.
- Designing homes for lease that maximize space, light, views and include luxurious and energy saving amenities.

These goals are embodied by a transparent, carefully articulated curtain wall that gives the building a thin and planar structural aesthetic. To add variation, the building's southern facade is slightly angled. Subtle textural modifications include operable windows, balconies, quilted and recessed stainless steel of varying color, and fritted reflective semi-opaque panels which create a lively enclosure with a sense of movement. This quality is further enhanced when sunlight and the sky reflect off the building's facade. This reflection is particularly appealing in winter, when the reflection of the Northwest's dull grey, rainy sky appears blue in the curtain wall's glass.

The base of the building also encourages movement and urban life. There, double-storyed retail spaces, featuring large, wood windows share space with the building's commercial and residential lobbies. The retail spaces have large, transparent facades in order to make a strong connection to the street. European-style café dining is anticipated along the building's south side; glass canopies provide weather protection for outdoor diners and pedestrians.

Street trees punctuate the base of the building, creating yet another layer of pattern and scale in concert with paving patterns adjacent to on-street parking. Two separate entrances and public lobbies for both the office and residential units have setbacks from the sidewalk, creating additional gathering and meeting places.

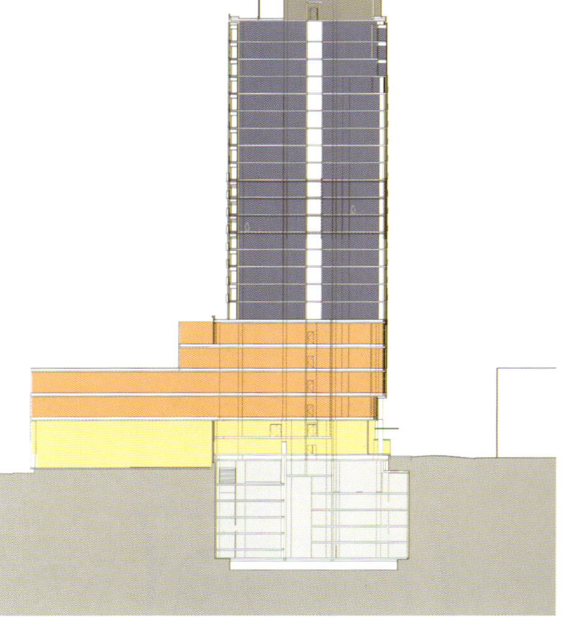

CAPITAL GATE

CAPITAL GATE大厦

RMJM 建筑事务所

项目地点	阿拉伯联合酋长国，阿布扎比市	Project Location	Abu Dhabi, UAE
总楼面面积	53 100平方米	Total Built-up Area	53,100 ㎡
建筑倾斜度	18度	Inclination	18°
建筑高度	160米	Building Height	160 m + helipad (4.5 m)
楼层	35	Floors / Storeys	35
建筑类型	综合用途(商业办公楼、住宅及五星级酒店)	Building Type	Mixed-use (Offices , Residence and Hotel)
建筑设计及顾问	RMJM建筑事务所	Architects & Main Consultant	RMJM

由国际建筑事务所RMJM设计、ADNEC（Abu Dhabi National Exhibitions Company）建造及发展的CAPITAL GATE，位于阿联酋首都阿布扎比，被列入吉尼斯世界纪录，成为全球最倾斜的人工大楼。

CAPITAL GATE向西倾斜18度，是比萨塔倾斜度的4倍。健力士世界纪录委员会于今年1月进行评估，当时大楼的玻璃幕墙部分已建至35层，楼高达160米。

对于CAPITAL GATE破纪录的创举，ADNEC主席Sheikh Sultan Bin Tahnoon Al Nahyan表示：CAPITAL GATE是阿布扎比的地标建筑，而吉尼斯世界纪录的肯定将使其成为全球最伟大的建筑之一，象征着阿联酋地区的未来。

CAPITAL GATE的首12楼是垂直于地基之上，而12楼以上的楼层逐层向西面超出300毫米到1 400毫米，形成倾斜的塔楼。

大楼的其他创新特色包括世界首次采用预拱度核（Pre-cambered core）的建筑技术，该技术使用了1.5万多立方米的钢筋混凝土，当中更添加了1万吨钢，令核心变得更坚固。建筑师在设计大楼时，核心刻意偏离中心位置，当大楼向上兴建灌注压缩混凝土时，渐渐增加了地台的重量，使核心移回垂直位置。

设计与兴建CAPITAL GATE的过程充满挑战性。但有赖经过精心组织及紧密安排的伙伴合作，亦多得各方人士在参与这个项目时作出的重大承诺与付出的努力，我们在过程中遇到的各种问题都得以迎刃而解。ADNEC及我们的伙伴都很享受参与创造这个建筑美感与功能兼备的标志性项目的机会。ADNEC主席Sheikh Sultan Bin Tahnoon Al Nahyan补充道。

整个CAPITAL GATE项目预计将于今年年底完成。届时会容纳即将进驻的五星级酒店——凯悦CAPITAL GATE（Hyatt Capital Gate）以及占地2万平方米的办公大楼。

凯悦国际集团西南亚总监Peter Fulton先生称：对于进驻CAPITAL GATE我们感到十分骄傲。集团很幸运可以在中东地区最出色的建筑项目中发展酒店。我们希望通过与ADNEC之间的合作，配合这座标志性建筑，为旅客提供独特和地道的酒店住宿体验，成为旅客来到阿布扎比的必然之选。

CAPITAL GATE由国际知名建筑事务所RMJM设计，位于阿布扎比首都中心，毗邻ADNEC（Abu Dhabi National

Exhibitions Company)。RMJM中东地区总裁Nick Haston先生表示：这项健力士世界纪录不但肯定了RMJM在中东地区40年来设计世界级建筑的经验，更奠定了CAPITAL GATE及阿布扎比于世界建筑发展及尖端建筑学上的重要地位。RMJM很高兴可以为这个成就导航。

CAPITAL GATE除了抗地心吸力的倾斜和它的预拱度核（Pre-cambered core）之外，以下是其他令其成为全球伟大建筑项目的元素：

每个单位都是独一无二的：玻璃幕墙和室内角度户户不同。

CAPITAL GATE的外壳采用斜肋构架（Diagrid）设计，有如强壮的骨骼般支撑整座大楼，它可以承受地台所有重量，更令大楼毋需使用柱子及横梁。曾采用同样设计的世界名厦包括：纽约赫斯特塔楼（Hearst Tower），伦敦瑞士再保金融大楼、（Swiss Re Building又称 The Gherkin）及北京中央电视台大楼。

CAPITAL GATE的玻璃幕墙为双层玻璃，大楼内的废气在排出之前，预先在双层玻璃中空部分被冷却，起了节能的作用。

阿联酋第一座大楼率先使用的低辐射玻璃幕墙，它可以在维持室内凉快感觉的同时降低外来强光，并保持玻璃的通透感。

19楼的特色是不锈钢网状设备，兼具设计元素，它可以在太阳照射未到达大楼之前已经阻隔百分之三十的热量。此设备向大楼的南面覆盖，避免大楼受猛烈阳光直接照射。

19楼的游泳池拥有开扬无边际景观。

离地面80米高的空中茶座从大楼外墙悬臂伸出。

大楼的地基采用密集的网状钢筋，在深达30米的地底使用的桩柱多达490支。

RMJM Strata is responsible for the landscape design around the tower, whereas RMJM's interior design studio, RMJM ID has been assigned the task of interior design for all office levels and the basement as well as back-of-house areas for both the office and hotel floors.

RPW Design is the interior designer for the hotel guest rooms, lobby areas and the spa floor at the Hyatt at Capital Centre hotel that is located within CAPITAL GATE.

The original pre-concept design uses the buildings unique form, shape and profile as a reference point for the landscape design, which responds and articulates the tower's leaning form by extending an arrival plaza and water feature out underneath the curve of the glazed façade above. The plaza consequently becomes an abstract representation of the roof plan on the ground plane. The water feature, a cascading reflection pool, defines the extent of the plaza and animates the arrival sequence.

Entrances for the hotel and offices are connected by a shared surface route with pedestrian links to the surrounding gardens and terraces and to the neighboring grandstand.

STRATA strives to make every design unique, respond to its context and have a quality which can only be found in that scheme i.e. creating a sense of place. Therefore the shape and form of the building has been transposed into the landscape design. The articulation of the dynamic shapes and forms is therefore unique to this building.

The landscape design has been documented to detail design by another design consultant. STRATA is currently reworking the podium layout following site boundary changes. We are currently in Scheme Design and intend for Detail Design to be completed in May / June 2010.

Construction of the landscape design is due to be completed in the third quarter of 2010.

The constraints of the site, together with the required internal access / circulation and the elevated height of the podium, have meant designing access for vehicles from ground level outside the site up to the podium level and port cochere has been a very challenging exercise. However STRATA strives on design challenges and has developed a response which enhances the original design.

Secondly re-working a scheme developed to a high level of resolve by another consultant has presented STRATA with the challenge of ensuring the inherent qualities of this work is

not lost through this phase of the CAPITAL GATE landscape re-design.

Sparkle and astrophysics. Imagine the earth or a meteoroid cut in half to reveal an interior never seen before. It would radiate warm colours to the core, starting with light yellow on the perimeter and finishing with strong red in the centre. RMJM ID's idea on this basis is to take reference from the colour graduation across the sphere diameter but using a cool colour palette that is both elegant and more suitable for interior applications in warmer climates. Thus RMJM ID has planned light blue veins of colour graduating through to darker blue veins in the white stone finishes of the tower lobbies. In light of value engineering requirements on the project, RMJM ID have supported a similar treatment in different material selections for example we have proposed the use of textured effect back painted glass as an appropriate alternative and cost effective option.

Lightness and pattern. The glazed skin allows the interiors to be washed through with natural light on all levels with minimal dark areas due to the curved plan which eliminates more typical sharp corned shadow play. Lobby patterning is primarily based on triangle shaping which takes reference from the skin diagrid itself. Shadows from the diagrid fall in unique way and at varying intensity, due to structure orientation, across curved floor plates and interact with the triangular patterning on both wall and floor surfaces to create unusual interior ambience.

The interior design work is already underway. The tower was topped out towards the end of 2009 with interior fit out commencing almost immediately. Work will continue through this year on all office and hotel levels.

Vertical sensations. The building challenges the laws of gravity both externally and within. The atrium provides a view to the sky and draws natural light into the very core of the building. The atrium follows the curved profile of the building elevation and plans to create unique drama with shape fluidity and overlap.

Straight and curve. Everything inside this iconic tower represents a design challenge in the sense that there are no rectangular interior spaces. Various applications and finishes to all surfaces follow curved lines, the outside plan shape of walls and ceilings, the elevation walls of lift shafts and the vertical curving of the skin grid. In addition every single floor plate has a slightly different shape from the level below or above. In summary, straight horizontal and angled line patterns are working to ever changing plan perimeter curves from floor to floor.

The Sky lobby and entrance reception on the 18th floor will have spectacular views of the surrounding city and coastline. Luxurious and uniquely shaped guest rooms starting from the 20th floor will provide a sense of flotation, where the convex curve of the structure seems to make the building disappear below on the North West elevation. Equally on the opposite side of the building which facing South East, there will be even more dramatic bedroom views and gravity play with flotation where the 19th floor cantilevered swimming pool seems to be suspended in mid air when viewed from above.

Guestrooms:

• The open plan nature of the room with interaction between the bath area and the guest living area, the combination of hard rough natural stone and marble with the contrast of the zebrano millwork combined with the grass effect panels and soft colour pallet leads to a very calm and relaxing space.

• The large egg shaped bath in which one can relax, will make the hectic business of the day feel a million miles away when one steps into the room.

• The suites with luxury infinity baths at the perimeter of the room, enable guests to relax and take in the view.

• Luxury fittings, fixtures and furnishings have been used throughout.

Public areas

• The main atrium chandelier is a bespoke design for ADNEC.

• On the ground floor we have an open plan food emporium, which will enables guests and visitors to ADENC and Hyatt to pick and choose from a variety of food outlets.

• The 18th Floor open tea lounge where visitors can enjoy afternoon tea whilst taking in the views of the Grand Mosque and its surrounding areas.

• The restaurant on the same floor with an inclined perimeter glass wall/ façade allows visitors to take in the view of the whole of Abu Dhabi city.

• At its outermost edges, the space will feel like it is floating above the ground.

• The Spa located on level 19 with 4 single private treatment rooms and a double treatment room, has been designed to encapsulate the calm and quietness of the desert.

The work on the interior design commenced in January 2007, right from the early stages of the project and is due to complete upon the opening of the hotel .

The interior design of the guest rooms has a calming, relaxed feel to them. Also, standing at the outer periphery of the room, one can get a feeling of floating as one cannot see the ground directly below one's feet.

This was an extremely difficult challenge when RPW were first handed the GA's (General Arrangement) for the building; potentially each room, shape and size could have been different. However, we were able to produce a design for the guestrooms which standardised elements of design within each guestroom, particularly in the wet areas, along with standardised modular design of joinery. This enabled RPW to deliver a package which offered the client some form of value-saving in modular design, removing the notion of 100 percent bespoke design in each room. However, that being said, the walls in each room are never parallel to each other, and this aspect will offer the repeat guest an experience of staying in a "new" room on each visit.

The New Visitor Center at Red Rock Canyon

拉斯维加斯红石公园新游客接待中心

线和空间建筑事务所/Line and Space

Floor Plan

Legend –

1. Visitor Entry Plaza
2. Transition Space / Entry
3. Temporary Exhibits / Multipurpose
4. Information Desk
5. Arrival Experience
6. Panorama Window
7. Classroom
8. Classroom Patio
9. Gift Shop
10. Transpired Solar Wall
11. Outdoor Amphitheater
12. Earth Pavilion
13. Tortoise Habitat
14. Tortoise View Deck
15. Fire Pavilion
16. Air Pavilion
17. Four Elements Exhibit
18. 360° View Deck
19. Cliff Walls
20. Water Walk
21. Natural Habitats
22. Desert Spring Exhibit
23. Desert Ecosystem Exhibit
24. Water Harvesting Storage
25. Earth Berm
26. Photovoltaic Array (60 kw)

0' 20' 40' 80'

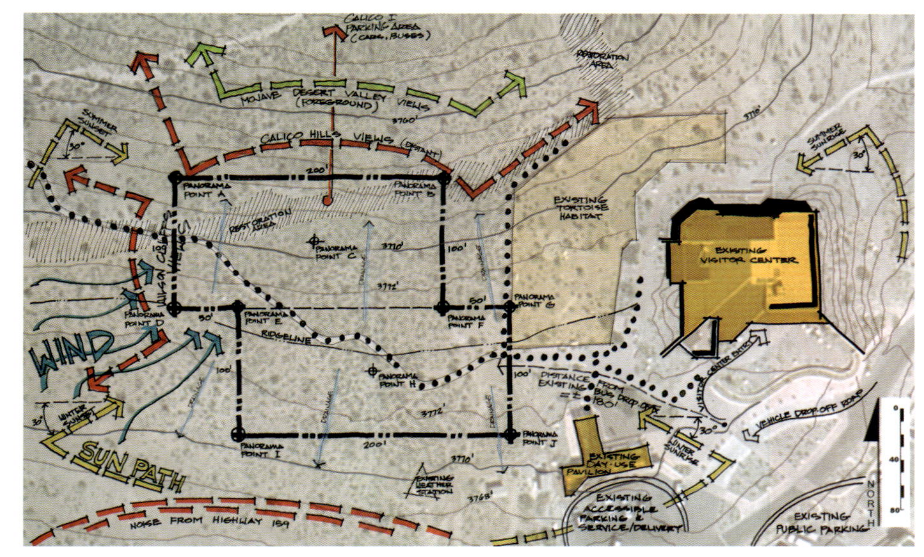

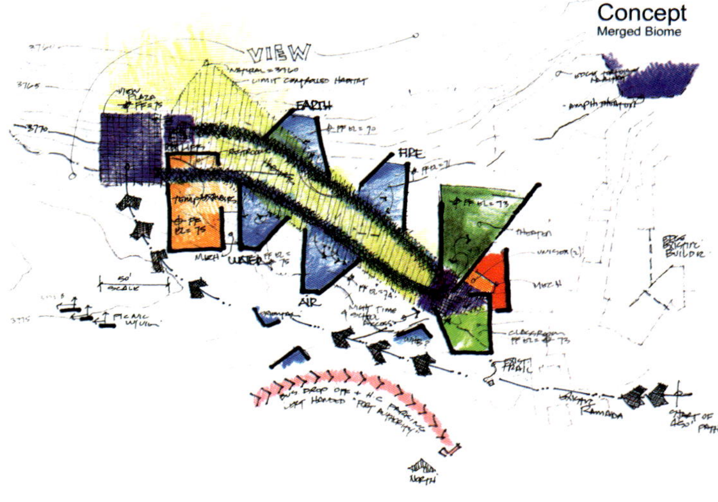

莫哈韦沙漠拥有特殊的生物种类，200 000平方英尺的红石峡谷国家级保护区称得上是一个奇观。

设计师意图每年使1 000 000的赞助人获益，其中蕴涵了无限的价值。在传统的透视面和录像的场景下，创造抽象的雕塑展品来增加邻近自然氛围的参与性。

本项目的概念是将相关科学、艺术和文化介绍给游客，可以增加他们在红石峡谷的经历以及启发他们去开发真正有意义的东西。

许多节约型资源的想法被运用到了参观者中心，包括对渗透型太阳能集热器的第一次公共使用，但是迄今为止，最大的能源节约还是来自于对整个设计项目的再思考和将展品从装有空调的室内重置到全日的日照和微气候中。

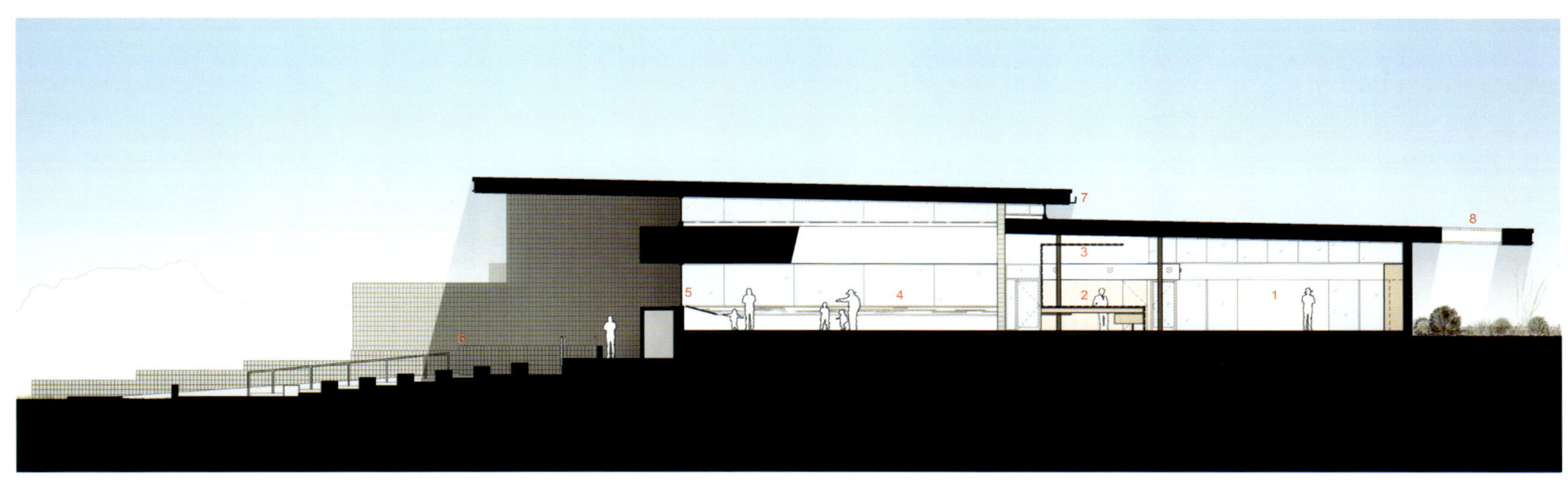

0' 8' 16' 32'

Within the special biome of the Mojave Desert, the 200,000 acre Red Rock Canyon National Conservation Area stands out as a place of wonder.

Our intent is to educate as many of 1,000,000 patrons a year, as possible, about the value of this seemingly barren desert. In lieu of traditional dioramas and videos, abstract sculptural exhibits are created to engage the participant in lessons focused upon adjacent natural phenomena.

The concept here is to submerge the visitor to the relevant science, art and culture that will enhance their experience in Red Rock Canyon, and then strongly encourage them to explore the real thing.

Many resource-conserving ideas are incorporated into the Visitor Center, including the first institutional use of transpired solar collectors, but by far the largest savings come from re-thinking the design program and relocating exhibits from air conditioned interior space to fully day lighted passive and active tempered micro climates.

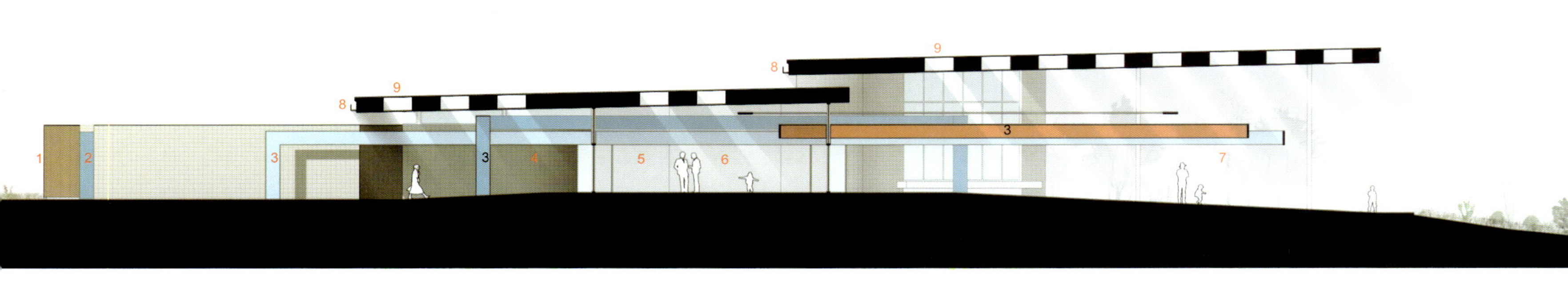

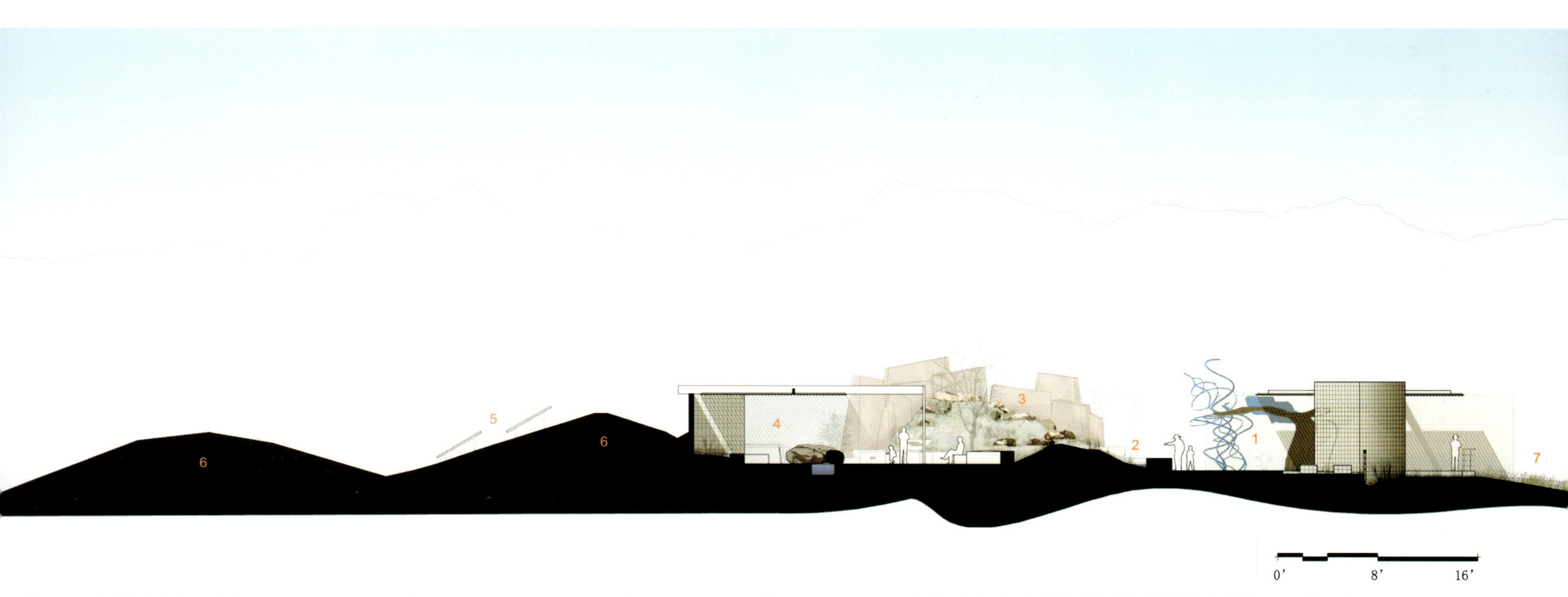

6
5
6
4
3
2
1
7
0'
8'
16'

FIRE

Eco-Wine Pavilion

生态葡萄酒馆

Michael Jantzen

Michael Jantzen的生态葡萄酒馆是该系列中的另一个设计构想。这个设计试图找到一种全新的、激动人心的方法将替代能源和存储系统整合到商业建筑当中。生态葡萄酒馆可在温带气候下作为酒厂和功能房来使用，例如葡萄酒品尝室或者特殊活动的举办场所。其结构由一块预制玻璃和钢铁组件构成。一系列的拱形钢铁和水平支撑物被玻璃面板包裹着，其中一些能够自由打开和关闭，以调节整个室内空间的自然通风。去掉玻璃后，该葡萄酒馆便会显露出隐藏起来的另一个样子：能够露天使用的葡萄酒馆。大量的钢和玻璃结构被两种不同半径的弯曲钢质翼片所覆盖。其中一些翼片被安置在最接近太阳的方向，其上面覆盖的光伏材料能够生成电，供室内使用。其余钢质翼片被漆成了和光伏材料相同的颜色。这些翼片奇妙地遮盖了室内结构。酒馆所有的额外用电均来自于安装在其附近的一个大型风力涡轮机。在风力涡轮机的基础上，内置的圆形长凳可以围绕着酒馆的支柱而建。这个设计的美学灵感来自于格子架上缠绕而生的葡萄藤。

Michael Jantzen's Eco-Wine Pavilion is another design proposal in a series, which explores ways to integrate alternative energy gathering and storage systems into commercial architecture in a new and exciting way. The Eco-Wine Pavilion is designed for use in a temperate climate at a winery and functions as a wine tasting and special events facility. The structure is formed from a set of pre-fabricated glass and steel components. A series of steel arches and horizontal supports are clad with glass panels, some of which can be automatically opened and closed in order to control natural ventilation throughout the interior space. The pavilion could also be constructed without the use of glass for an open-air shaded version. The majority of the steel and glass structure is covered with a series of curved steel panels formed with two different radii. Some of these panels, which are located in the appropriate orientation relative to the sun, are covered with a flexible photovoltaic material that generates electricity for use inside of the pavilion. The rest of the curved panels are painted with the same color as the photovoltaic material. These panels also shade the interior of the structure. All of the additional electricity needed to power the pavilion is generated with a large vertical axis wind turbine, which is mounted near the structure. At the base of the wind turbine there is a circular bench built-in around the perimeter of the support column. The aesthetic inspiration for this design comes from the symbolic image of grape vines growing over a trellis.

E-Solar Visitors Center

E-Solar游客中心

Michael Jantzen

这项为e-Solar游客中心进行的设计，其灵感来自于e-Solar公司提供给Michael Jantzen的一些原理图，这些原理图显示了太阳光如何通过一系列的镜子反射到位于中心的接收塔上。接收塔通过镜子将太阳的热量集中起来，使水变为蒸汽，从而用这些蒸汽来发电。在Michael Jantzen的设计中，游客可以爬上一座欢庆楼梯，到达距地面40英尺的观景台。登上观景台后，人们可以利用望远镜来查看发电厂的细节部分。一台实时摄像机会为那些不想爬楼梯的游客展示中心的内部。该平台由一些大型的黄色翼片支撑，它们象征着被镜子反射到收集点上的太阳光线，基层结构上也有类似的翼片。处于大型接缝间的游客中心底层翼片看起来就像一个矩形网格（指反射镜的形状）。象征着太阳光线的翼片创造了一个大型的（1 800平方英尺），兼具了展览、办公、会议等功能的封闭空间。这个封闭的空间在其中心的地方有一个巨大的天窗（依旧象征着太阳）和一个较小的天窗，以照亮中心的入口。封闭空间里的大型遮光玻璃门将面对电厂而建。中心的室内通风将通过能预先制冷空气的高架阴影和由太阳能热泵驱动的空调系统及空气热交换机来实现。游客中心所有的电力都将由太阳能电站来供给。此外，雨水也将被收集起来进行再生利用。围绕着封闭空间而建的宽敞的开放式纳凉空间将被用作公共活动和私人活动的场地。最具可持续性的可用材料都将运用在e-Solar游客中心的建设当中。

Michael Jantzen's design for this e-Solar visitors center is inspired by schematics supplied to him by the e-Solar company, which illustrate how the sun rays are reflected off of a series of mirrors and onto the top of a central receiving tower. Heat from the sun reflecting off of the mirrors and focused onto the receiving tower converts water into steam, which is then used to generate electricity. In his design, the visitors can climb on a large celebratory staircase to an observation platform located about 40 feet above the ground plane. On the platform, a telescope will be installed to view specific details of the power plant. A real- time video camera may also be added to display the view inside of the visitors center for those who do not want to climb the stairs. The platform will be supported by large yellow panels that symbolically represent the rays of the sun, which have been reflected by the mirrors to a receiving point above the observation platform. Similar panels located at the base of the structure symbolically represent the rays of the sun as they originally intersect with the mirrors. A rectangular grid (referring to the shape of the mirrors) is graphically depicted on the floor plane of the visitors center by seams between large tiles. The symbolic sun rays panels shade a large (1,800 square foot) enclosed space below used for exhibits, offices, meeting rooms, etc. This enclosed space has a large

skylight at its center (symbolically referring again to the sun) and a smaller skylight that illuminates the entrance to the visitors center. Large shaded glass doors will be built into the enclosed space facing the power plant. Ventilation will be supplied to the interior space from under the elevated structure, where the air is naturally pre-cooled by the shade of the visitors center, and by air conditioning operated by solar powered heat pumps and air to air heat exchangers. All of the energy needed for the visitors center will be supplied by the solar power plant, and rain water will be captured for use at the center and recycled. A large open shaded space around the enclosed space will be used by the public and for special private events. The most sustainable materials available will be used in the construction of the e-Solar visitors center.

eSolar
eSolar
eSolar 
eSolar

Xixi Leisure Center

西溪休闲中心

徐甜甜

The tiny Thumbelina woke up very early in the morning, and when she saw where she was she began to cry bitterly; for on every side of the great green leaf was water, and she could not get to the land.

项目地点	浙江，杭州
项目功能	休闲
建筑面积	6 300平方米
客户	杭州西溪国家湿地公园三期工程有限公司
设计单位	DnA _Design and Architecture
设计时间	2008/1~2009/7
施工时间	即将进入施工阶段
建筑师	徐甜甜

Project Location	Hangzhou, Zhejiang
Program	Leisure
Building Area	6,300 m²
Client	Hangzhou Xixi National Wetland Park Phase III Engineering Co., Ltd
Design Company	DnA _Design and Architecture
Design Milestones	2008/1—2009/7
Construction Milestones	About to Enter Construction Phase
Architect	Tiantian Xu

西溪休闲中心是坐落在西溪国家湿地公园的12个艺术和文化建筑群的单体建筑之一。这个建筑群是由杭州市政府委托给12名中国建筑师的集群项目。杭州以其悠久的历史，丰富的文化和西湖景观而众所周知，而西溪艺术和文化建筑群则恰恰促进了其文化及旅游业的发展。

休闲中心延续了原有湿地的路径和形态，营造出由地面上的休闲功能空间和下沉的活动池塘组成的连续、开敞的空间形态。在二层，特色水疗室则成为相对独立的空间，就像漂浮在水上的睡莲一样，娱乐空间从它下方经过。在这样的意境中，一根根纤巧的支柱如竹林般立起，支撑起荷叶状的屋面。

这种荷叶的组织和连接方式使人们从户外的步行道通过中间的各层荷叶到达屋顶上的小池塘，从而最终在屋顶观赏到这片湿地的全景。

在荷叶间的逗留会勾起人们似曾相识的感觉，好像我们祖先千百年前的一个幻想，经过这么多年的紧张的心理和生理上的变化，我们，作为人，仍然可以轻如一片鸿毛，敏捷如一条祥龙，在这充满微风和熏香的自然环境中冥想。

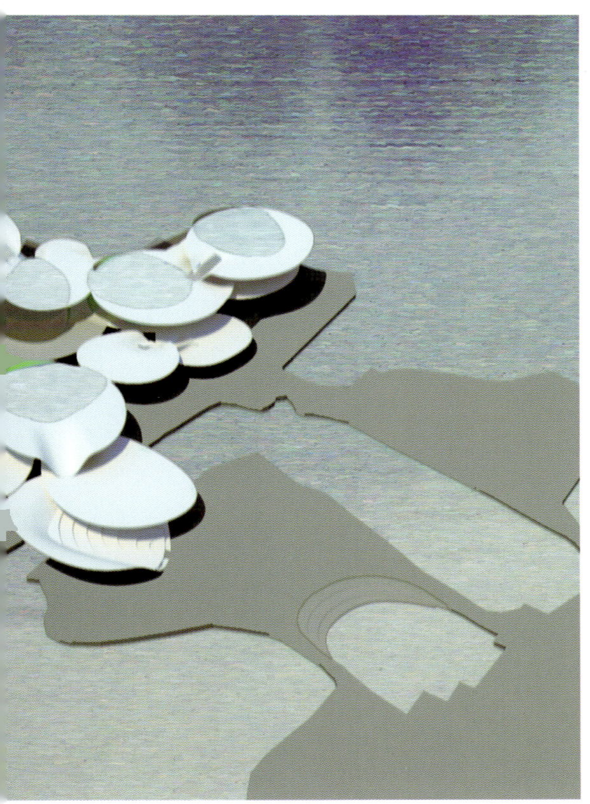

Xixi Leisure Center is one of the 12 buildings located in an art and culture compound in Xixi National Wetland Park, a cluster architecture commissioned to 12 Chinese architects by Hangzhou city government. Hangzhou has been well known for its long history, rich culture and the West Lake landscape. Xixi art and culture compound is part of development to stimulate cultural tourism.

The leisure center will house leisure function as open and continuous circulation on ground level and sunken into activity pools imitating paths and ponds of wetland topography, while on second level, spedicalized SPA rooms become rather individual spaces, like the thumbnail floating in water lily leaves, carried by leisure circulation underneath.

In such an atmosphere, tender and bulk columns are applied like a bamboo forest supporting the volumes and leaves.

The format and organization of leaves leads an outdoor promenade from the paths up to roof terraces discovering small water lily ponds, eventually unveiling the wetland to a panoramic view from above.

A pause on these leaves might evoke a déjà vu, a fantasy traced back thousands of years ago from our ancestors' believes, that after years and years of intense mental and physical training, we, human beings, are able to pause upon water, light as a fur, alert as a dragonfly, meditating the breeze and aroma in our nature.

3 Pubic

公共

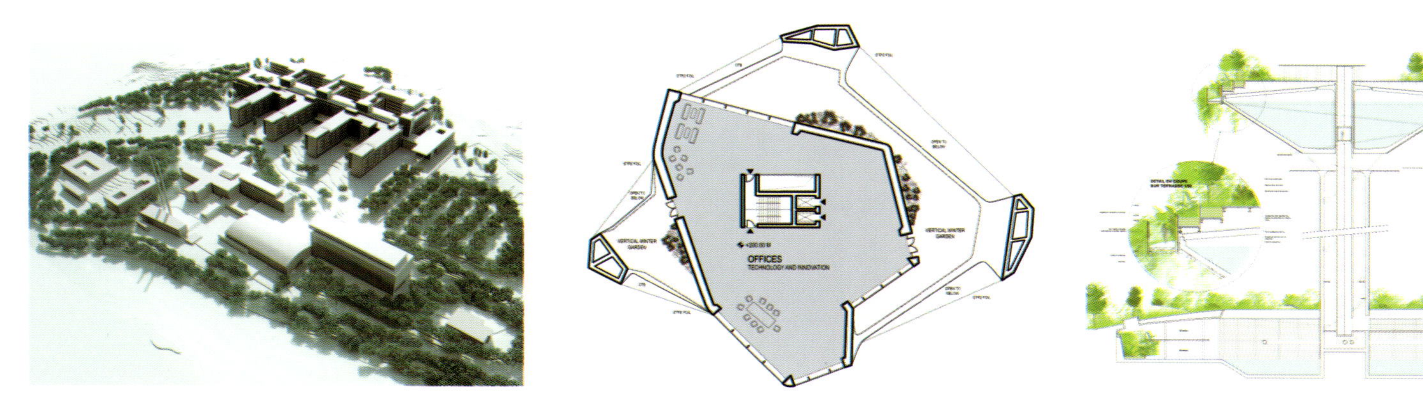
VERTICAL WINTER GARDEN
VERTICAL WINTER GARDEN
OFFICES
TECHNOLOGY AND INNOVATION

Akershus University Hospital
阿克斯胡斯高等学院校医院

Arkitektfirmaet C. F. Møller

项目地点	挪威，奥斯陆
项目时间	2000年~2011年
项目面积	137 000平方米（118 000平方米新建）
设计单位	Arkitektfirmaet C. F. Møller
客户	Helse Sør–Øst RHF
合作方	Multiconsult AS, SWECO AS, Hjellnes COWI AS / Interconsult ASA, Ingemannson Technology, Nosyko/Erstad og Lekven
景观设计师	Bjørbekk & Lindheim AS, Schønherr Landskab A/S
艺术家	Troels Wörsel, Gunilla Klingberg, Mari Slaattelid, Knut Henrik Henriksen, Jan Christensen, Tony Cragg, Birgir Andrésson, Petteri Nisunen, Tommi Grönlund, Julie Nord, Per Sundberg, Vesa Honkonen, Janna Thöle–Juul, Kristine Halmrast, Mikkel Rasmussen Hofplass
摄影师	Torben Eskerod, Arkitektfirmaet C. F.，Møller

Project Location	Oslo, Norway
Project Milestones	2000-2011
Project Area	137,000 m^2 (118,000 m^2 newbuild)
Design Company	Arkitektfirmaet C. F. Møller
Client	Helse Sør-Øst RHF
Collaborators	Multiconsult AS, SWECO AS, Hjellnes COWI AS / Interconsult ASA, Ingemannson Technology, Nosyko/Erstad og Lekven
Landscape Architect	Bjørbekk & Lindheim AS, Schønherr Landskab A/S
Artist	Troels Wörsel, Gunilla Klingberg, Mari Slaattelid, Knut Henrik Henriksen, Jan Christensen, Tony Cragg, Birgir Andrésson, Petteri Nisunen, Tommi Grönlund, Julie Nord, Per Sundberg, Vesa Honkonen, Janna Thöle-Juul, Kristine Halmrast, Mikkel Rasmussen Hofplass
Photographer	Torben Eskerod, Arkitektfirmaet C. F. , Møller

SYKEHUSVEIEN
Parkeringsplass
VITTENBERGBEKKEN
NORDBYHAGAVEIEN
TÅBYVEIEN
GAMLEVEIEN
NN
FB
S1
S2
S3
S4
EP
TS
UK
BUK
AMM
KA
T

8
1
9
19
18
10
5
3
19
11
18
19
12
3
4
6
12
18
19
7
3
12
19
13
18
19
2
16
19
17
14
15

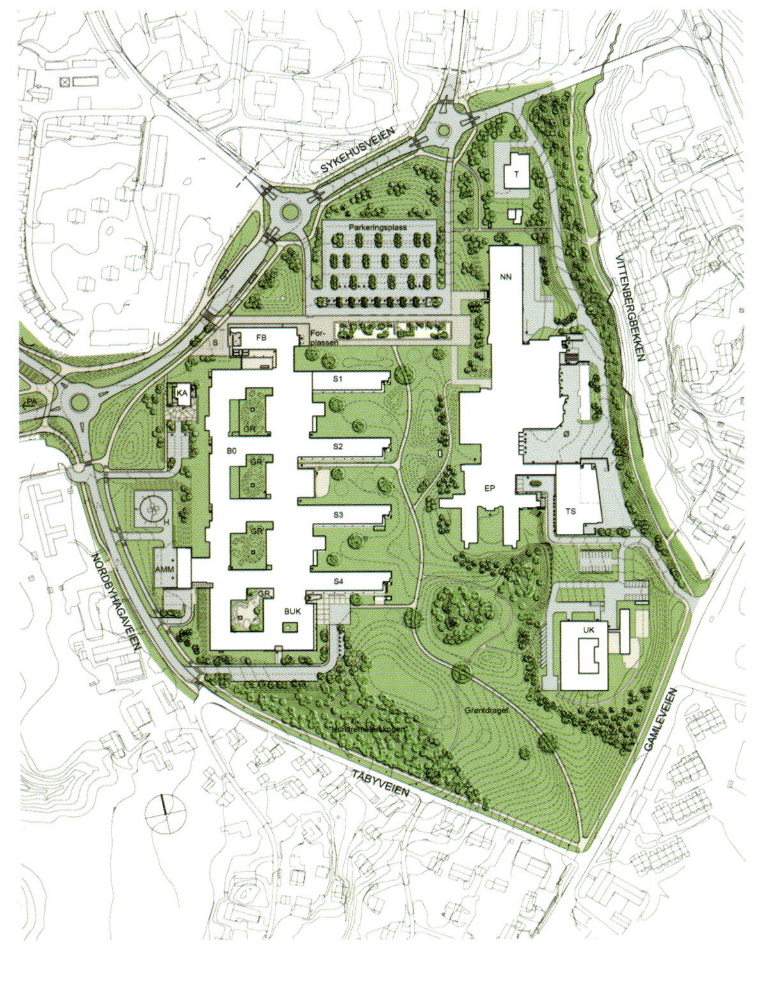

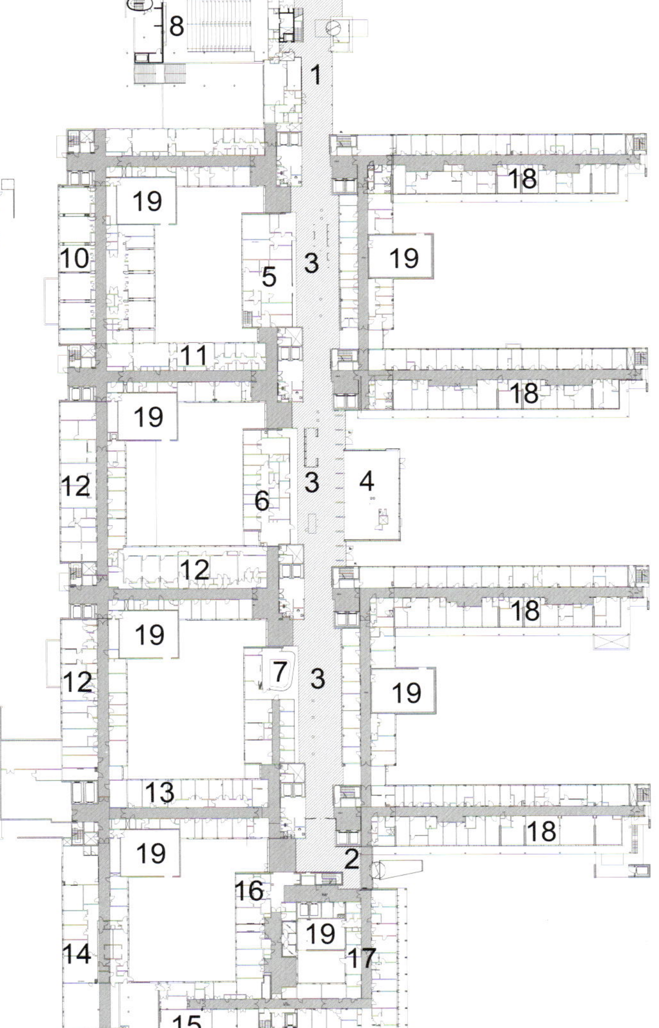

1 st floor

阿克斯胡斯高等学院校医院坚持高度可持续发展的设计理念，充分利用当地的物质资源和地热能，为医院提供了85%的热能及40%多的总能源消耗。各种近距离的服务设施、分工明确的组织和包括机器人在内的现代科技的广泛运用为工作人员提供了更多时间在病人身上。

再生能源

该医院总能源消耗约为20GWh每年，相当于1300个家庭所消耗的能源。再生能源的利用需通过地热交换系统实现，结合基岩的热容量，所剩的热能量（例如来自太阳能、人体、技术设备、冷却和通风设备）可以储存进350个深200米的能源井。

随着时间的推移，现代化的医院产生了大量的剩余能源，特别是在通风系统和一些技术密集的地方。一个主要医院的其他部分也需要大量的热量，而存储和恢复系统就是解决该能源问题的有效方案。

BTES系统是欧洲最大的系统之一，它能提供加热所需能量的85%以及包括冷却在内的超过40%的医院总能耗。二氧化碳的排放量同原先的医院相比减少了50%多。

就规模而言，一块两公顷的土地被用来容纳这些能源井。总之，大约有150千米的管子铺设228个能源井，这些能源井内有250 000升的防冻剂。在BTES上方的田野将继续种植庄稼。

这些能源井在储存冷水机剩余热量的同时，也可以用来隔绝对外界的噪声污染。传统的冷水机需要有一个放在建筑楼顶上的笨重的冷却塔，用大型风扇来排放暖气。

环保材料

医院采用的所有建筑材料都是健康有益的，有利于保持室内空气清新。医院的建造广泛地采用了木材和石头等当地的物质材料，为医院营造了一种安全如家的氛围。

建造期间，首先坚持的原则就是在建造过程中减少废弃物浪费，资源循环利用和节约能源。

The Akershus University Hospital is a highly sustainable design, making use of locally sourced materials and geo-

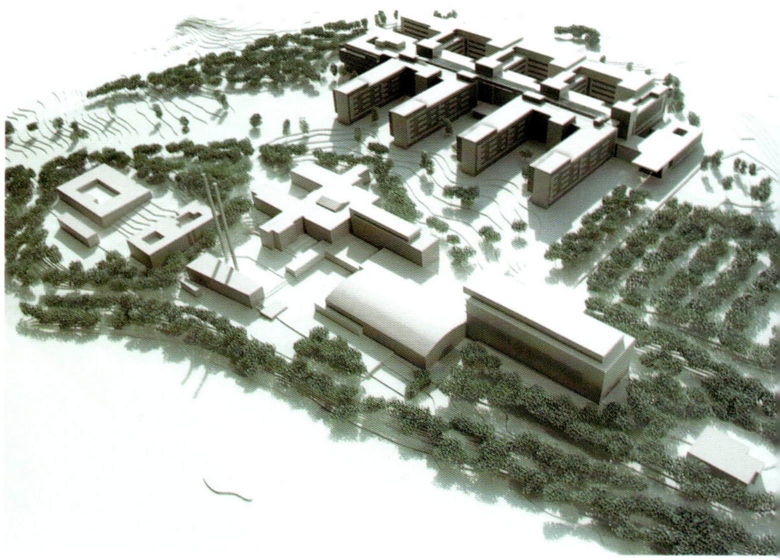

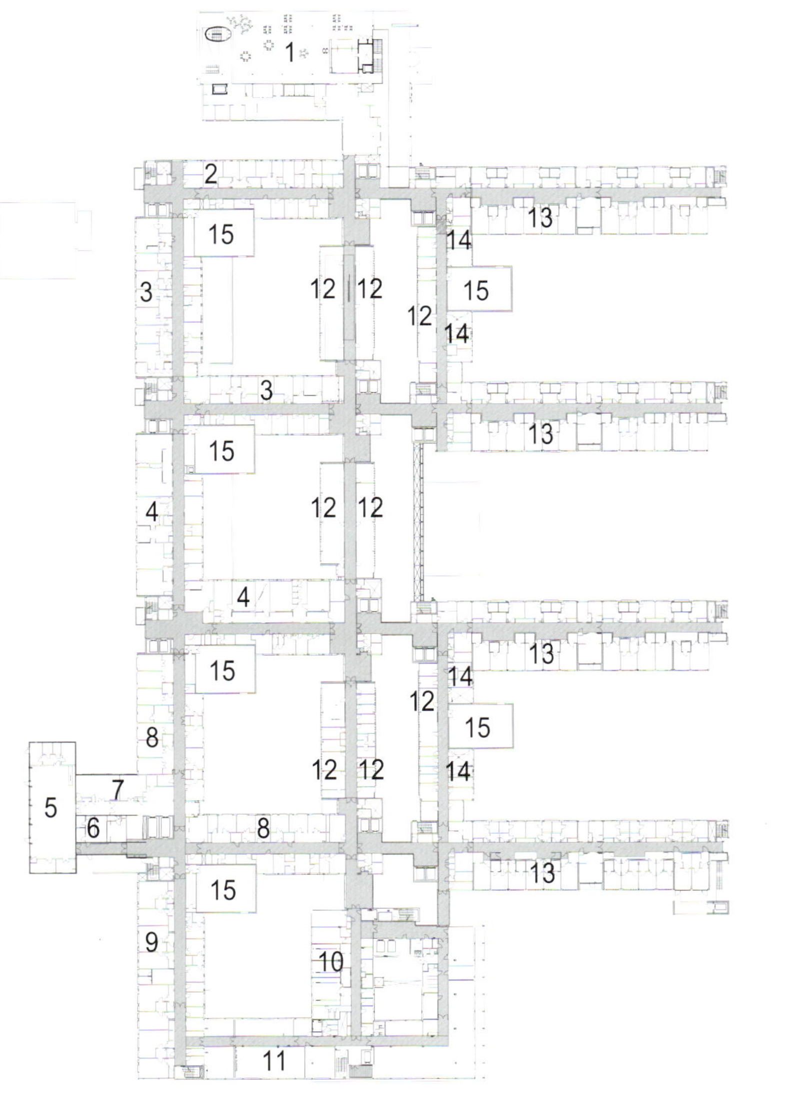

2 nd floor

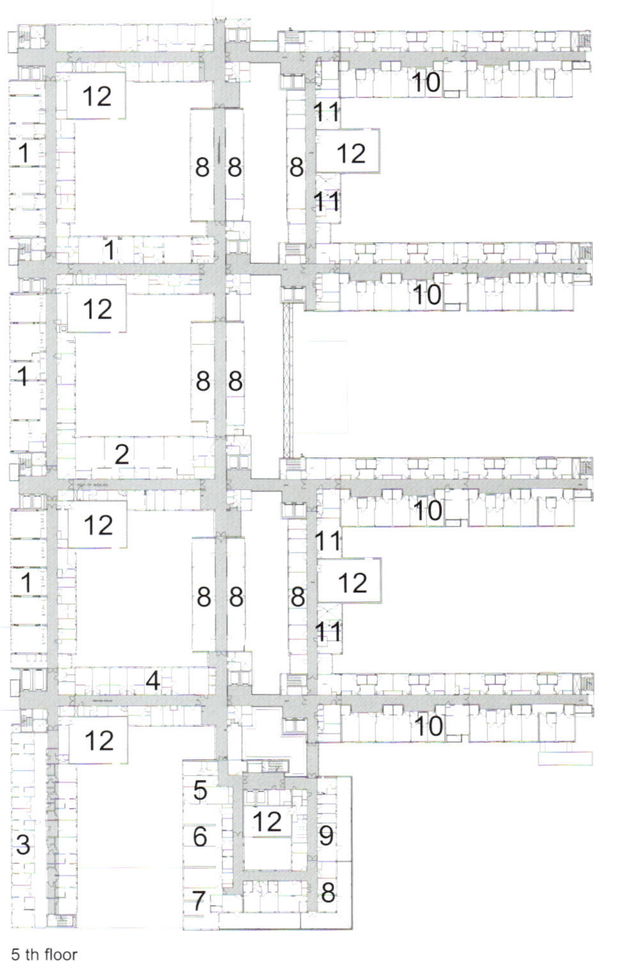

5 th floor

thermal energy to provide 85% of the hospital's heating and more than 40% of the total energy consumption. Short distances between functions, a clear organisation and extensive use of modern technology including robotics give staff more time for patients.

Renewable energy sources

The total energy consumption of the hospital is approx. 20 GWh/year, similar to the consumption of 1,300 single family houses.The use of renewable energy is based on a ground-heat exchange system, combined with thermal storage capacity in the bedrock, where surplus heat (for instance from solar gain, people, technical equipment, cooling and ventilation plants) can be stored in 350 energy wells drilled to a depth of 200 m.

Modern hospitals generate a large amount of surplus energy over time, particularly in ventilation systems and extra technology-intensive spaces. When there is also a large need for heating in other parts of a major hospital, a storage and recovery system is an effective solution to the energy problem.

The Borehole Thermal Energy Storage (BTES) system is one of the largest of its kind in Europe, and it generates 85% of the energy used for heating, and covers over 40% of the total energy consumption in the hospital, including cooling. This reduces CO_2 emissions by over 50% compared to the former hospital's performance.

To illustrate its size, an area covering two hectares of land has been used to contain the wells. All in all, around 150 kilometres of pipe have been laid to 228 wells that are filled with some 250,000 litres of antifreeze solution. Above the BTES, the field will continue to be used for cultivation of crops.

By using wells to store surplus heat from the chiller, it is also possible to spare the external environment from noise. Traditional chillers require bulky cooling towers on the roof of the building, with large fans to remove the warm exhaust air.

Healthy materials

All materials used in the hospital are healthy materials with a good indoor climate performance. Locally sourced materials such as wood and stone are used extensively, which helps create a homely and secure atmosphere as well.

In the construction phase, very high priority was given to reduce waste and recycle materials, as well as reducing the energy consumption in the construction process.

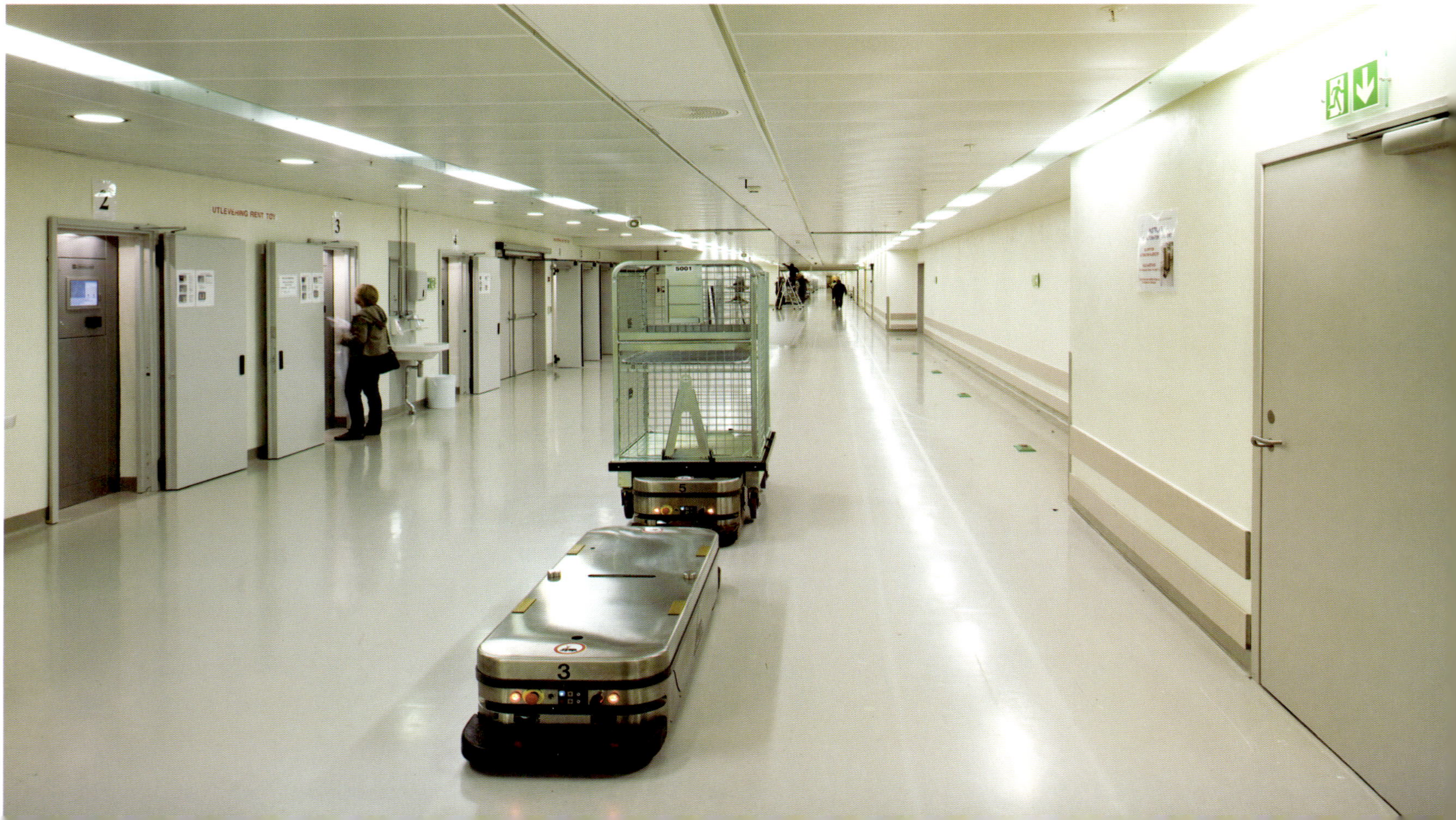

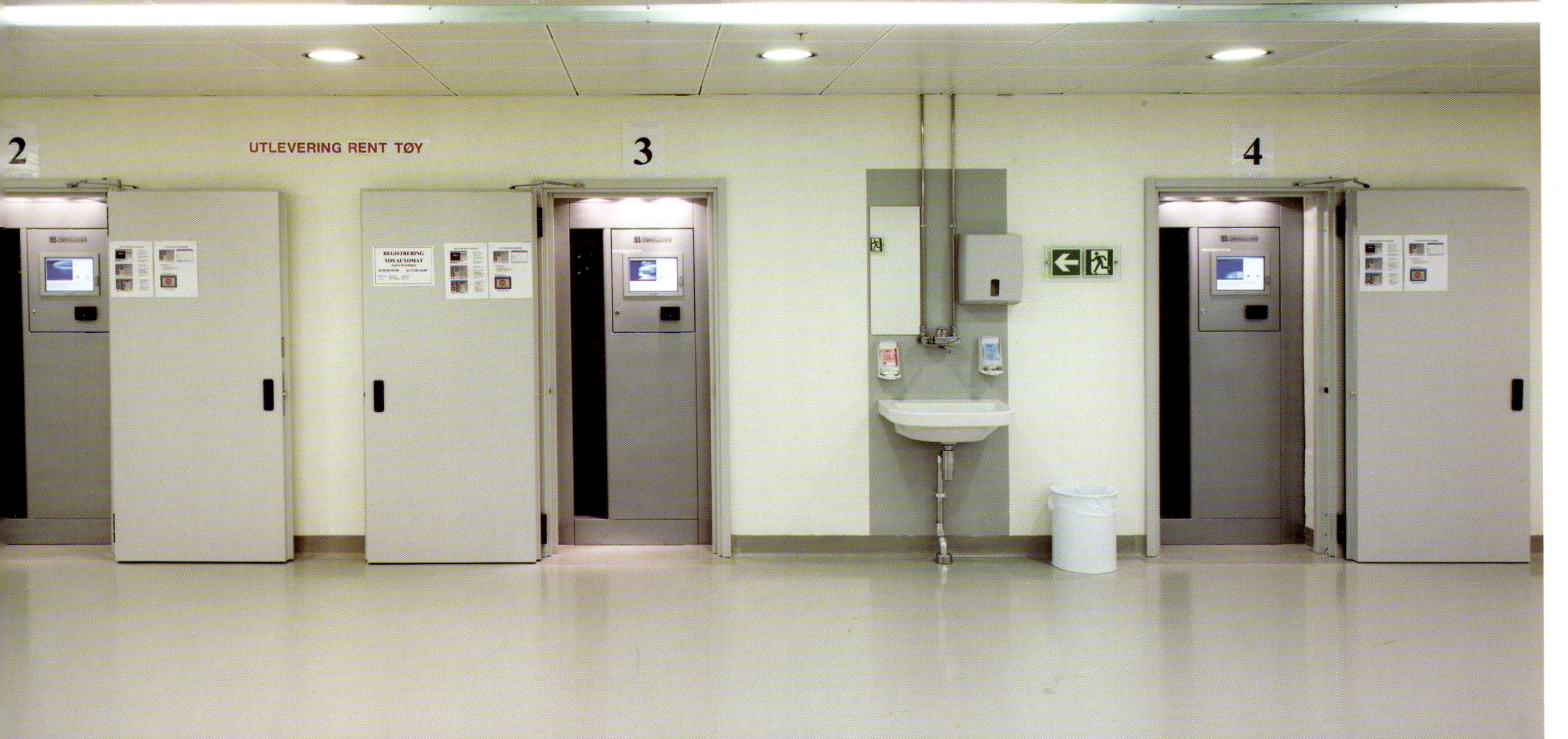
2
UTLEVERING RENT TØY
3
4

Cheongna City Tower

青罗城市大厦

Tom Wiscombe 、Takeshi Masuyama、Josh Sprinkling、Alina Grobe、Jon Anderson、Seton Beggs

项目地点	韩国，仁川
项目时间	2008年
建筑类型	瞭望休闲塔
项目面积	50000平方米
客户	Korean Land Corpor
设计团队	Tom Wiscombe 、Takeshi Masuyama、Josh Sprinkling、Alina Grobe、Jon Anderson、Seton Beggs
工程师	Buro Happold、Matt Melnyk

Project Location	Incheon, Korea
Project Milestones	2008
Building Type	Observation and Leisure Tower with Sunkeng
Project Area	50,000 m²
Client	Korean Land Corpor
Design Team	Tom Wiscombe , Takeshi Masuyama, Josh Sprinkling, Alina Grobe, Jon Anderson, Seton Beggs
Engineer	Buro Happold, Matt Melnyk

A. Macrotermes michaelseni
B. Macrotermes subhyalinus
Direction of air movement
Fungus combs
Nursery galleries
Royal cell
1 m

FERRY PIER
Viewing Platform
KIDS PARK
Retail
Food Court
SUNKEN GARDEN
TOWER LOBBY
TERRACE RESTAURANT
PROMENADE
Terrace
Tower Zone
CENTRAL PARK
EURO PARK
AQUA PARK
Siteline

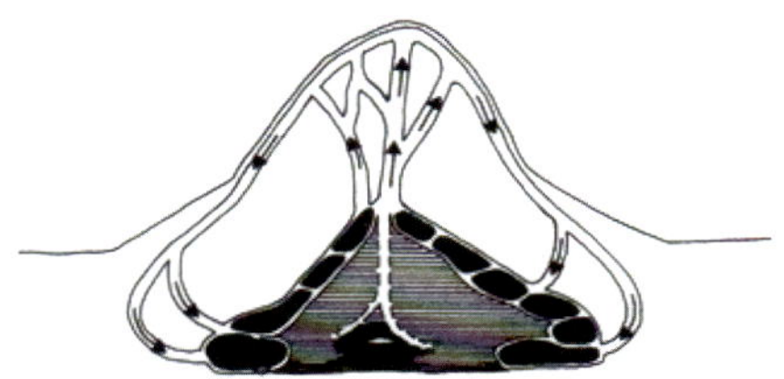

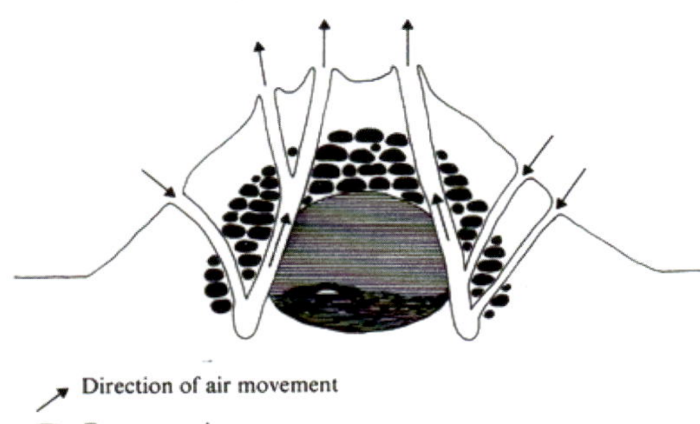

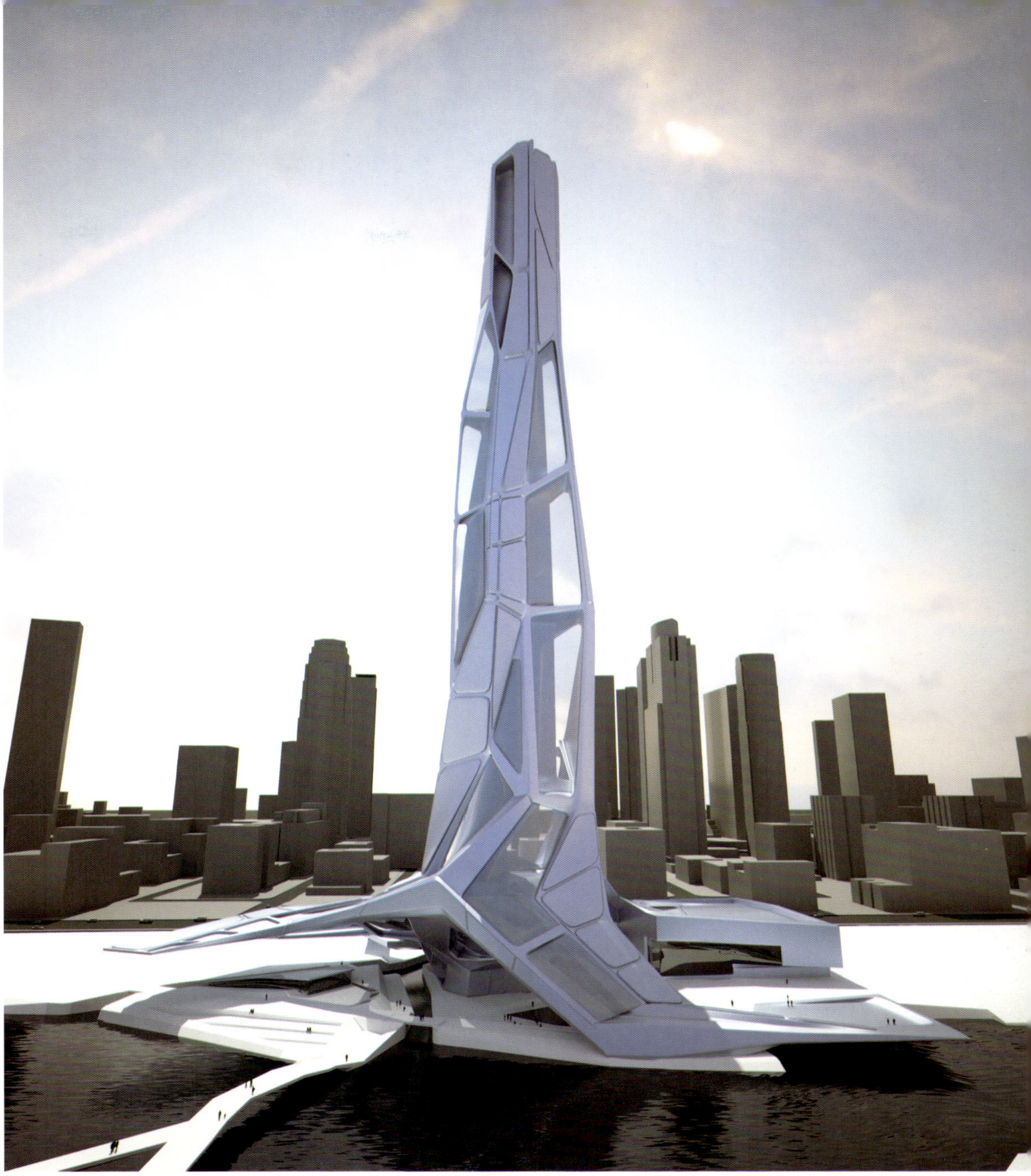

塔

青罗城市大厦的设计旨在创造一个具有创新空间性和结构性的能源生产设备，这种设备将会成为未来IFEZ青罗地区的象征符号。本项目位于东西走向的主人行道和南北走向的主要人工水道的交叉口，它将成为城市活动中心和该区域一个新的聚集地。塔高400米，在此塔上可以尽览海洋风景，它位于仁川机场的东边和Mt. Geyang的西边。

该塔的较低层拥有艺术和设计展示空间、集会和学术空间、礼品店、酒吧等休闲空间来举办各种休闲文化活动。在该塔的中层每隔50米就有一个公共天台；还有一个商业脊柱，它包括对各种科技公司和文化机构开放的办公展示空间。而塔的最高层有天文观测台，名厨汇集的季节性高端餐厅、各哨点及观测甲板。

作为能量装置的“垂直的冬季花园”

该塔的外层由双壳结构建造而成，形成了一个“垂直的冬季花园”。它可以容纳各种观测甲板，植物园和休闲活动场所。晚上，从飞机上或者是在其周边看塔，它变得五光十色，灯火通明。

“垂直的冬季花园”是一个在外观上用ETFE膜隔开的自然通风的缓冲地带。在冬天，它被用作温室来收集热能向建筑物里散发热量；夏天，它是热绝缘体，保护建筑物的内部空间不受热侵害。上升的热空气聚集在顶部空间，在空气流出之前通过换热器和微型燃气轮机补充能量。

另外，在夏天“冬季花园”则被用作存储冷空气的设备，这些冷空气是通过晚上开放的位于塔顶的大型风力摄入器收集而来的，根据大型被动夜间冷却塔的模式设计而成。最后通过大楼外部集成的可逆式循环加热系统可以提供辐射加热和冷却。它通过从地热源热泵获得

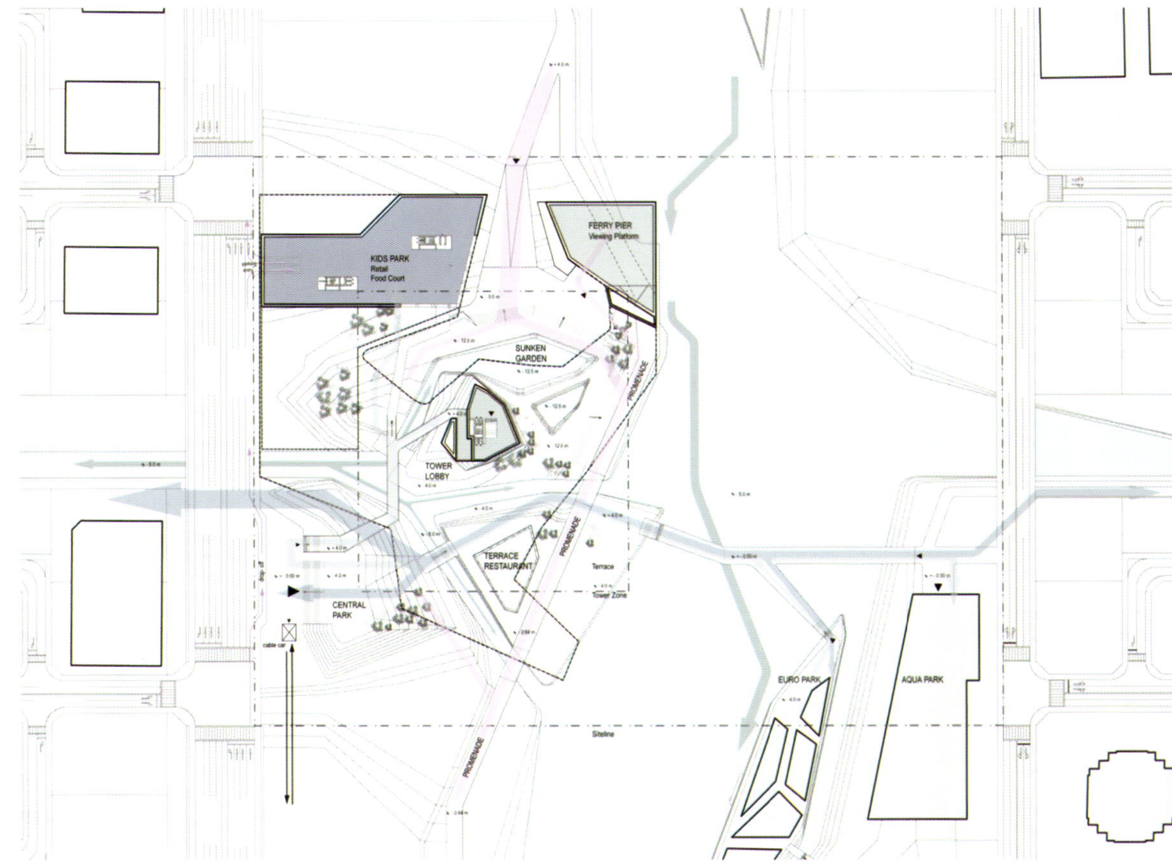

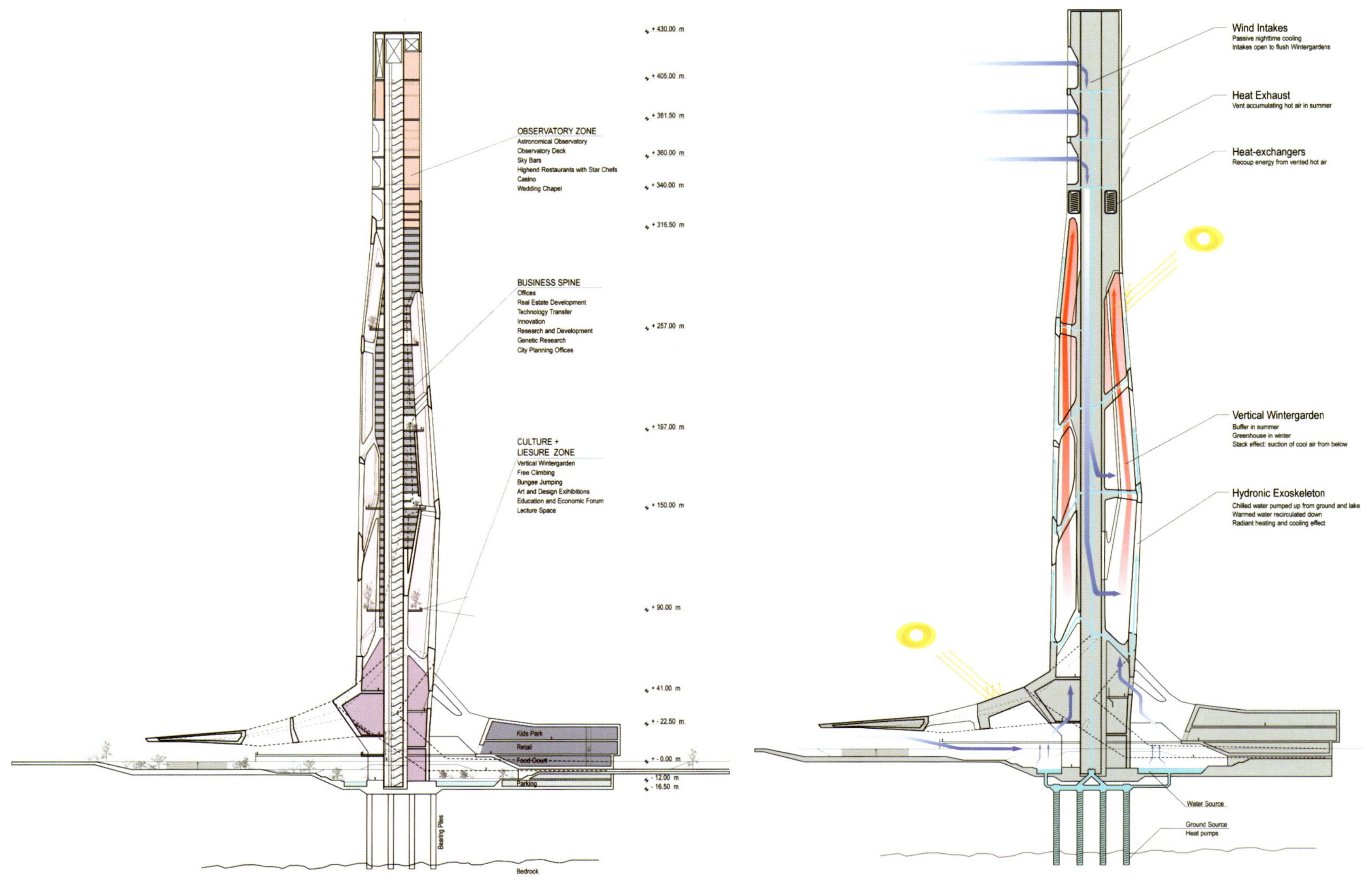

冷热气流，但有时也用湖盆的可利用冷水。

广场和屋顶

垂直塔和水平底座的类型是一建筑上个经典的问题。自高层建筑出现以来，就涌现出了许多解决如何使塔和地面及水平方向的建筑物相连接的方法，但这些方法大多太理想化或是太具投机性。为创造出一种流动性同时也可保护下方空间，本方案更偏重于塔和其水平范围内的连续性。三个不同功能的建筑在此范围内：塔的东边是渡轮平台，北边是零售业建筑，南边是广场屋顶。

该广场具有集城市化与自然化于一体的地形特点，包括茂密的植被和具有异国情调的鲜花、密布的水道和水池及风景各异的阳台。另外该广场内还有咖啡馆、餐厅，各种文化活动和服务性的设施，是一个很繁华、有活力的城市化场所。在这里，游客可以尽情地放松，与朋友一起畅谈，还可参与各种休闲活动。它既是一个休息场所，也是一个充满活力的城市化场所。

创新性外部构造

该塔的结构采用了钢外壳而非传统的结构性核模型。三股主结构脊沿外立面交织，并随着深度、宽度以及不停的转动而变化，以配合垂直和横向外力以及设计团队制定的几何规则。与龟壳的内部构造相似，这些主结构柱共同交会形成一个由硬壳构造和框架—表皮构筑类型的混合体。整个结构形态在建筑表面、表面浮凸以及支架之间形成了平滑的梯度，这恰恰是EMERGENT事务所一直在关注的。

CATIA，modeFRONTIER及ROBT

参数化软件，人口集成软件及结构分析引擎被运用于一个自下而上的非线性工程化过程中。这个过程就如大自然中物竞择生，优胜劣汰的自然规律一样。最后的解决方案通过不断地检测直到搜索范围一点点缩小才得以通过。

CATIA在参数约束下被用于设定观测系统，并考虑到特殊的行为范围。ModeFRONTIER则根据随机的非线性原理被用到种群变异的生成中。这些突变通过ROBOT自动加工，而形成了突变的基本结构评估，以此来评估它们在建筑上的潜能。而这些“幸存者”又进入modeFRONTIER中新一轮的筛选，由此形成一个反馈环。

本项目的关键是多目标的优化过程，在任何一个单独的搜索中都没有100%的最佳目标，它更倾向于是一个多元化独立取向的生态方式，能够在高水平的组织中营造出一种错综复杂的趋势。以下是该方案的三个边界条件：

Ⅰ. 翘曲稳定性响应（自转的）

Ⅱ. 弯曲响应(深度)

Ⅲ. 地基响应（转换）

这种工作方式重新界定了工程的工作流程，同时潜在地界定了建筑师和工程师在设计过程中所扮演的角色。我们需要自觉避免的危险是一种优越感，它是一种效率的原本需求，而不是一个生成的设计过程，能够创造出有效率的建筑种类。

The tower

This design for the Cheongna City Tower is based on creating an innovative spatial and structural energy production device which will become an operational symbol of the future for the IFEZ Cheongna region. Located at the intersection of the main pedestrian passageway from east to west and the main artificial waterway from north to south in Lake Park, the Tower is intended to be a hub of urban activity and a new destination for the region. It is 400 m tall and offers views of the ocean, the Incheon Airport to the west and Mt. Geyang to the east.

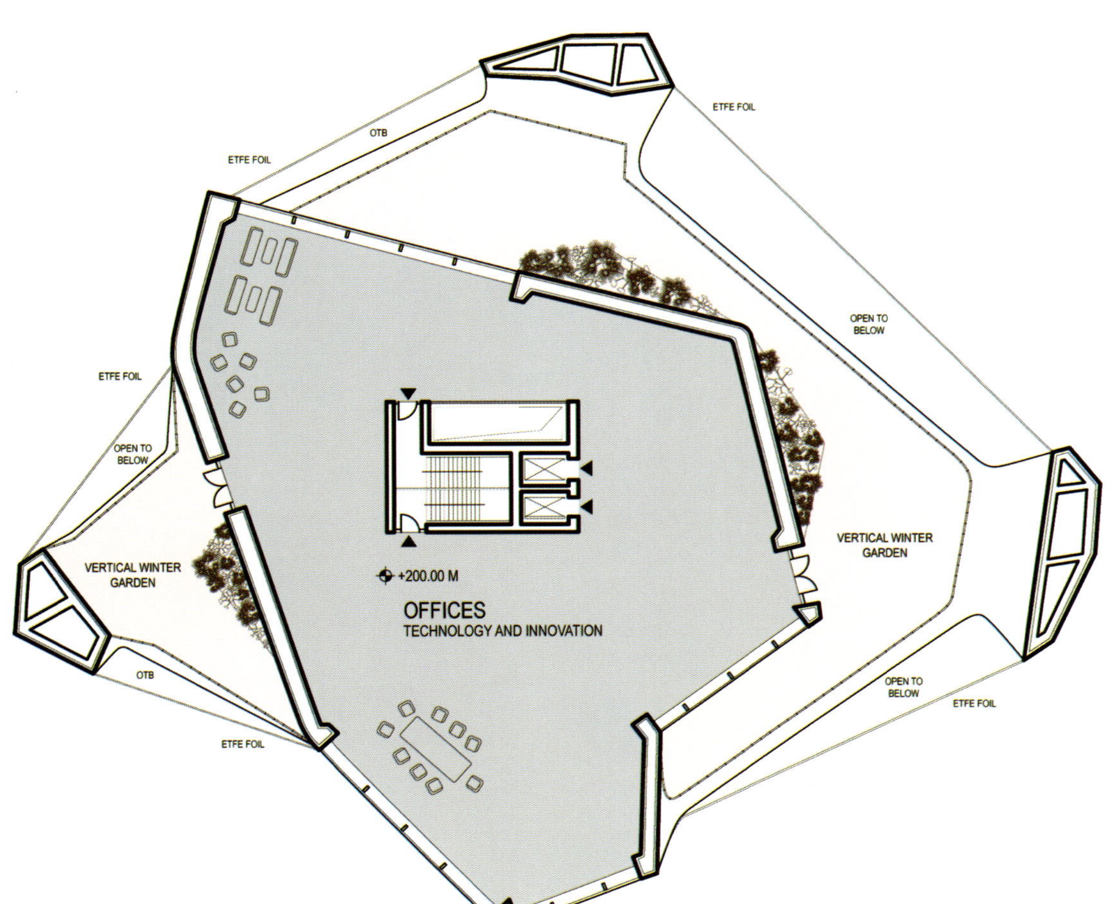

The lower levels of the Tower contain various leisure and cultural activities such as art and design exhibition spaces, an assembly and lecture space, gift shops, and bars. The mid-levels of the Tower contain public Sky-Terraces every 50 m as well as a Business Spine which contains showroom office space for various technology companies and cultural institutions. The upper levels of the Tower contain an astronomical observatory, a seasonal high-end restaurant with star chefs, and various lookout points and observation decks.

The Vertical Wintergarden as energy device

The envelope of the Tower is based on a double-shell construction which creates a Vertical Wintergarden. It houses various observation decks, botanical gardens, and leisure activities. At night it glows in changing color and light, visible from landing aircraft as well as the surrounding territories.

This Vertical Wintergarden is a naturally ventilated buffer zone isolated with ETFE foil on the exterior. It operates as a greenhouse in the winter, collecting heat energy to be circulated throughout the building and as a thermal insulator

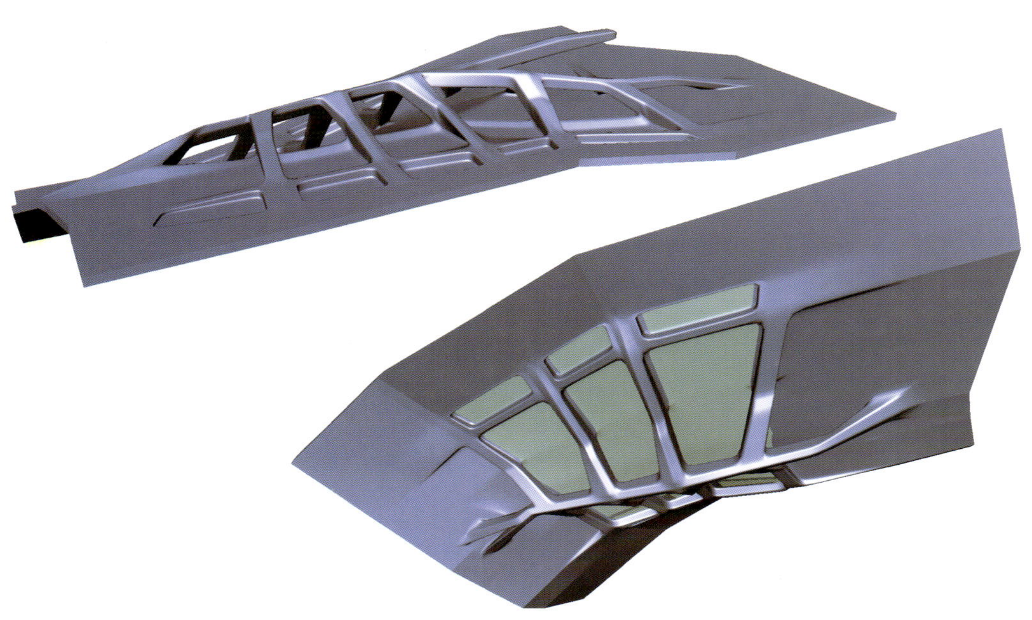

in the summer, protecting the internal spaces from thermal gain. Rising hot air in the space accumulates at the top where it is passed through heat exchangers and micro-turbines which recoup the energy before the air is exhausted.

In addition, the Wintergarden is used in summer as a storage device for cool air which is collected at the top of the tower through large wind intakes which are opened during the night. This is based on the model of passive nighttime cooling towers literalized at a mega-scale. Finally, a reversible hydronic system is integrated into the exoskeleton of the building which provides radiant heating and cooling. It is supplied with cooler or warmer fluids from geothermal source heat pumps but it also utilizes the available cool water from the lake basin on site when possible.

The plaza and the roof

The typology of vertical tower and horizontal plinth is a classic architectural problem. Since the birth of the high-rise, there have been many approaches—both utopian and opportunistic—to resolve how towers connect to the ground as well as to horizontal buildings. This proposal biases continuity between the Tower and its horizontal extensions in order to create a fluid and protected space beneath. These extensions adapt to perform three different functions: ferry platform to the east, retail building to the north, and Plaza Roof to the south.

The Plaza is characterized by urban and natural topography, including lush vegetation and exotic flowers, a network of waterways and pools, and hardscape terraces. Interspersed are cafes, restaurants, cultural events, and the attendant amenities. The space becomes a complex, lively urban space where visitors can relax, meet friends, and engage in leisure activities. It is both a sanctuary and a vibrant urban atmosphere.

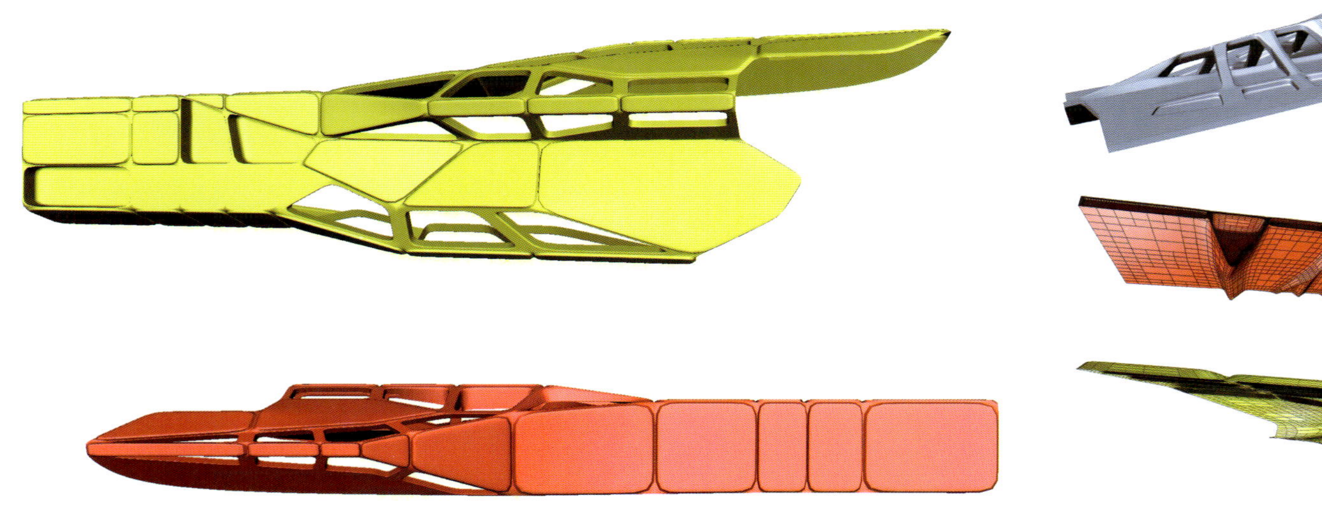

I. SHELL-TO-BEAM

II. SURFACE-TO-PLEAT

III. SURFACE-TO-ARMATURE

Evolutionary exoskeleton

The structure of the Tower is based on a steel exoskeleton rather than a traditional structural core model. Three main structural spines weave along the facades, varying in terms of depth, width, and rotation in response to vertical and lateral forces as well as geometrical rules set by the design team. Similar to the inside of a turtle's shell, these spines are merged together to form a hybrid of monocoque and frame-and-skin construction types. The structural morphology becomes that of smooth gradients between surface, surface relief, and strand which is a continuing interest of EMERGENT.

CATIA, modeFRONTIER, and ROBOT

A combination of parametric software, population generating software, and structural analysis engines is used in a bottom-up nonlinear engineering process. This process is similar to natural selection in nature where populations of mutations are generated and then fitness tested by environmental forces. Successful solutions are then bred and tested until the search eventually narrows.

CATIA is used to set up the geometry within parametric constraints, allowing for a particular range of behavior. ModeFRONTIER is then used to generate populations of mutations based on a stochastic, non-linear method. These mutations are then processed automatically through ROBOT which performs a basic structural evaluation of the mutations, which are then evaluated for their architectural potential. The "survivors" are then propagated into a new generation in modeFRONTIER and so on, establishing a feedback loop.

The key here is a multi-objective optimization process, which does not have the goal of 100% optimum in any single search but rather a more ecological approach of multiple individual tendencies producing complex, emergent tendencies at higher levels of organization. For this project three boundary conditions are used:

I. Spine Buckling Response (Rotation)

II. Spine Bending Response (Depth)

III. Spine Footing Response (Translation)

This way of working redefines engineering workflows and potentially, the roles of architect and engineer in the design process. The danger here—something we are consciously avoiding—is an anemic understanding of optimization as a reductive quest for efficiency rather than as a generative design process which can result in architectural species characterized by both efficiencies and excesses.

Castle in the Sky

空中城堡

RAMDAM团队

项目地点	意大利，拉蒂纳
项目	水楼的修复
客户	Foundation Wilmotte & City of Latina
基地面积	9300平方米
建筑面积	3000平方米
建筑师	RAMDAM团队

Project Location	Latina, Italy
Project	Rehabilitation of a Water Tower – Water Consciousness Trail & Offices
Client	Foundation Wilmotte & City of Latina
Site Size	9,300 m^2
Building Area	3,000 m^2
Architect	RAMDAM

水塔细长的结构正好适合这个原本是水资源管理的区域。此水塔是城市之外的制高点。它被打造成现代的休闲项目，成为了我们这个时代生态问题的一个标志。本项目中公共空间得到了充分的利用。

渗透： 2 000平方米的休闲活动区基于半隐藏的水塔。靠近水库打开了第一个由街道通往地下的通道，将生气带入到水池中。

蒸发： 栏杆由高反光的不锈钢包裹，使得水塔的底部仿佛不存在一般。与地面脱节，使得水塔变得十分不可亲近。参观者通过电梯才能使其与地面之间产生联系。

挥发： 参观者于此，仿佛与外界脱节，独立于世界之外，在顶端有一个平台，是一个原用于庆典活动的顶楼空间，并有用于跳舞的舞池。

降落： 一条坡道环绕水库从雨篷顶部穿过。它指引游客在一个高层次的通道上通过。在这个旅途中，穿过云层，从另一个视角欣赏这座城市。水塔变成了在天空中的城市。

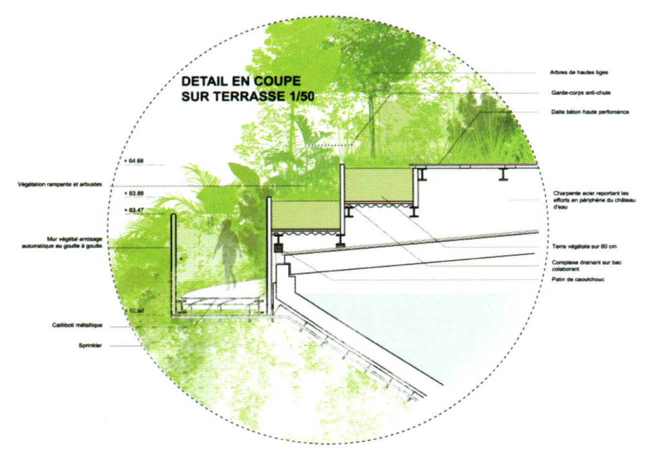

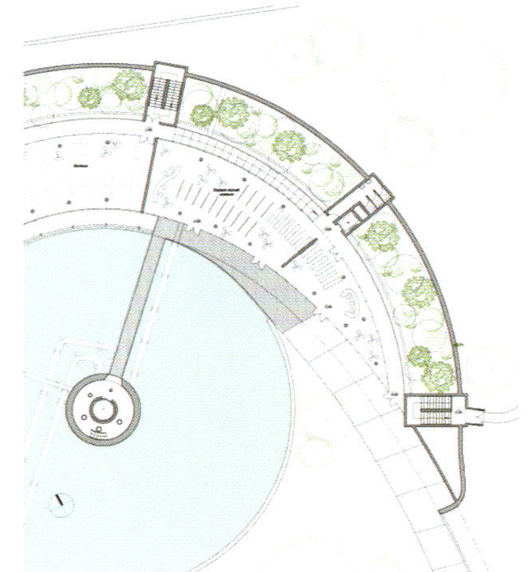

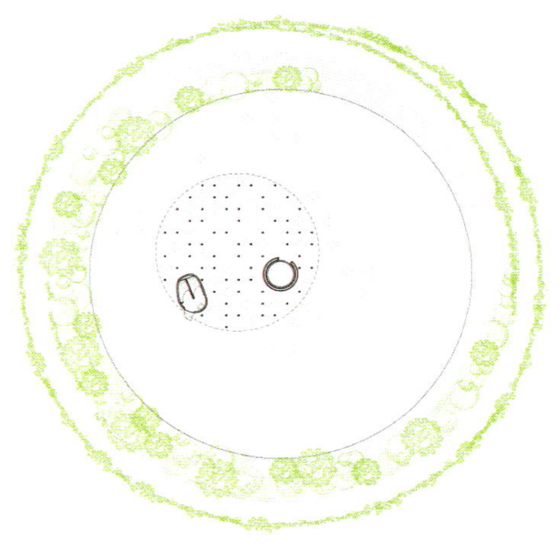

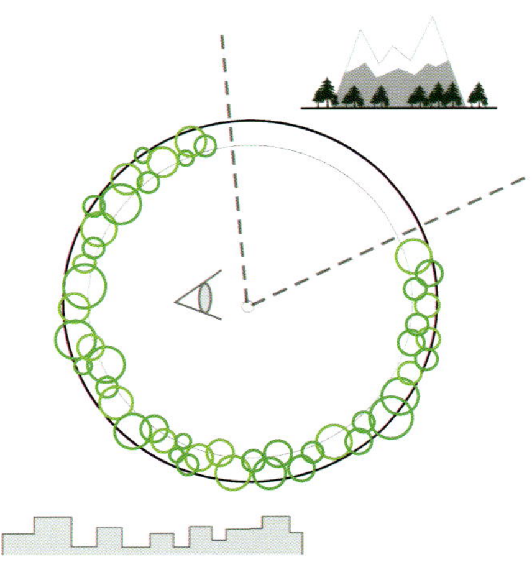

The slender architecture of the water tower dominates a territory that is historically devoted to water management. The reservoir is a summit outside of the city. It is enriched by a contemporary recreational programme, and it becomes the symbol for the ecological issues of our time. The public space finds itself amended.

Infiltration : 2,000 square metres of tertiary activities constitute the semi-buried base of the water tower. Access to the reservoir opens up the first walkway from the street to the underground that brings life around a pond.

Evaporation : The shaft, dressed in highly reflective stainless steel, makes the foot of the reservoir appear not to be there. Disconnected from the ground, it takes on a volatile and inaccessible aspect. The visitor is invited to reconnect with it via an elevator.

Vaporisation : Disconnected from the outside world, the visitor discovers a terrace at the summit, an original rooftop space for festivities with the spirit of a dance hall.

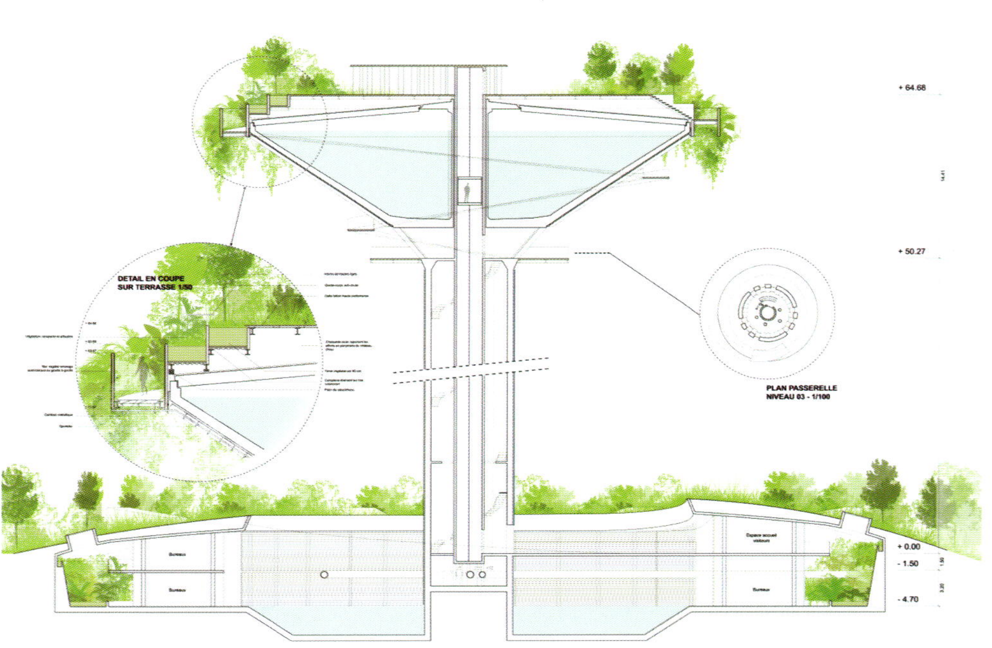

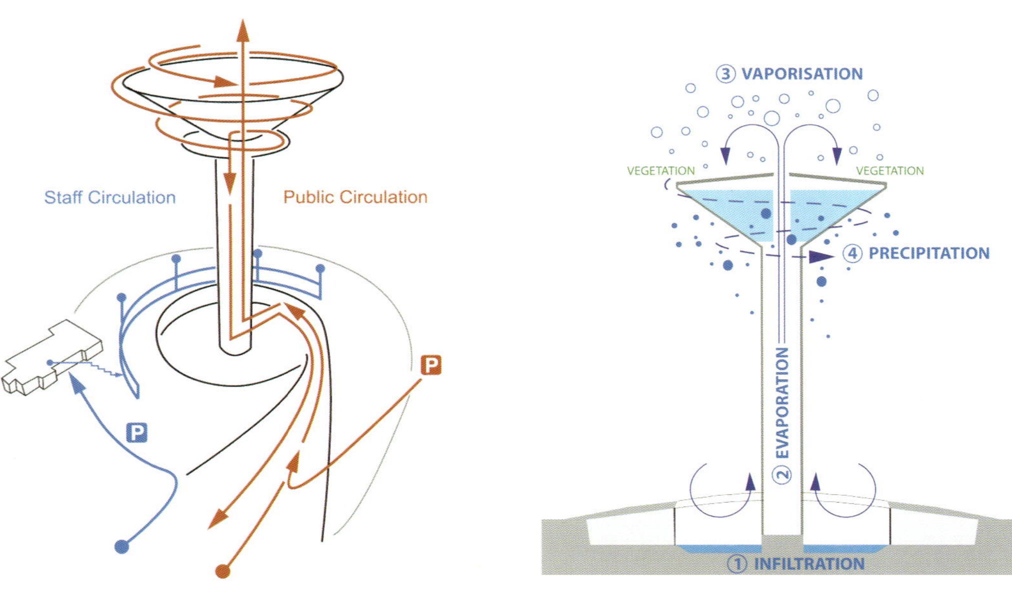

Precipitation : A ramp threads its way round the reservoir through the canopy. It guides the visitors on a high-level walkway. During this journey through the clouds, they see the city from another perspective. The water tower has become the city up there.

Mojave Discovery Center

莫哈韦发现中心

线和空间建筑事务所/Line and Space

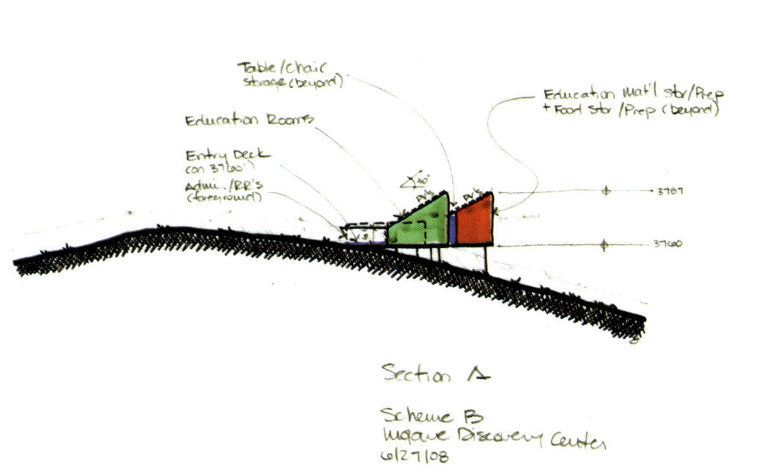

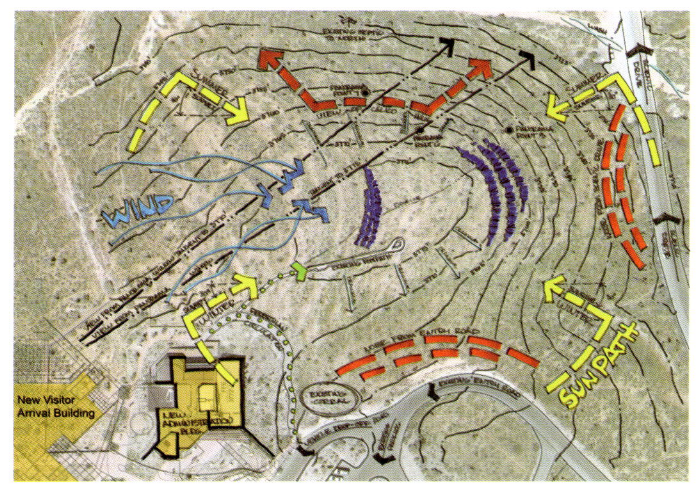

红石峡谷莫哈韦发现中心通过对莫哈韦沙漠环保系统不断增长的知识和理解来进行管理和设计。为了强调教育儿童对沙漠环境尊重的使命，新大楼的设计元素被综合起来并演示了土地与生活圈并不是被完全隔离开的。其中的设备被帮助人们获得身临其境的、亲身体验，在那里可以为当地学校的五年级分担交流课程和实验，来培养与自然环境的一种长久的关系。大楼将作为一个物理意义上的样板，通过可利用的户外空间、提供充足的庇荫、雨水收集以及产生自己的能量和利用自然和持久耐用的材料来阐释如何在沙漠中生存和驻留。

The Red Rock Canyon Mojave Discovery Center is dedicated to instilling stewardship by increasing knowledge and

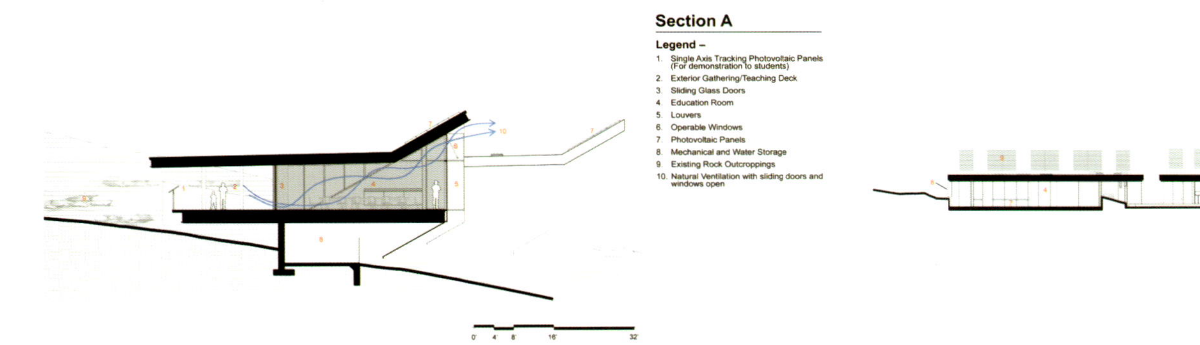

Section A

Legend –

1. Single Axis Tracking Photovoltaic Panels (For demonstration to students)
2. Exterior Gathering/Teaching Deck
3. Sliding Glass Doors
4. Education Room
5. Louvers
6. Operable Windows
7. Photovoltaic Panels
8. Mechanical and Water Storage
9. Existing Rock Outcroppings
10. Natural Ventilation with sliding doors and windows open

Section B

Legend –

1. Staff Workroom (beyond)
2. Catchment for Water Harvesting
3. Entry Bridge
4. Exterior Gathering/Teaching Deck
5. Sliding Glass Doors (beyond)
6. Single Axis Tracking Photovoltaic Panels (For demonstration to students)
7. Operable Windows
8. Vertical Louvers
9. Photovoltaic Panels

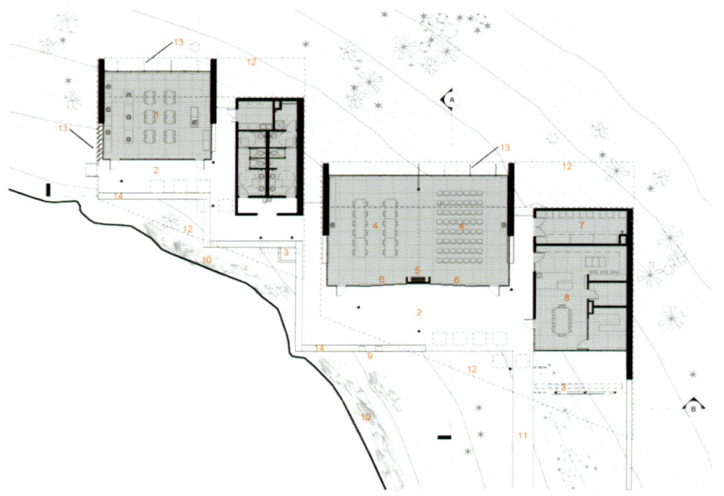

Floor Plan

Legend –

1. Ecology Lab
2. Exterior Gathering/Teaching Deck
3. Catchment for Water Harvesting
4. Education Room
5. Folding Partition
6. Sliding Glass Doors
7. Table/Chair Storage
8. Instructor's Workroom
9. Single Axis Tracking PV Panels (For demonstration to students)
10. Existing Rock Outcroppings
11. Elevated Walkway
12. Edge of Roof (Above)
13. Vertical Shade Screens
14. Concrete Bench (Typical)

understanding of the Mojave Desert ecosystems through a unique experiential discovery program. To emphasize the mission of teaching children respect of the desert environment, design elements of the new building are integrated and expressed to demonstrate that land and life ethic are not separable. The facility is designed to provide an immersive, hands-on experience where the 5th graders from the local school district partake in interactive lessons and experiments designed to foster a lasting relationship with the environment. The building will stand as a physical example of how to exist and conserve in the desert by extending usable outdoor space, providing ample shade, harvesting rainwater, generating its own energy, and using natural and durable materials.

Solar City Tower
太阳能瀑布塔

瑞士RAFF建筑设计公司

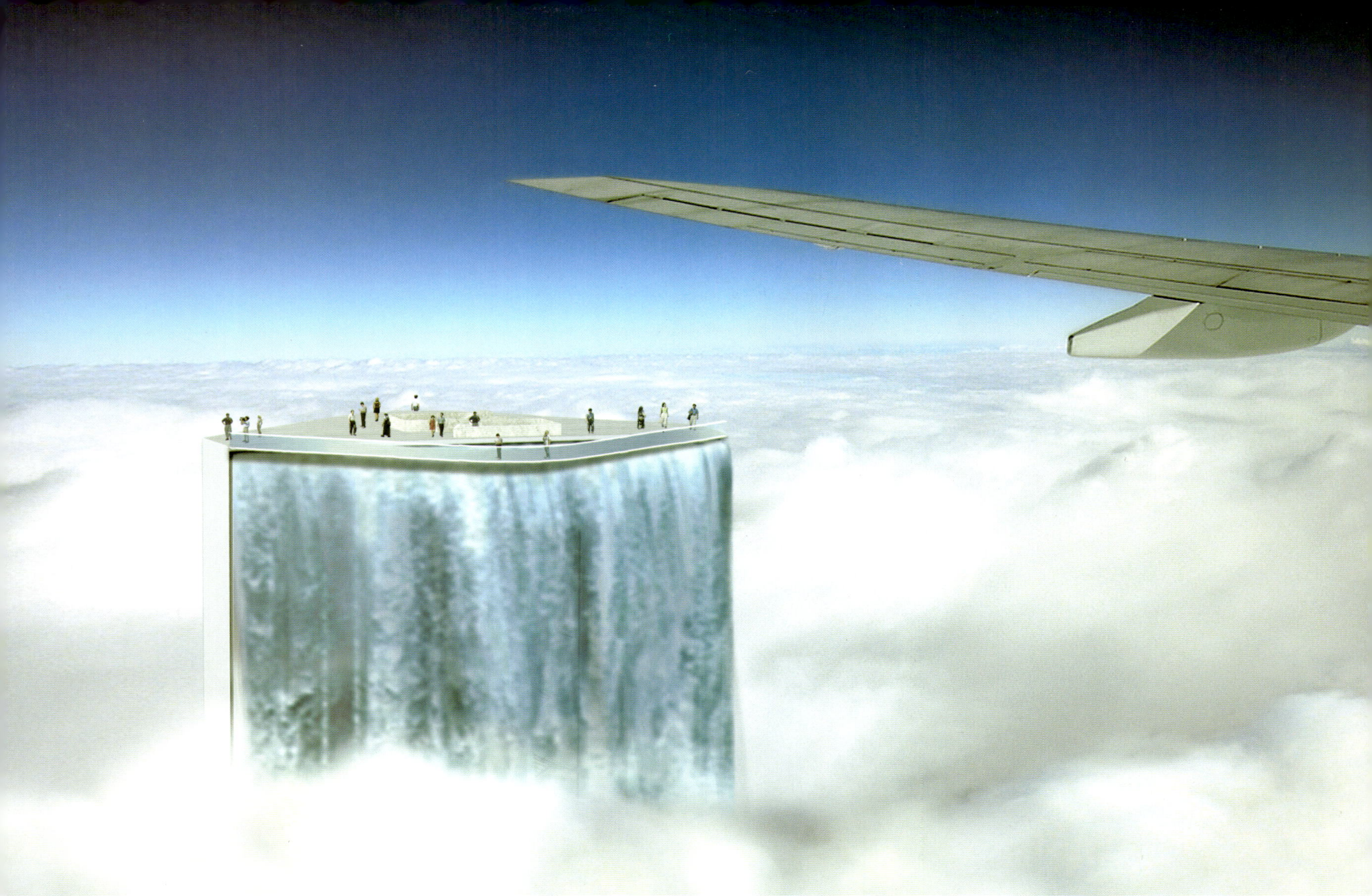

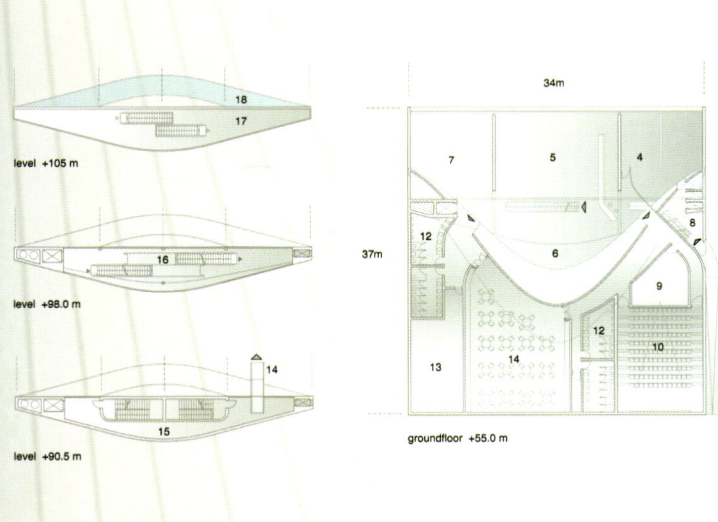

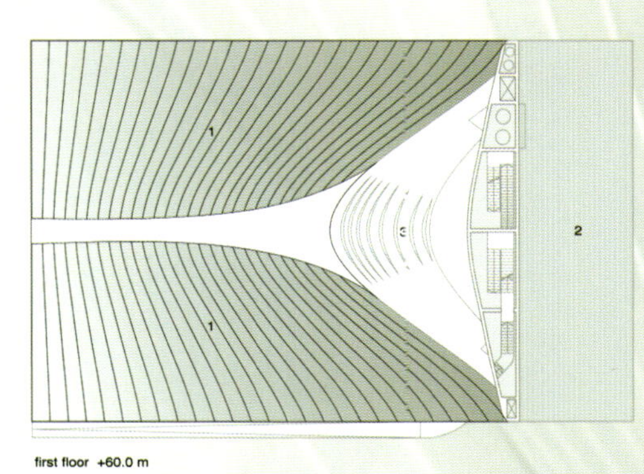

1	solar energy plant	920 sqm
2	water platform	390 sqm
3	amphitheatre	120 sqm
4	administration	100 sqm
5	cafeteria	339 sqm
6	entrance / info	123 sqm
7	shop	112 sqm
8	entrance staff	65 sqm
9	maintenance	52 sqm
10	auditorium	138 sqm
11	M.U.S.	197 sqm
12	lavatories	98 sqm
13	storage	81 sqm
14	bungeejump	6 sqm
15	observation deck 1	127 sqm
16	observation deck 2	127 sqm
17	urban balcony	127 sqm
18	skywalk	78 sqm
	TOTAL	1770 sqm

在设计任务书的说明中，这个项目坐落在里约热内卢Cotonduba岛的海湾，面对着Copacabana海滩。Cotonduba岛承担着里约热内卢空中门户的作用，外来飞机均在此着陆。因此该岛希望建造一个醒目的纵向建筑，以形成类似于航标的作用；同时，也希望为里约热内卢申办2016年夏季奥运会在城市形象上增加足够的说服力。所以，设计的挑战在于拿出一个具有地标作用的塔楼建筑方案，以此欢迎所有来里约热内卢的访客。当然，这包括无论是乘飞机而来，或者是从海上乘船而来的所有访客。

设计理念

本案的目标在于引起人们对传统地标建筑含义的重新思考。它并不仅仅强调一种夸张的外形，更多的是面对即将到来的后石油时代，在生态意义上探寻建筑的实际使用效果。这其实反映了未来社会发展的诉求，也是该项目最为核心的一个理念。本案对传统意义上的建筑的突破在于，试图营造一个“类机器制造”的建筑或城市：它能够有效利用自然资源，为里约热内卢的城市及市民使用自然资源提供能源。我们希望获得一个具有政治感染力的国际奥林匹克标语。继里约热内卢在1992年举办过地球高峰会之后，这座城市不断探寻以可持续发展方式构建城市，参与创建全球宜居环境的方法和途径。而这个项目也许可以成为“零碳奥运”发展历程中的一个最有益的探索。

设计方法

该项目包括一个太阳能发电厂，白天发电供给奥运村，剩余的电量则通过海水泵送到大厦内的蓄水池得以储存起来；到了晚上，海水重新被释放出来，并借助涡轮机为城市供电。总之，产生的电能在为该建筑照明之余，还能用于城市供电。从某种意义来说，这个带有发电厂的塔楼简直就是一个令人叹为观止的自然奇观，就像一个从天而降的城市瀑布是自然力量的象征；也必将反映出生活在城市中的人们，对于周围那秀丽景色的回应。

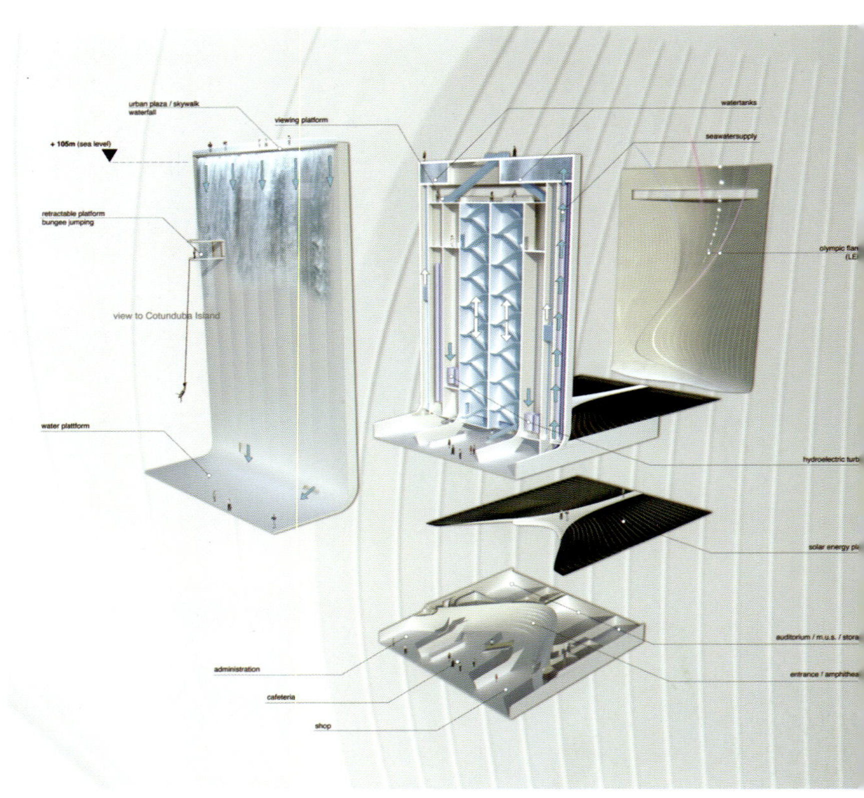

功能布局

通过一个位于海平面60米高度之上的城市广场，行人由此进入建筑；在穿过一个阶梯状的门厅之后，可来到位于一层的入口处。所有的入口区域以及阶梯状门厅都可以充当供人们社交与集会的场所，而该项目对公共空间的重视与营造由此也可见一斑。餐厅和商店就位于“瀑布”的下方，因此具有绝佳的景致。升降电梯可以将游客带到观望台以及建筑最顶端的城市广场。而穿过门厅，就可以找到管理处。可以说，内部的动线正是通过这个专用的出入口以及升降电梯组织起来的。不完全开放的半公共空间位于建筑的后部，因此对它们的使用具有一定的独立性和私密性。还有一个用来蹦极的可回收平台，被设置在海平面以上90.5米处。在海平面以上98米处，有一个观望台用来为人们远距离观景提供便利。城市平台被设置在塔楼的顶部，距离海平面105米处。游客在这里能够以360° 视角观赏到周围的景观，穿过玻璃围栏还可以直接接触到水幕。

The project under consideration should be located in the bay of the city of Rio de Janeiro on the Cotonduba Island, which is the obligatory approach for aircraft landing and will comprise a vertical structure which will seek to become a symbol for those arriving in Rio, creating an image potent enough to enable Rio to triumph in its bid to host the summer Olympics in 2016. The challenge, therefore, consists of designing an observation tower which will become a symbol welcoming all those who visit Rio de Janeiro, whether they arrive by air or sea.

Concept

The aim of this project is to ask how the classic concept of a landmark can be reconsidered. It is less about an expressive, iconic architectural form; rather, it is a return to content and actual, real challenges for the imminent post-oil-era.This project represents a message of a society facing the future; thus, it is the representation of an inner attitude. Our project, standing in the tradition of "a building/city as a machine", shall provide energy both to the city of Rio de Janeiro and its citizens while using natural resources. We hope to attain an international Olympic message with a political appeal. After hosting the United Nation's Earth Summit in 1992, Rio de Janeiro will once again be the starting point for a global green movement and for a sustainable development of urban structures. It will perhaps even become a symbol for the first zero carbon footprint Olympic Games.

Approach

The project consists of a solar power plant that by day produces energy for the city and the Olympic village respectively . Excessive energy will be pumped as seawater into a tower. By night, the water can be released again; with the help of turbines, it generates electricity for the night. The electricity produced can be used for the lighting of the tower or for the city. On special occasions, this "machine

building" turns into an impressive wonder of nature: an urban waterfall, a symbol for the forces of nature. At the same time, it will be the representation of a collective awareness of the city towards its great surrounding landscape.

Organisation

Via an urban plaza located 60 meters over sea level you gain access to the building. Through the amphitheatre, you reach the entrance situated on the ground floor. Both entrance area and amphitheatre can serve as a place for social gatherings and events. The public spaces are also accessible from this point. The cafeteria and the shop are situated beneath the waterfall and offer a breathtaking view. The public elevator takes the visitor to the observation decks and the urban balcony. The administration offices can be reached directly from the foyer. Its inner circulation is organised by an own entrance and the elevator. The semi-public spaces are located in the back area of the building; thus, they can be used separately. A retractable platform for bungee jumping is located on level +90.5. Long distance observation can be done from the observation deck on level +98.0. The urban balcony is situated at the top of the tower 105 meters above sea level. Here the visitor has a 360° view of the landscape and can experience the waterfall while walking over the glass sky walk.

Culture Education

文化｜教育

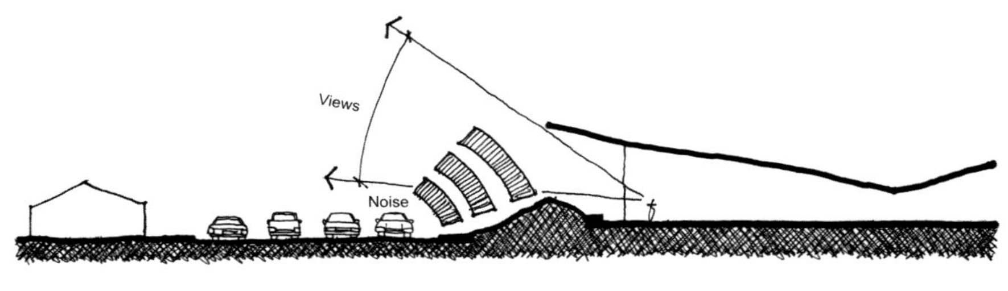
Views
Noise

MAS Museum Aan DE Stroom

MAS博物馆

Neutelings Riedijk Architects、Rotterdam、The Netherlands

项目地点	比利时，安特卫普
建造时间	2006年10月
竣工时间	2010年2月
项目面积	地板面积 19 557平方米 户外建设 11 415平方米
建筑设计	Neutelings Riedijk Architects、Rotterdam、The Netherlands
结构设计	Bureau Bouwtechniek、Antwerp、Belgium
施　　工	ABT België、Antwerp、Belgium
建筑物理分析	Peutz bv ingenieuze adviseurs、Mook、The Netherlands
装置设计	Marcq & Roba、Brussels、Belgium
消防安全	IFSET International Fire Safety Engineering Technology、Asse、Belgium
总承包商	THV MAS、Antwerp (Interbuild, Willemen, Cordeel)
摄　　影	Sarah Blee、Antwerp and Scagliola/Brakkee、Rotterdam、The Netherlands
照片版权	Neutelings Riedijk Architects and City of Antwerp

Project Location	Hanzestedenplaats 2000 Antwerp Belgium
Construction Start Date	October 2006
Completion Date	February 2010
Project Area	19,557 m2 Floor surface 11,415 m2 Outdoor construction
Architectural Design	Neutelings Riedijk Architects、Rotterdam、The Netherlands
Structural Design	Bureau Bouwtechniek、Antwerp | Belgium
Constructive Design	ABT België、Antwerp、Belgium
Building Physics	Peutz bv ingenieuze adviseurs、Mook、The Netherlands
Installation Design	Marcq & Roba、Brussels、Belgium
Fire Safety	IFSET International Fire Safety Engineering Technology、Asse、Belgium
General Contractor	THV MAS、Antwerp (Interbuild, Willemen, Cordeel)
Photography Building	Sarah Blee、Antwerp and Scagliola/Brakkee、Rotterdam、The Netherlands
Copyrights Photography	Neutelings Riedijk Architects and City of Antwerp

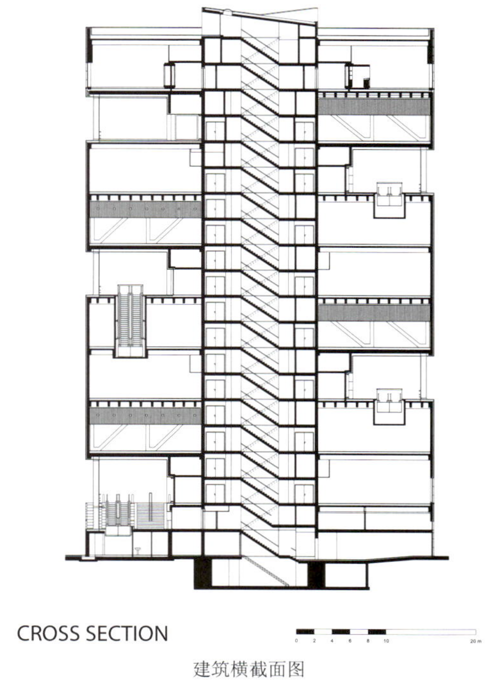

CROSS SECTION

建筑横截面图

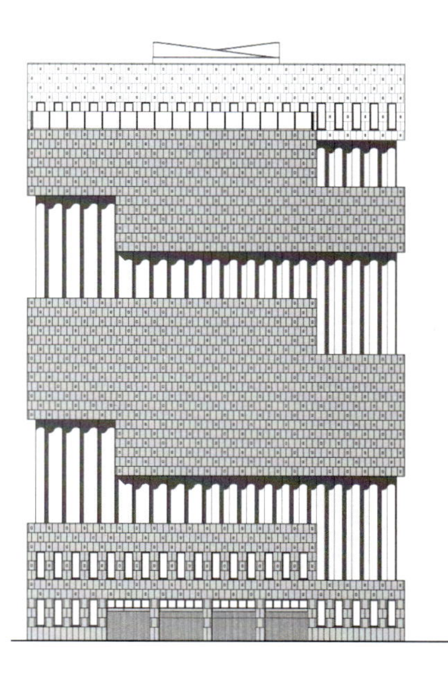

WEST FACADE

西立面图

新的MAS (Museum aan de Stroom Museum by the River)坐落于Eilandje地区，位于旧城市中心的港口。Eilandje名称的由来也是因为这个地区四周被河流所环绕，当桥被架起的时候，这里其实是一个岛屿。

地理位置

MAS博物馆位于Eilandje中心的旧街区。该区域正是安特卫普中心最重要的城市复兴项目所在地之一，并作为一个新兴的城市正在全速发展。

大楼

MAS被设计为高60米的塔楼。十个巨大的天然石块累积起来昭示着历史的沉淀感，充满沧桑，象征着一直以来祖先们的历史演进。它就像是位于旧港区中心装满历史的仓库。塔楼的每一层都围绕中心旋转成直角，创造了一个巨大的螺旋形楼梯。螺旋楼梯空间由波纹玻璃表面包裹而成，形成了一个公共城市展览空间。

全天然的石头表面

建筑的表面由来自印度Agra100厘米x60厘米的全天

然红色砂岩石覆盖。由四面红色的随机图案组成，以此来打破建筑大面积的表面区域，每一面都是不同的颜色。墙、楼梯以及画廊的天花板则都由相同的红色砂岩面覆盖，以此来强调其具有雕塑般风味的体积。广场和建筑本身也由同样的岩石构成，这样一来原本复杂的形式被组合成了一个共同体。红色的砂岩都经由手工刻画并且未经抛光，在岩石的能见面创造了一种调压设计，使得整个建筑呈现一种强烈的表面存在感。

恒温博物馆大厅

严格稳定的博物馆气温维持在22摄氏度，而55%的相对湿度是在博物馆大厅通用的。大厅位于封闭的大楼中，没有窗户，这样一来可以避免能源的流失。这也与博物馆最初的设计吻合，使这些具有历史感的陈列物不会在阳光下曝光。

展厅中多变的可调节气温

展厅由巨大的透明展示玻璃组成，这样一来就为参观者提供了最大的城市景观。由于对流通区域并没有十分严格的要求，因此设计师选择了一个多变的可调节气温系统：室内的气温随着不同的季节而有所变化。12摄氏度将作为画廊冬天室内的最低温度，而在夏天的最高温度则为30摄氏度。

博物馆大厅与画廊的能量平衡

设计师考虑到了高能耗需求的博物馆大厅与低能耗需求的画廊之间的能量平衡，在春秋季节通过利用博物馆大厅以及画廊之间的排气来平衡，这样一来博物馆大厅的余热可以供给画廊。相反地，在寒冷季节画廊则像温室一样可以捕捉热能，并为博物馆大厅提供如温室般的效应。

作为能量缓冲区的画廊

画廊作为能量缓冲区，在这里石头材质可以随着季节时间的变化而变得温暖或是凉爽。石头材质的性能可以调节太阳能的缓冲。地板上设置有含水系统用来输送冷热水。这个系统使得大楼南北区域不同的温度得以平衡。

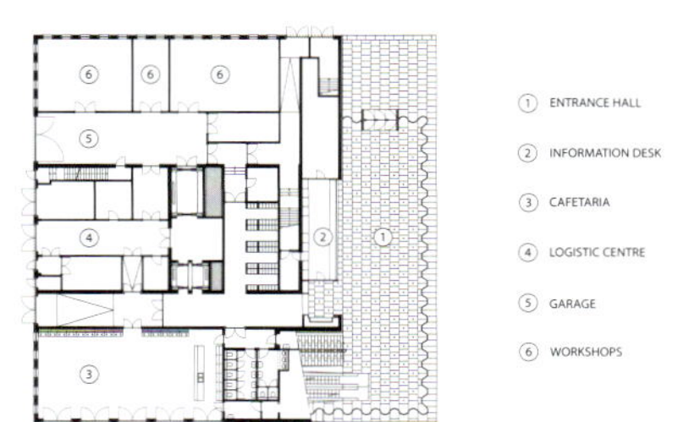

GROUND FLOOR

总平面图

LEVEL 1

一层平面图

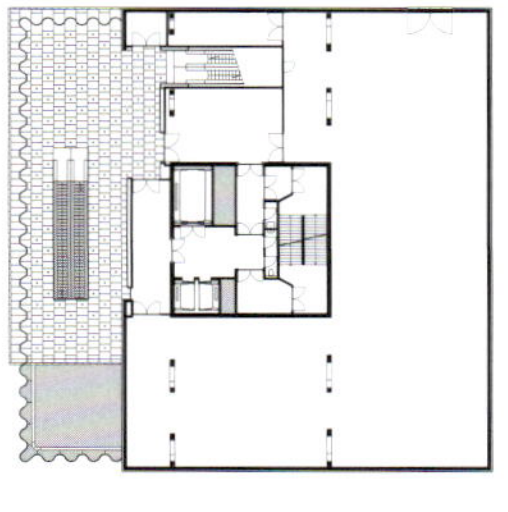

LEVEL 2

二层平面图

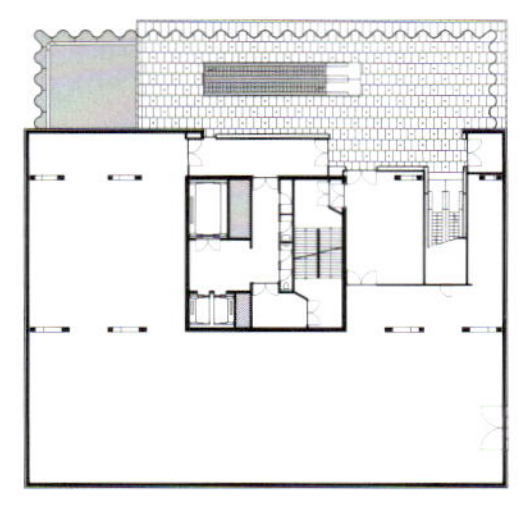

LEVEL 3

三层平面图

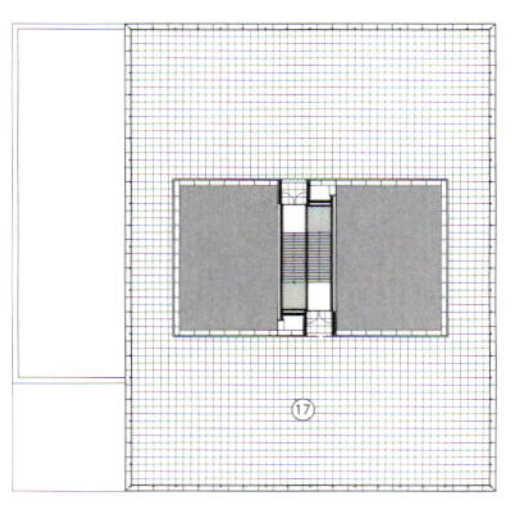

⑰ ROOF TERRACE

ROOF PLAN

屋顶平面图

冷却水处理

MAS将会由来自Bonapartedok的水进行冷却，那里的水温通常低于空气温度。通过利用区域水与室外气温之间的差别，用水泵以及换热器来冷却通风。通过这种方式，附近的天然能源以一种可持续的方式被搜集起来。

集中设置

整个加热、冷却和通风装置都被集中设置在一起。单独的空气处理器则被安置在每一个楼层，这样一来博物馆每一个大厅的气温都能够被很好地调节。

The new MAS (Museum aan de Stroom Museum by the River) is located in the centre of the district called Eilandje, the old harbour district by the centre of the old city. The name Eilandje stems from the fact that this area was surrounded by water on four sides, so that it actually was an island when the bridges were up.

The location

The MAS (Museum aan de Stroom) stands between the old docks in the centre of the Eilandje. This old harbour area is the most important city renovation project in the centre of Antwerp and is in full development as a new dazzling city district.

The building

The MAS is designed as a 60m high tower. Ten gigantic natural stone trunks are piled up as a physical demonstration of the heaviness of history, full of historical objects that are the legacy of our ancestors. It is a storehouse of stacked history in the middle of the old harbour docks. Every storey of the tower has been rotated a quarter turn, creating a gigantic spiral staircase. This spiral space, in which a facade of corrugated glass is inserted, forms a public city gallery.

Natural stone facades

The facades are covered with 100x60 cm red sandstone panels from Agra in India. A random pattern with four shades of red has been made in order to break up the large facade surface areas, whereby no more than two panels of the same colour are adjacent to each other. Walls, floors and ceilings of the galleries are covered with the same red sandstone panels, which emphasises the sculptural nature of the volume. The plaza and pavilion are also constructed of the same stone, so that the complex forms one single unit. The red sandstone is hand carved and unpolished, creating a relief design on the visible side of the stone, thereby giving the building an accentuated tactile appearance.

A stable climate in the museum halls

A strict stable museum climate of 22 degrees Celsius and 55% relative humidity prevails in the museum halls. The halls are located in the closed building components that are best insulated without window displays, thus limiting loss of energy. This coincides with the basic assumption of the museum that the historic objects may not be exposed to daylight.

Variable semi-controlled climate in the gallery

The character of the gallery is transparent due to the large glass displays that provide a maximum view on the city. Since no special museum requirements are imposed on this circulation area, a variable semi-controlled climate has been chosen; the indoor temperature fluctuates depending on the season. A temperature of about 12 degrees has been selected as the lower limit for the indoor temperature of the gallery in the winter, and an upper limit of about 30 degrees in the summer.

Energy balance between the museum halls and the gallery

The balance between the museum halls with a high energy demand and the galleries with a lower one is utilised by exhausting the ventilation air from the museum halls via the galleries in the interim seasons, so that the residual heat from the halls heats up the gallery. Inversely, in the cold periods the gallery functions as a conservatory that captures the solar heat and preheats the ventilation air for the museum halls.

The gallery as energy buffer

The gallery functions as an energy buffer, whereby the stone masses of the galleries are heated or cooled depending on the season and the hour of the day. The inertia of the stones regulates the buffering of the solar energy. A water-bearing system has been installed in the floor which can transport both warm and cold water. This system makes it possible to compensate for the differences in temperature between the north side and the south side of the building.

Cooling with dock water

The MAS will be cooled by means of the dock water from the Bonapartedok, which always has a lower temperature than the air. The difference in temperature between the dock water and the outside air is utilised by means of pumps and heat exchangers to cool the ventilation air. In this way primary energy is collected from the immediate surroundings in a sustainable manner.

De-centralised set-up

The entire installation for heating, cooling and ventilation is set up de-centrally. A separate air conditioning unit is set up on every storey so that the climate for each museum hall can be regulated separately.

Surry Hills Library and Community Centre

莎莉山图书馆及社区中心

Francis-Jones Morehen Thorp

社区综合体	公平，愿望和可持续性
设计总监	Richard Francis–Jones
设计单位	Francis–Jones Morehen Thorp

Community hybrid	Equity, aspiration and sustainability
Design Director	Richard Francis-Jones
Design Company	Francis-Jones Morehen Thorp

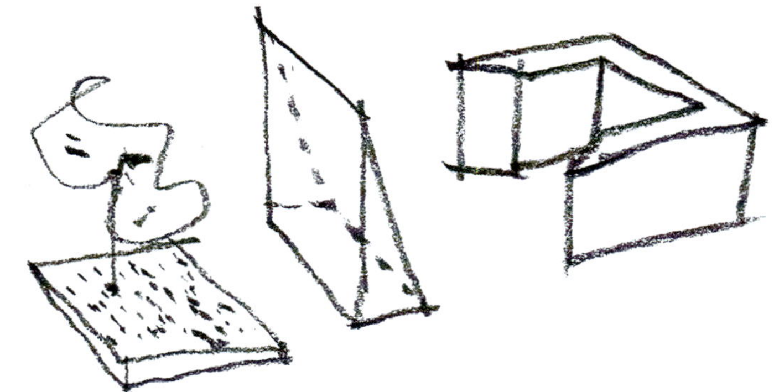

本案位于莎莉山的心脏地带，也是悉尼郊区人口较多的地区。社区里居住的人群在年龄、收入和文化背景方面都是多样化的。建筑所在的环境也是多元的，已有的公寓、排屋、商店，商业和工业场所尺度不一，多为维多利亚式风格。基地的条件很受限，测量结果为25米乘以28米，三面临马路，向东有莎莉山的主干道王冠大街，向南、向西各有两条住宅区街道。

形成项目方案概要时，我们与这个活跃的社区进行了密切的商议，从中产生了方法的关键——社区想要一个人人共享的设施，而不只是一个单纯的图书馆、社区中心或儿童中心，所以，在一个地方，一个建筑里同时实现这些功能，就显得尤为重要。这样，整个社区的人们才能以不同的使用方式，真正地共享这座建筑。同样重要的是，建筑要能够代表和反映所在社区的价值观。

为了响应以上的需求，我们开发了一种新的方式。对悉尼的公共建筑来说，它是全新的，不是一个已有众多先例的单纯的建筑类型，而是一个混合型的公共建筑。它集多种功能于一身：图书中心/资料中心，社区中心，幼托中心。多种功能被整合成一个规模适中的，人人可用的建筑。

“透明度”成为了多个层面上的建筑主题，带来一个引人入胜的、受欢迎的建筑，形成开放的、易于使用的公共景观。重要的是，这座建筑并不只是透明，或仅仅暴露其中的设施，还代表和体现了社区的价值观——易用，开放，透明，可持续。这些价值观，正是人们普遍的愿望。

从我们早期的研究中，产生了四个形式因素：一个新的、简单的开放空间和平台，一个带有棱柱体玻璃的中庭环境，一个悬挂式的U型木架，一个过渡性的门厅空间。

在南侧，柯林斯街封路转化为一个适度的、带有一个上升玻璃平台的公共停车场。这个新空间延伸了建筑的功能，强化了它作为公共空间的特点。

锥形的玻璃中庭响应了该设计颇具雄心的可持续发展目标，带来了有层次的透明感和设计的品质感。玻璃棱柱系列创造出了开放的、透明的立面，使得建筑就像一个开放的玩具房子，标注出新的开放空间，由此，在社区中心举行的活动都能被外面的人看到。这鼓励着人们走进来参与。

U型木架围绕着棱柱形的中庭环境，朝向南面和新的小停车场。木架中固定的部分结合了自动化的百叶窗系统，它过滤并控制阳光，调节景观。这个木架被抬离地面，以创造出建筑的透明度和易用性。

门厅是一个较低的过渡空间，调节着建筑与相邻的商店的比例，创造出一个亲切、透明的入口。悬挂着的屋脊剖面呈云彩状，把日光带进室内空间，并延伸至街道上方，标明了建筑的入口。

景观中庭成为中心位置的象征，并清晰地界定了新建筑与公共场所。视线穿过棱柱体立面，建筑的功能显而易见。

图书馆位于地面及地下一层，包括可供借阅的3000

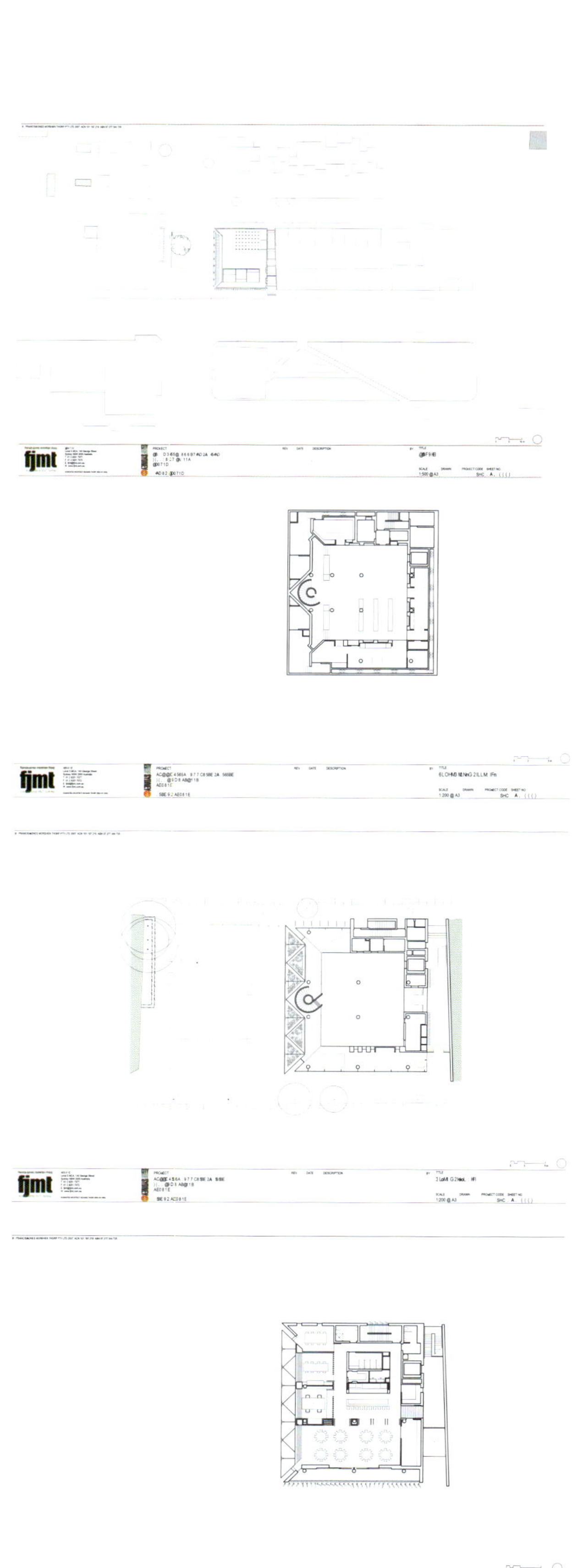

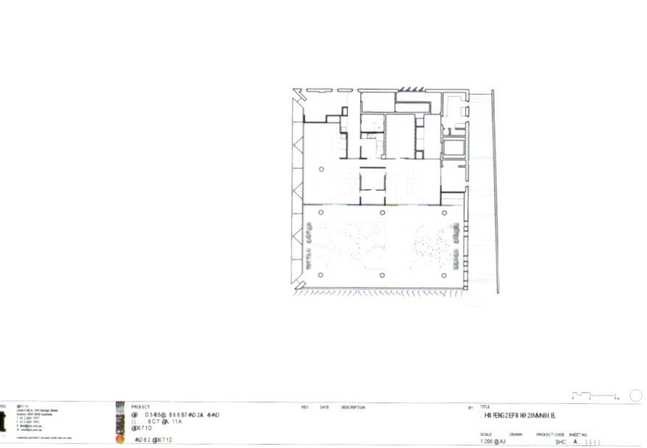

余册图书、地方志、一些参考资料和公共计算机。位于一楼的社区中心包含一个能容纳125人的功能厅和相邻的走廊、会议室、商用教学厨房、邻里中心办公室和一些便利设施。在二楼，幼托中心能够容纳两组共26个儿童（两组儿童分别为1~2岁，2~5岁），还包括一个带有自动遮阳顶棚的户外游戏场地。

本项目其中一个关键目标，在于建立澳大利亚城市建筑可持续型、环境友好型设计的新标准。本建筑实施了多项创新设计，并寻求它们与建筑整体的统一，探索这些系统在表现力上的潜力。在景观中庭中，这表现得最为明显。

景观中庭里一系列三角形、锥形的通风井把户外的洁净空气带入室内，并对它进行冷却。同时，使用植物来进行生物过滤，以减少污染物质。我们对花园里的植物进行特殊化的挑选，把植物放进玻璃围成的装置中。自然光通过这些不同层次的玻璃体和花园，被过滤后流入室内。由于使用独特的空气质量系统，以自然植物为过滤体，空气在建筑底部自然冷却。

在环保节能方面，我们在设计中应用了一组热迷宫，作用是过滤室内空气并调节室内温度。此外，创新之举还包括阳光跟踪式的木制百叶窗系统，自动的遮阳织物，混合式通风设备，大面积的光伏矩阵，地热冷却孔，绿屋顶。

电脑化的智能建筑管理系统（BMS）能够自动监测和控制内部环境，全天候调整通风、遮阳百叶窗以控制热负荷、光线和遮阳。在有需要时，系统能够开关灯具。BMS还能监测和记录电、水压系统，使建筑环境效率达到最大化，并能够辨认系统错误。

自开放以来，中心受到了当地社区的欢迎。对所有年龄、分属于不同社会团体的人来说，这都是一个受欢迎的社区场所。它除了提供设施，还体现着如下信息：让人们公平地取得信息和资源，是建造一个社区共同体的本质。

This project is prominently located in the heart of Surry Hills, an inner-city suburb of Sydney whose community is characterised by a diversity of age, income and cultural backgrounds. The architectural context is also diverse: residential apartments, terrace housing, shops and commercial/industrial premises vary in scale though their architectural style is predominantly Victorian. The site is very constrained, measuring just 25 metres by 28 metres and bounding on three edges by roads: Crown Street, the main street of Surry Hills to the east and two residential streets to the south and west.

The project's brief is developed in close consultation with the very active local community. The key approach that emerges from these discussions is that the community wants a facility that everyone could share. Rather than only a library or a community center or childcare center, it becomes clear that it is important to have all of these facilities together in one building, in one place. In this way the building becomes a truly shared place where the whole community could meet and use in different ways. Important, too, is for the building to represent and reflect the community's values.

In response we develop what for Sydney is a new type of public building. It is not a singular typology, for which there are many precedents, but a hybrid public building that is many different things in one: a library/resource centre, community centre and childcare centre are all integrated into one modest building and accessible by all.

Transparency becomes an architectural theme at many levels, allowing an inviting and welcoming building that is accessible and open to public view. At the same time it is important that the building is not merely transparent, or only expose what is accommodated within, but that it represents and embodies the values of the community. Accessibility, openness, transparency and sustainability are key values as a general sense of aspiration.

From our early studies, four integrated formal elements emerge: a new simple open space and platform, a prismatic glass environmental atrium, a suspended U shaped timber form and a transitional foyer space.

On the southern edge, the Collins Street road closure is converted to a modest public park with a raised grass platform. This new space extends the function of the building and reasserts itself as a public place.

The tapered glass atrium evolves in response to the ambitious sustainability objectives of the project, and equally to the sense of layered transparency and the project's aspirational quality. The series of glass prisms create an open, transparent façade, akin to an open dolls house, and address the new open space so that all the different activities of the centre are visible and displayed, encouraging participation.

The timber U form embraces the prismatic environmental atrium and orients both towards the south and the new little park. The solid sections of this timber form are made of automated louvre systems that filter and control sunlight and view. This warm timber form is lifted above the ground to create transparency and accessibility.

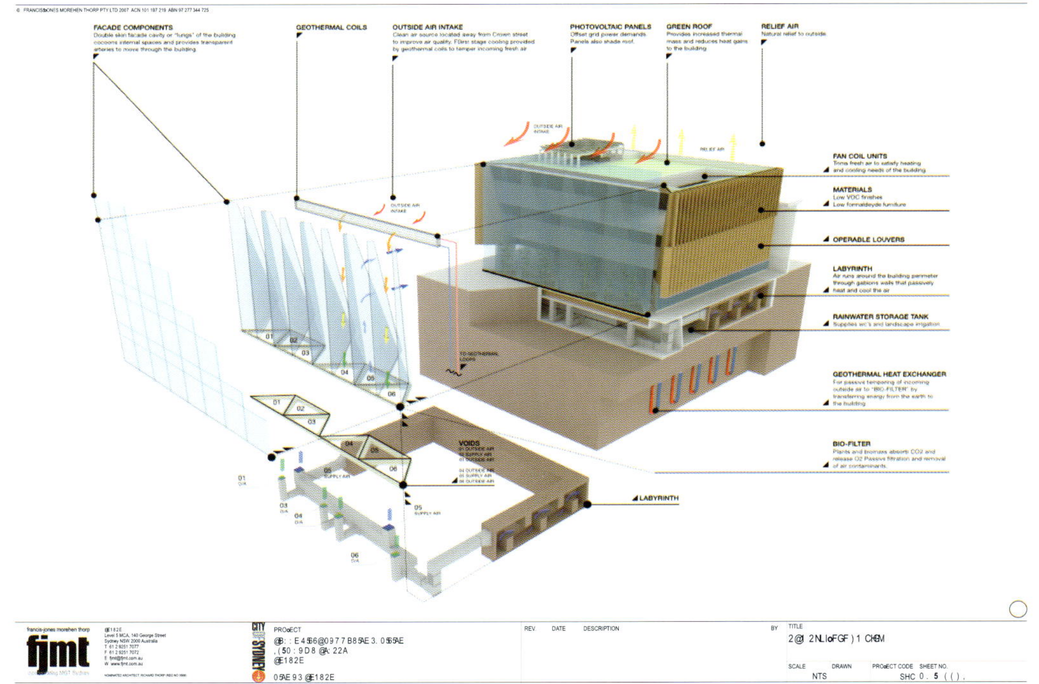

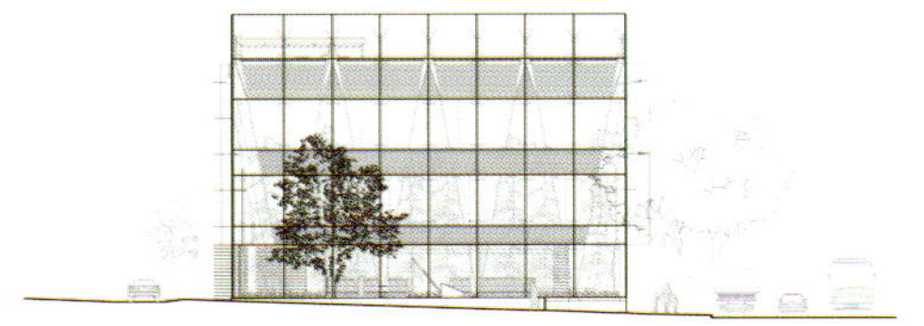

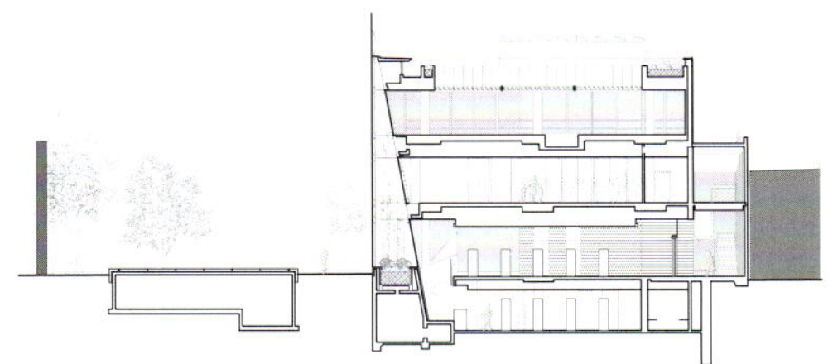

The foyer space is a lower transitional form that mediates the scale of the building against the adjacent shops while creating a welcoming, transparent entry. Suspended cloud-like roof profiles bring daylight into this space and extend out above the street to mark the entrance.

The environmental atrium has become emblematic of the center and clearly identifies the new building and public place. Looking from the new park through this prism façade, the functions of the building are apparent.

The library on the ground and lower-ground levels contains a diverse borrowing collection of approximately 30,000 items, local history collections, some reference material and public access computers. The community centre on level one comprises a function facility for 125 people and adjoining verandah, meeting rooms, commercial teaching kitchen, Neighbourhood Centre administration offices and amenities. On level two, the childcare centre provides accommodation for 26 children in two groups (1–2 and 2–5 years) and includes an outdoor landscaped play space with automatic shade roof.

A key project objective is to establish a new Australian standard of excellence for environmentally sustainable design in civic buildings. The building incorporates many sustainable design innovations and seeks to integrate these into the architecture and explore the expressive potential of such systems. This is most evident in the environmental atrium.

The environmental atrium's series of triangular, tapering airshafts draw in clean outside air and passively cool it. Experimental use of plants to bio-filter pollutants is integrated in the gardens of specially selected plants within these glass enclosures. Natural daylight is filtered through these layers of glass and garden and flows deep into the interiors.

The array of environmental initiatives intrinsic to the design also include a thermal labyrinth for passive filtering and tempering of the air, solar-tracking timber louvre systems, automated fabric shading, mixed mode ventilation, extensive photovoltaic array, geothermal cooling bores, green roof, rainwater collection and recycling, and sustainable material selection.

© FRANCIS-JONES MOREHEN THORP PTY LTD 2007 ACN 101 197 219 ABN 97 277 344 725

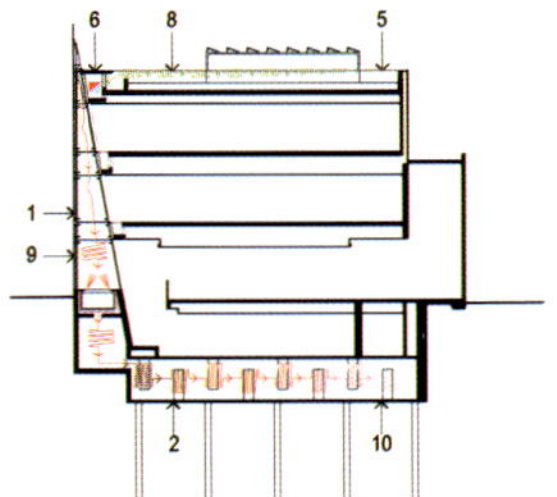

Intake Air Path

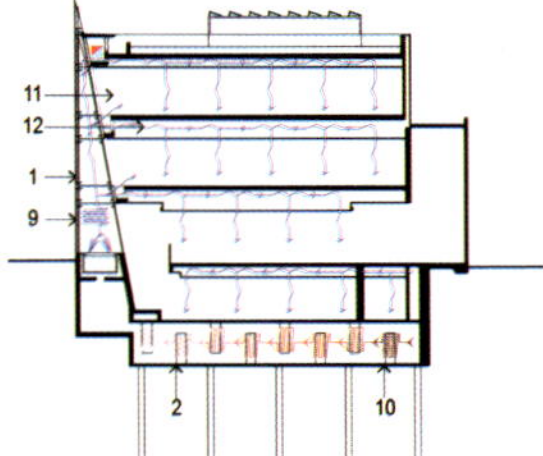

Supply Air Path

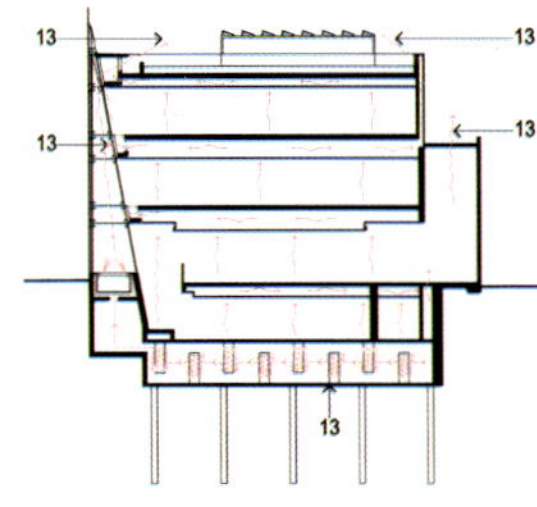

Night Flush

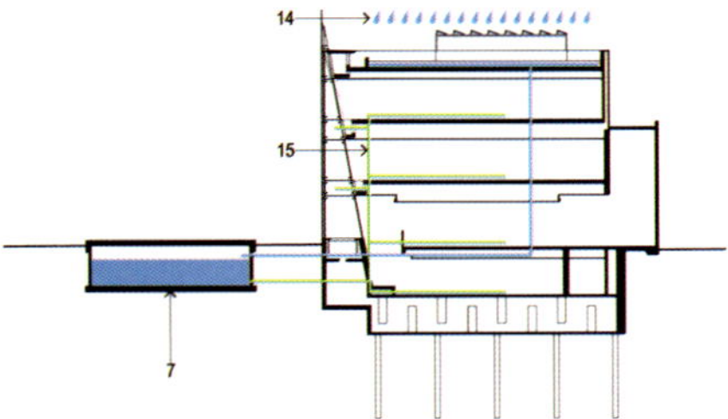

Onsite Water Collection

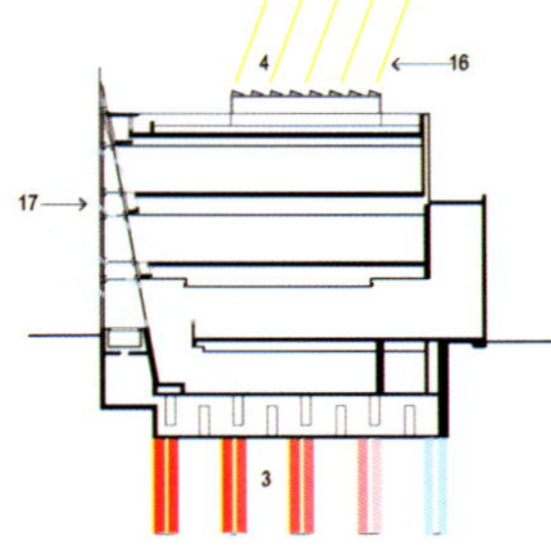

On-site Power Generation

LEGEND
1 - ENVIRONMENTAL ATRIUM
2 - THERMAL LABYRINTH
3 - GEOTHERMAL BORES
4 - PHOTOVOLTAIC CANOPY
5 - GREEN ROOF
6 - GEOTHERMAL CHILLERS
7 - WATER STORAGE TANK
8 - SUPPLY AIR DRAWN ACROSS GREEN ROOF
9 - AIR TEMPERED AND FILTERED THROUGH ATRIUM
10 - AIR TEMPERED AND FILTERED BY THERMAL LABYRINTH
11 - FILTERED TEMPERED FRESH AIR SUPPLY
12 - "VAV" UNITS TREAT AIR TEMPERATURE AS REQUIRED
13 - NIGHT FLUSH EXHAUST AIR PATH
14 - WATER CAPTURED ON GREEN ROOF AND DIVERTED TO RAINWATER STORAGE TANK
15 - RAINWATER RE-USED IN IRRIGATION AND SANITARY FLUSHING
16 - PHOTOVOLTAIC ARRAY OPTIMISED TO OFFSET GEOTHERMAL SYSTEM
17 - FILTERED SOUTHERN LIGHT THROUGH ATRIUM MINIMISES HEAT GAIN

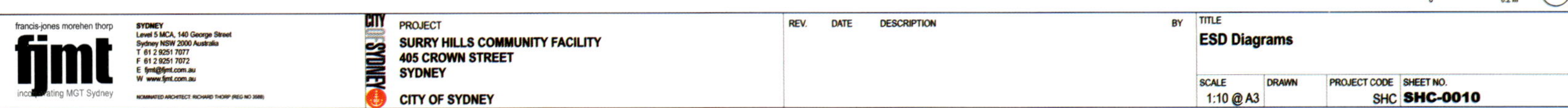

Computerised building management and control systems (BMS) automatically monitor and control the internal environmental conditions of the building, adjusting the ventilation and sunshade louvres throughout the day to control heat load, light and shade, and switching lights on and off when required. The BMS also monitors and records both electrical and hydraulic systems to maximise the environmental efficiency of the building and identify system faults.

The centre has been embraced by the local community since its opening. It is a welcoming community place for all ages and all social groups. It provides facilities that embody the values of equity of access to information and resources that are essential to build communities.

Ordos Art Museum

鄂尔多斯美术馆

徐甜甜

项目地点　内蒙古，鄂尔多斯
项目功能　美术馆
客户　鄂尔多斯市江源水务工程有限公司
设计时间　2005年8月~2006年4月
建造时间　2006年5月~2007年8月
场地面积　4200平方米
建筑面积　2700平方米
建筑师　徐甜甜，纪尧姆.奥布里，陈英男
摄影师　Savoye/周若谷

Project Location　Ordos, Inner Mongolia
Program　Art Museum
Client　Ordos Jiangyuan Water Engineering Co. Ltd
Design Milestones　2005.08—2006.04
Construction Milestones　2006.05—2007.08
Site Area　4,200㎡
Building Area　2,700㎡
Architect　Tiantian Xu, Guillaume Aubry, Yingnan Chen
Photographer　Savoye/Ruogu Zhou

建筑模型图

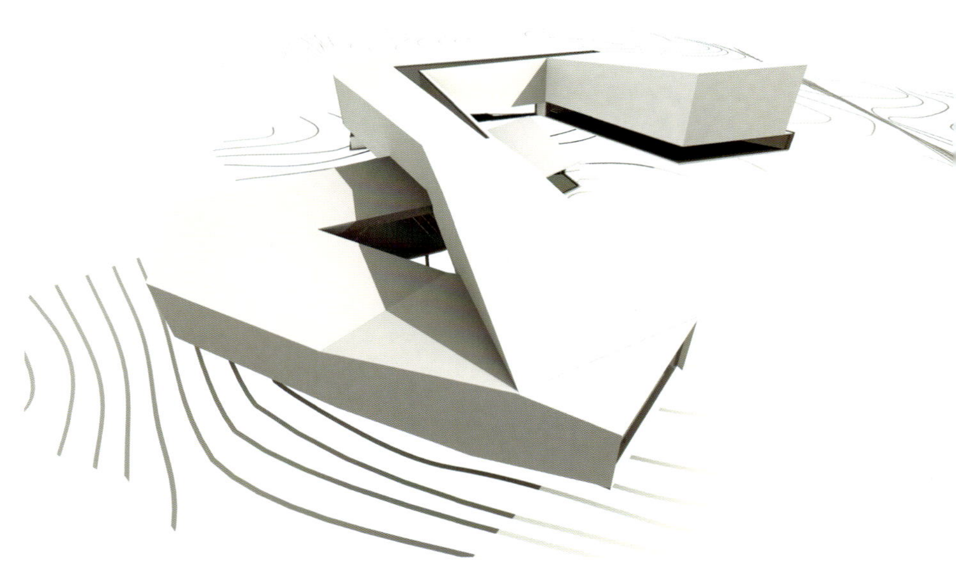

立面展开示意图

景观与背景

鄂尔多斯新区位于东胜与阿镇之间，距东胜22公里、阿镇3公里，规划控制面积155平方公里，建设面积32平方公里，到2020年人口达到30万。新区背靠青春山，濒临乌兰木伦河，具有开阔平坦的地貌地形，以及考考什那水库的水源保障，以市政府迁入为标志，将规划发展成鄂尔多斯市未来的政治、文化、科教、金融中心和技术产业基地。

然而2005年第一次到考考什那时，很难想象几年之内这里将会是一个容纳数十万人口的新城：四周荒无人烟，考考什那水库平静的水面倒映着湛蓝的蒙古天空，地势开阔平坦，偶尔有条形沙丘舒缓平展。这一片沉睡千年的荒原转眼灯红酒绿回归人世。

在这片原生态的景观中根据规划中成熟城市的条件设计美术馆，既能给建筑师带来自由发挥的可能，也是我们面临的最大的挑战：建筑跨跃时空，要以现有的生态景观作为起点和记忆，还要满足将来城市的功能需求。

在规划协调之后，美术馆选址定于沙丘之上，利用沙丘建设要求的低密度，又可以使美术馆占据高地，在未来形成城市里显著的公共建筑。

建筑

美术馆有两条功能主线：公共展览线路和内部资料线路。

展览的交通以线性展开，水平低矮的入口顺沿沙丘坡地的自然地形扭转，在高处则挑起远望考考什那水库，继而反转，以自身建筑作为结构承载，蜿蜒回落到

地面展厅——既是公共展览流线的结束高潮，或又向下延续开始内部资料流线——作为出口与入口相互对望，整个公共交通呈现一个"8"字型的连续线路。

建筑形体因此在沙丘上蜿蜒延展，与地形相互作用形成半围合的院落或广场。

这个流线空间一路上跌荡起伏，根据高度和周边地势形成大小高宽尺度不同的横截面。

沿路结合景观，空间和地势，或以尽端点式玻璃幕远眺水库，或以侧墙连续玻璃引入院落风景，或以顶部开窗围合展览内容。

光线也因此呈现多样化，光影和空间交织出特有的节奏。

在这个节奏里，建筑空间内的艺术情景和外部自然风景交替呈现，参观者的心理空间在艺术情境，建筑尺度和自然风光之间弹性转换。

城市

鄂尔多斯美术馆于2007年八月落成开馆，这时候的鄂尔多斯新区基础设施早已完成，市政建筑也已经落成大半。美术馆旁边正在施工的灰砖房子是艾未未设计的艺术家工作室；附近沙丘上一系列由艾未未率领的瑞士建筑师参与的文化类建筑也正在设计中。以这条"沙丘文化走廊"催生的创意文化园区围绕周围的平地，已经通过规划。周边地区的城市化不可避免，一个新的城市正在崛起。美术馆像撒下的第一颗种子，渐渐蔓延开来生成一片。

不过在这一刻，它还站在沙丘上，翘首期盼它的城市的生成。

Ordos Art Museum is the first building of Ordos' new civic center on a stretch of sand dunes along the lake that is dedicated as a "public corridor" with art and cultural facilities. This 29,000 sq.ft. of exhibition and research space is distributed within an undulating form with a central span lifting clear off the ground, suggesting a desert viper winding over the dunes.

The space is conceived as one uninterrupted room with a series of openings absorbing natural light while offering cinematic views of the raw surroundings; the art exhibition mingles with the natural landscape that becomes an integrated experience for viewers.

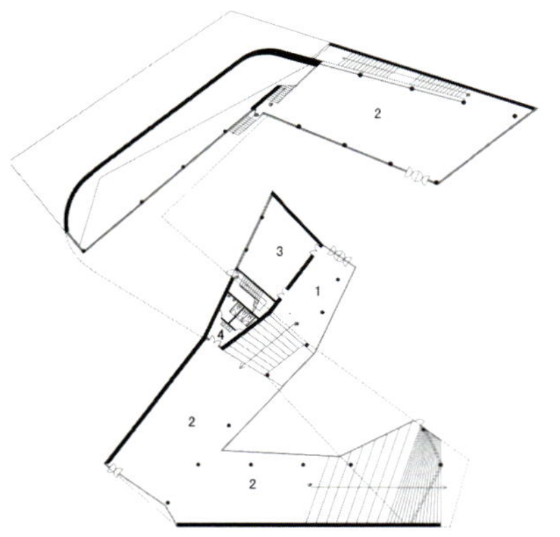

一层展厅平面
1st Floor Plan

1 大厅/Lobby
2 展厅/Gallery
3 办公室/Office
4 厕所/Toilet

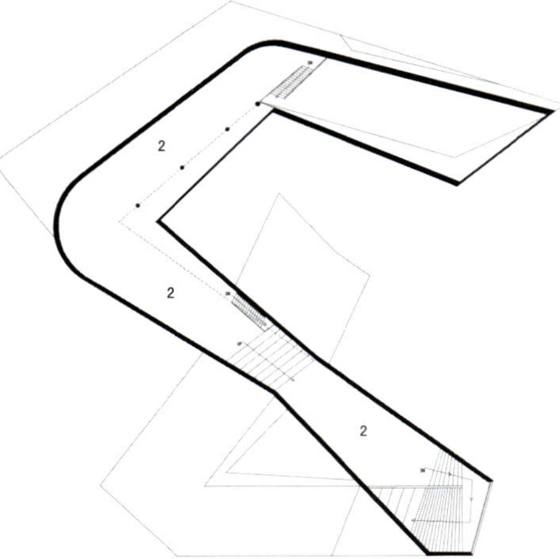

二层展厅平面
2nd Floor Plan

2 展厅/Gallery

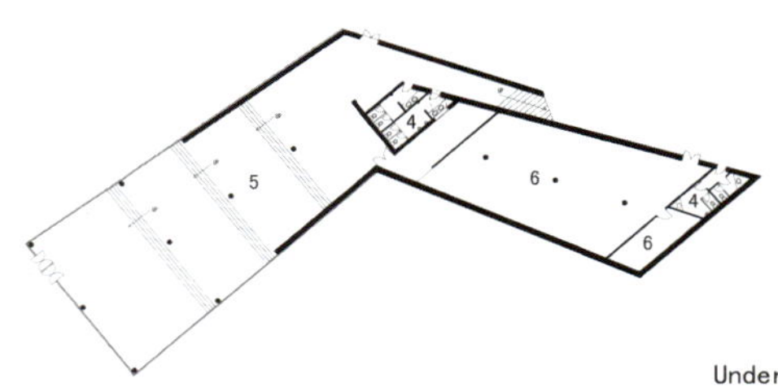

Underground Floor Plan
研究资料层平面

4 厕所/Toilet
5 研究资料/Research&Archive
6 贮藏/Storage

Yeosu Oceanic Pavilion2012

韩国丽水世博会主题馆

EMERGENT Tom Wiscombe, LLC in collaboration with KOKKUGIA

建筑类型　漂浮在海洋上的世博馆
客户　2012 韩国丽水世博会组委会
项目面积　6000平方米
设计师　EMERGENT Tom Wiscombe, LLC in collaboration with KOKKUGIA
设计团队
EMERGENT　Tom Wiscombe、David Stamatis、Chris Eskew、Brent Lucy、Graham Thompson、Zeynep Aksöz
KOKKUGIA　Roland Snooks、Pablo Kohan、Fleet Hower

Building Type　Floating Marine Expo Pavilion
Client　Yeosu 2012 Expo Committee
Project Area　6,000 m²
Designers　EMERGENT Tom Wiscombe, LLC in collaboration with KOKKUGIA
Design Team
EMERGENT　Tom Wiscombe, David Stamatis, Chris Eskew, Brent Lucy, Graham Thompson, Zeynep Aksöz
KOKKUGIA　Roland Snooks, Pablo Kohan, Fleet Hower

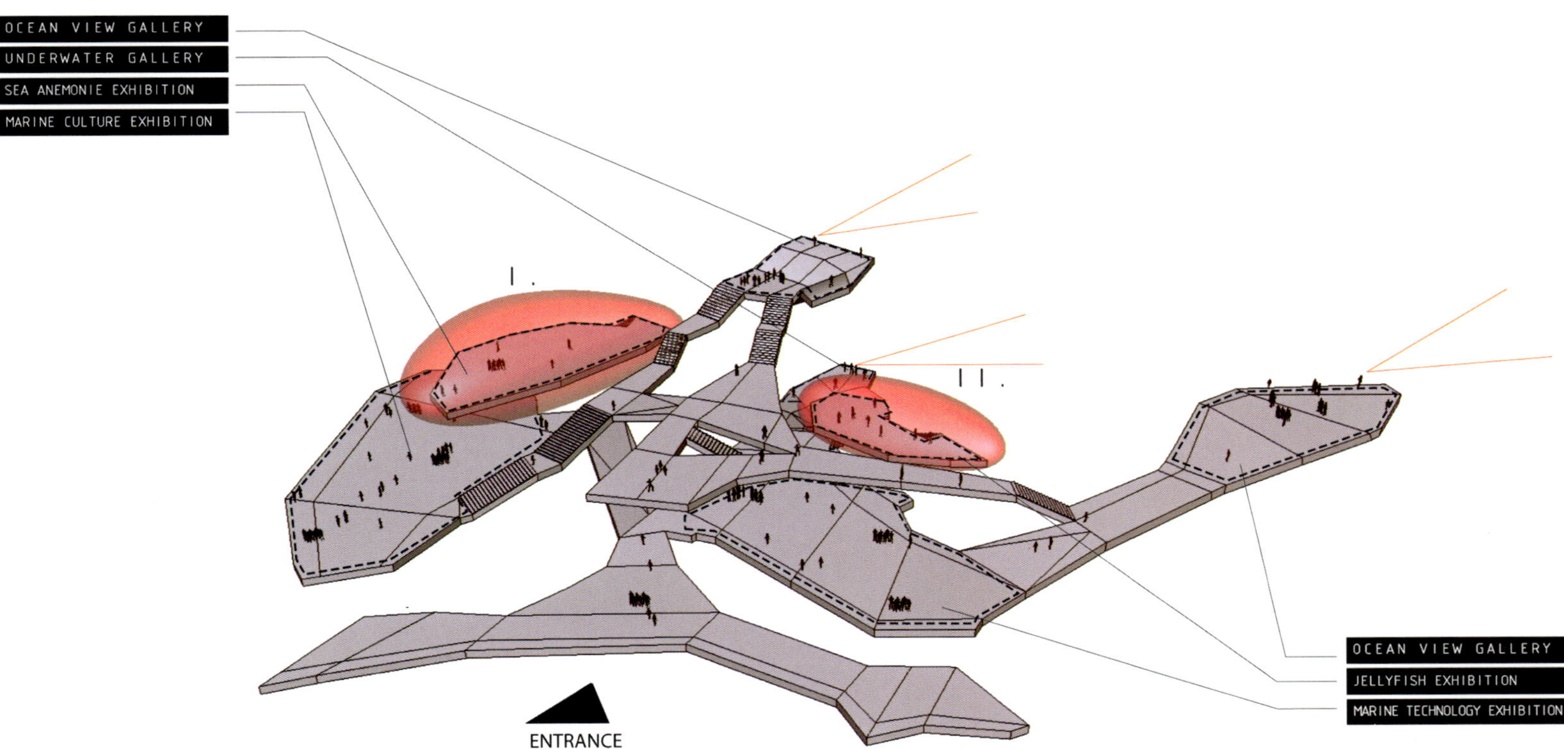

CONNECTIONS

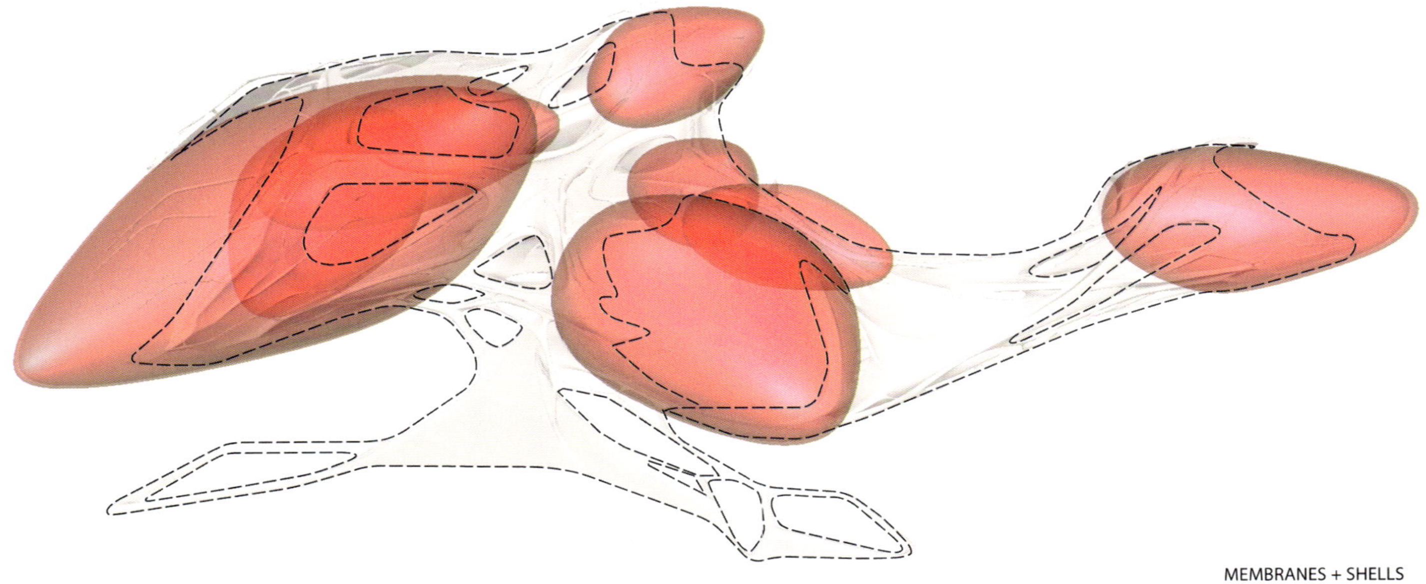

MEMBRANES + SHELLS

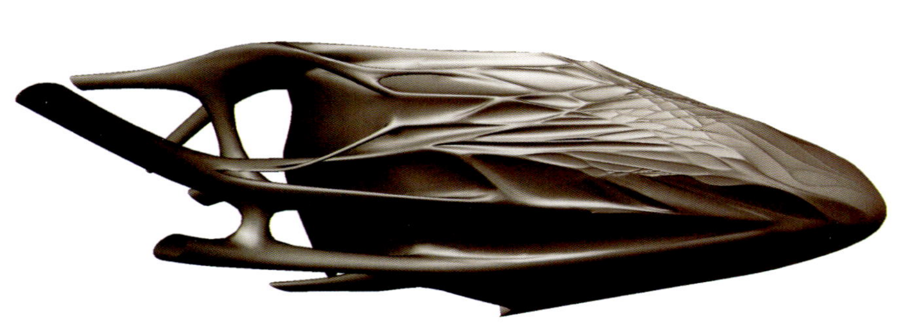

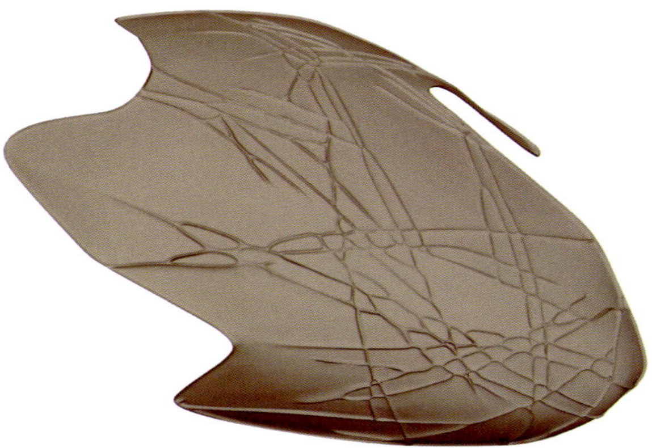

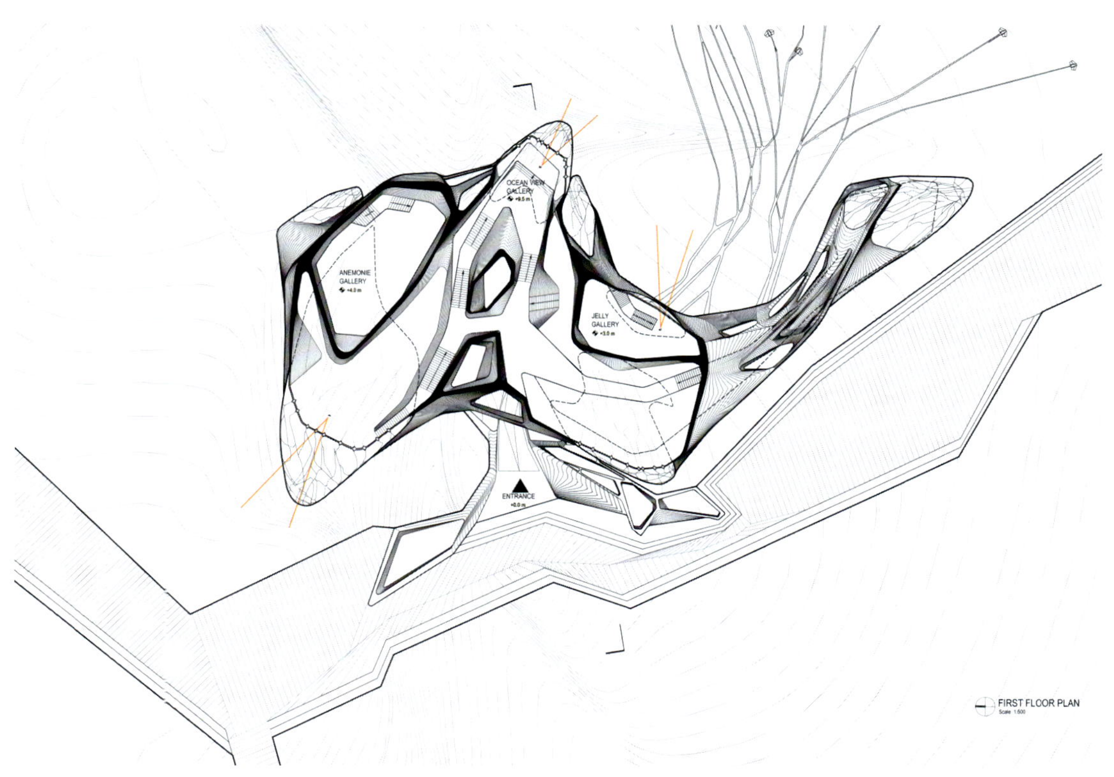

协作与背景

本方案由EMERGENT和KOKKUGIA合作设计，彼此之间分享心得，交流专业技能。方案的完成建立在持续不断的设计与脚本书写上，完成的经过是个大量计算的探索过程。它代表了一种松散、开放式的设计方式，相比于自我调整式的设计过程，更偏重于设计结果。

本方案将成为2012韩国丽水世博会的核心，是一个赞美海洋作为生物体、人类文明和海洋生物系统共生为主题的展馆。在此设计方案中，建筑和它本身的区域进入一个反馈环。正如环境孕育物种，不同的物种也会选择与其相适应的生存环境，因此建筑师应对建筑周围及其地下环境中的各种元素和能量进行积极地再组织。

构造与颜色

本项目是基于软膜泡沫及硬式外壳建造而成的。这两种系统具有反映表面清晰度的特点，尤其是对其物质性而言。然而，这些特点由于变化融合，逐渐变得多余了。

构成结构钢度的深褶和特大电枢与纤维复合材料外壳有关，而优质的双褶气垫承载条板扩散开来使拱形的ETFE膜材更加稳固。微型电枢越出膜壳结构，在系统中创造出了结构化的、具有装饰感的连续性。

颜色可以用来在视觉上强调结构行为的转换（例如，莫霍克呈现出橙色或黄色而特大电枢则倾向于紫色或粉红色）。然而颜色梯度并非具有100%的指示性，也并非完全准确；它们相互关联同时也互相干扰。在综合性的生态特征上，颜色不再让位于形式，在整体的生态特征中变得至关重要。

Collaboration and context

This project is the result of a collaboration between EMERGENT and KOKKUGIA, intended to capitalize on both shared sensibilities as well as individual expertise. It is an exploration of messy computation in the sense that the project is the result of moving in and out of the realms of designing and scripting. It represents a loose, open-ended way of working that biases effects over self-

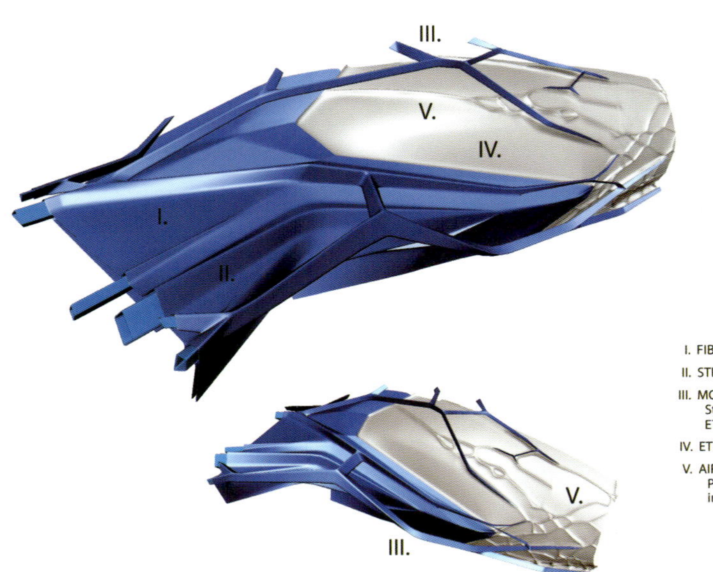

justifying processes.

The Pavilion is intended to be the centerpiece for the Yeosu 2012 Expo, a space which celebrates the ocean as a living organism and the co-existence of human culture and ocean ecosystems. In our design proposal, the building object and its territory enter into a feedback loop. The role of the architect is expanded to include the active re-organization of matters and energies around and underneath the building, where the species selects its environment as well as the environment selects its species.

Tectonics and color

The building is based on an aggregation of soft membrane bubbles merged together with a hard monocoque shell. The two systems are characterized by patterns of surface articulation which are specific to their materiality. Nevertheless, features tend to migrate, hybridize and become redundant.

Deep pleats and mega-armatures that create structural stiffness are generally associated with the fiber-composite shell, while fine, double-pleated Air-beams spread over and stabilize the vaulted ETFE membranes. Micro-armatures (a.k.a. Mohawks) transgress thresholds between shell and membrane, creating structural and ornamental continuity between systems.

Color is used to visually intensify transformations in structural behavior (for instance mega-armatures tend towards purple/pink while Mohawks tend towards orange/yellow). Nevertheless, color gradients are neither 100% indexical nor are they completely smooth; they are coherent yet glitchy. No longer secondary to form, color becomes critical in an overall ecology of features.

ADEPT House

ADEPT主题屋

MVRDV.Rotterdam、ADEPT.Copenhagen

项目地点	丹麦，腓特烈斯贝
建筑师	MVRDV.Rotterdam、ADEPT.Copenhagen
景观建筑师	Soren Jensen
工程师	ImitioWinnie、Ricken Copenhagen
合作建筑师	Max Fordham、Ducks Sceno、ADEPT团队

Project Location	Frederiksberg, Denmark
Architect	MVRDV. Rotterdam, ADEPT. Copenhagen
Landscape Architect	Soren Jensen
Engineer	ImitioWinnie, Ricken Copenhagen
Associate Architect	Max Fordham, Ducks Sceno, ADEPT team

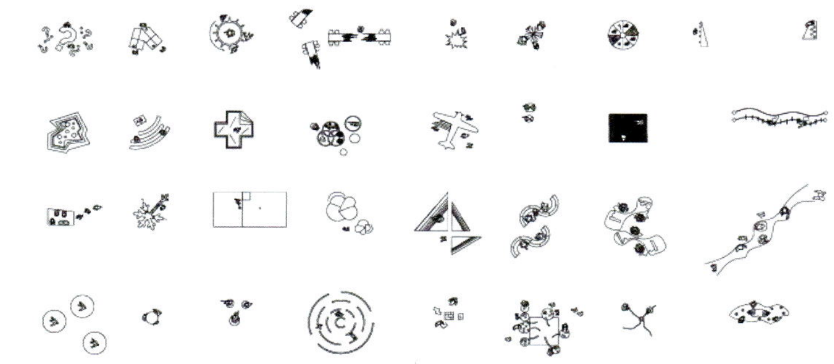

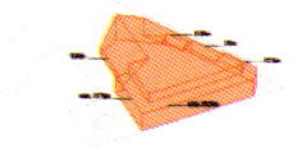
BUILDING ENVELOPE

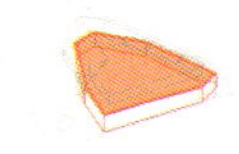
URBAN PROPORTIONS

COMMERCIAL PROGRAM

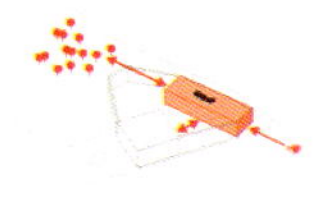
SPARK HOUSE 1

SPARK HOUSE 2

URBAN ENSEMBLE

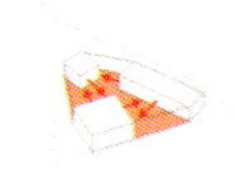
SPARK SQUARE

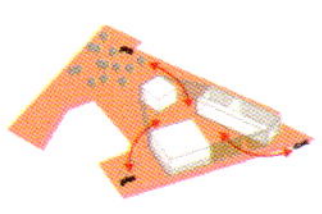
URBAN DYNAMO

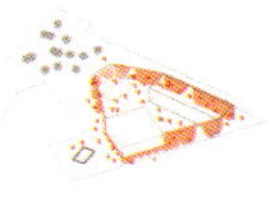
URBAN CURTAIN

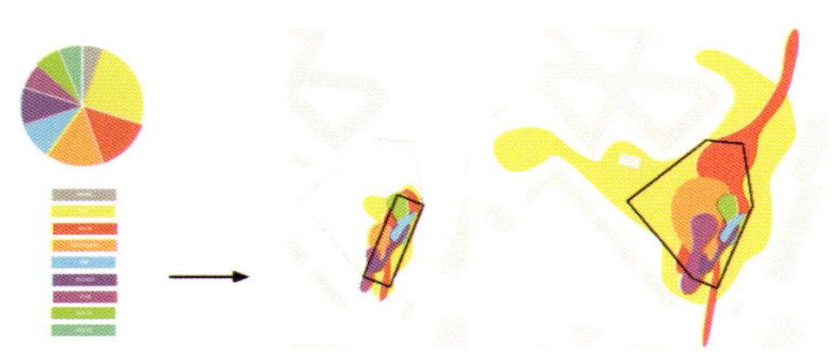

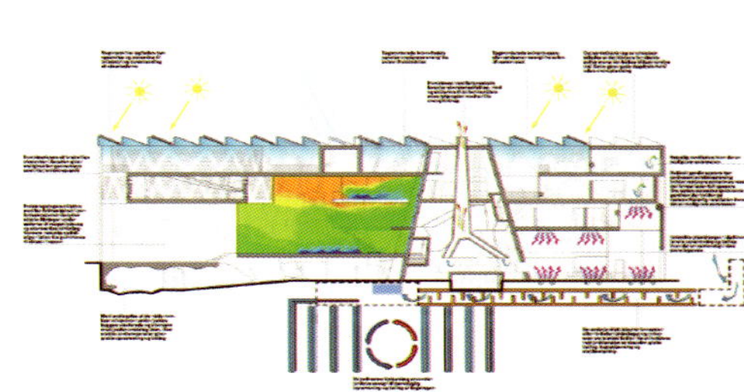

最近丹麦腓特烈斯贝市、丹麦文化与运动设施基金和Realdania宣布了MVRDV与ADEPT共同赢得了位于丹麦腓特烈斯贝的文化运动馆的设计竞赛。该建筑是一个新的城市类型体，融合了社区中心、展览和演出场地、运动场地、公园以及健康中心。文化运动馆的主旨在于为腓特烈斯贝市居民提供一个健康积极的生活方式。4 000平方米的建筑位于一个4 500平方米的公共公园中，是三个系列建筑中的第一个。第一阶段将于2015年完成，总预算为1 700万欧元。

文化运动馆的主要目的是为Flintholm社区所有年龄段的居民，提供一个富有朝气的、适合各种活动的聚会场地。健康、文化、休闲和教育理应完美地结合在一起，创造一种终极建筑体验。

主要建筑物，即文化运动馆本身，或者叫做Ku–Be，是一个长方形的玻璃体量，里面容纳了六个叠加起来的理想化功能元素体。体块间的空间称为灵活多变的“游戏区”，可以满足各种活动和相应空间流线。这六个叠罗的元素容纳着更多功能：剧院、保健区、餐饮区、禅区、学习中心和展览厅、健身和活动中心、健康中心以及管理办公区。剧场功能灵活，布置有不同的舞台和观众席；而且剧场所具有的大窗子可以使其成为一个户外的演出场所，大家可以从花园观看节目。该建筑是一个真正为使用者精心设计的多功能公共活动中心。

三个体量被“城市帘幕”所包裹，并形成花园的框架。这个帘幕具有极大的使用多变性，可以是艺术作品、自行车场、水景和灯光设施、表演场地，或者只是帘幕。

与SLA景观建筑师合作设计的公园同样具有多功能使用要求，可以作为有活力的吸引人的公共空间。室外景观与文化运动馆的室内各主题相适应，分为表演区、健康和运动区、安静区、联系区域以及为二期工程预留的空旷区域。

项目将分阶段实施。运动馆、花园和城市帘幕将首先实施。接下来会是商业建筑部分和运动馆的第二部分。气候和能源技术的运用均基于可靠的技术措施，例如太阳能板、自然通风、地下冷热蓄能技术，最终将产生一个高效的低能耗建筑。

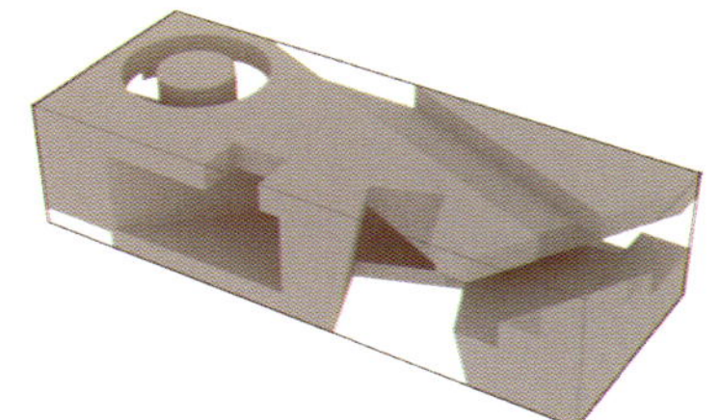

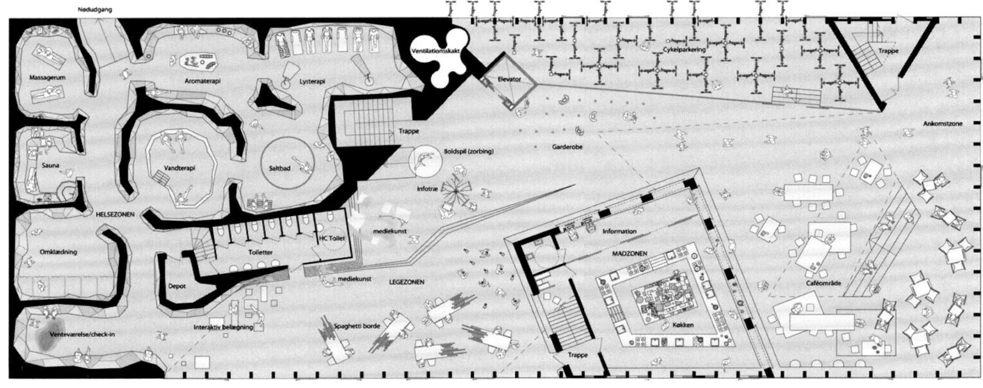

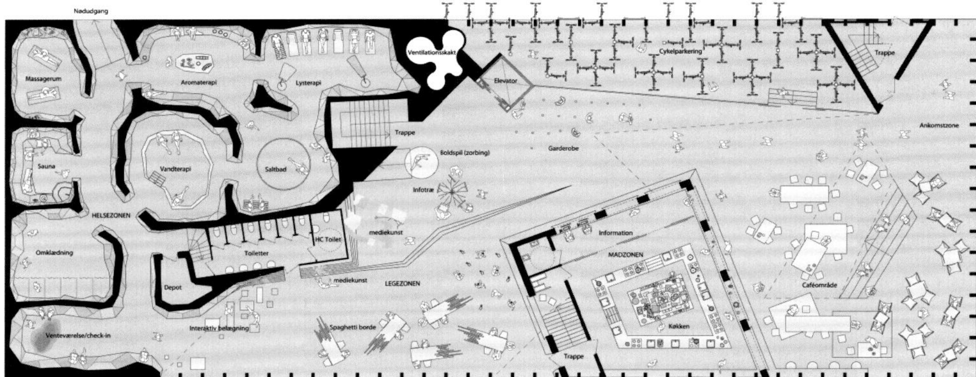

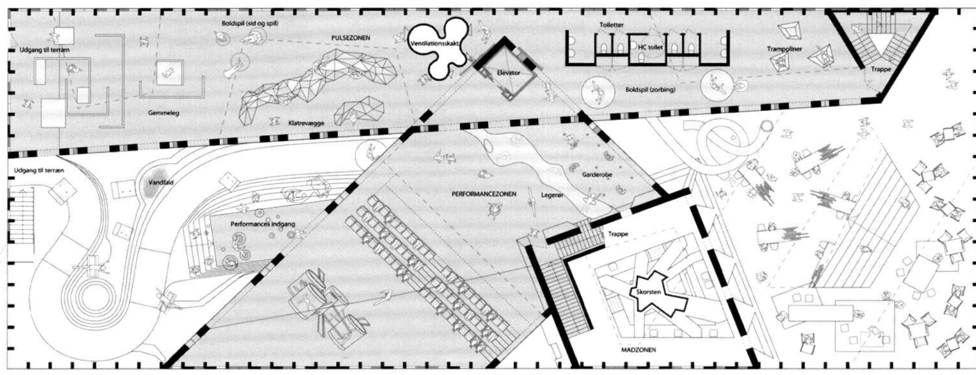

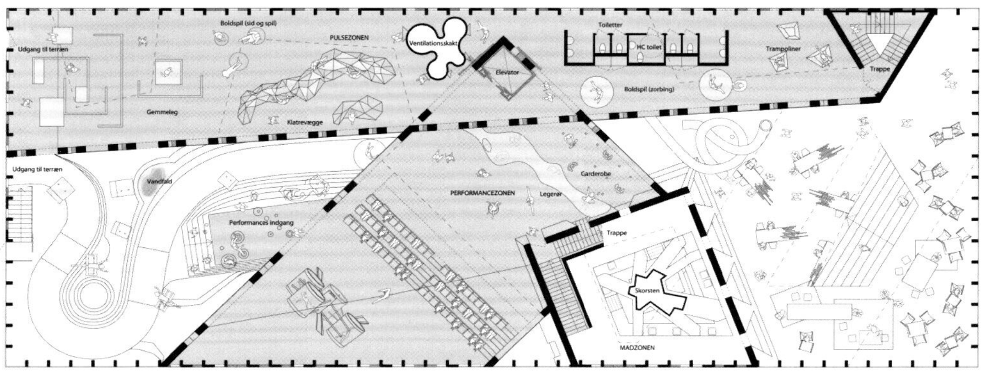

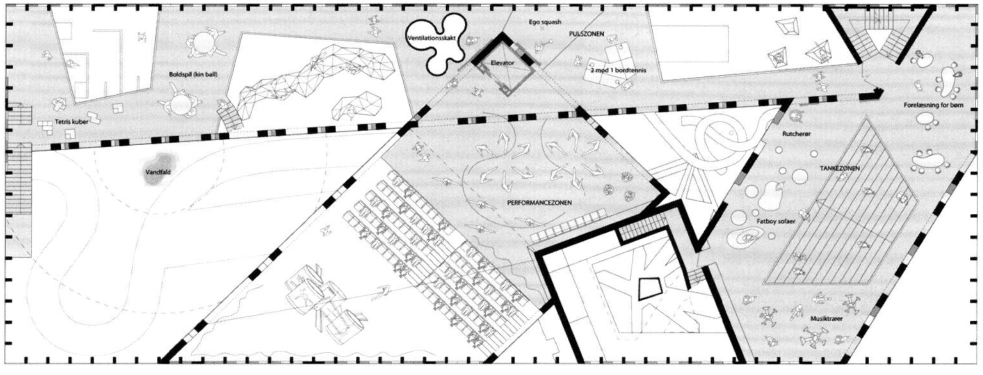

Today the City of Frederiksberg, Denmark, the Danish Foundation for Culture and Sport Facilities and Realdania announced MVRDV and ADEPT winner of the House of Culture and Movement competition in Frederiksberg, Denmark. The building is a new urban typology with its mix of community center, exhibition and performance, playground, park and health center. The House of Culture and Movement is aimed to engage the population of Frederiksberg in a healthy and active life style. The 4,000m² building is set in 4,500m² public gardens and is the first in a series of 3 buildings. The first phase is to be completed in 2015 and has a total budget of 17 million Euro.

The main ambition for the House of Culture and Movement is to offer the Flintholm neighborhood a dynamic meeting point for people of all ages taking part in a wide range of activities. Health,culture, leisure and education should smoothly blend together to create a spectacular architectural experience that will become a destination.

The main building, the House of Culture and Movement,or Ku-Be(Kultur-og Bevagelseshus)is a rectangular glass volume containing six stacked ideal programmatic elements. The space in-between can be programmed flexibly as a "play zone" with various activities and main circulation. The stacked elements hold more specific uses: a theatre, a health zone, a food zone, a zen area, a study centre and exhibition hall, a fitness and activity centre, a wellness centre and an area for the administration. The theatre is flexible and can be used in different stage and audience settings; in addition its large window allows it to be used as an open air theatre where the public stays in the garden. The building is a truly multifunctional public centre which engages its users.

The 3 volumes are wrapped in an urban curtain that acts as frame for the garden. It offers great flexibility and can be used for art projects, bicycle parking, water and light installations, performances and curtains.

The garden, designed in collaboration with SLA landscape architects, is fit for multiple uses acting as an activated, spectacular public space for the area. The landscaping follows the themes of the interior of the connecting zones and an empty zone reserved for House of Culture and Movement.

The project will be phased The House of Movement, the garden and the urban curtain will be first realizations. In later stages a commercial building and a second House of Movement will be added. Climate and energy technology is based on reliable technologies such as solar panels, natural ventilation and underground hot and cool storage resulting in a highly efficient low energy building.

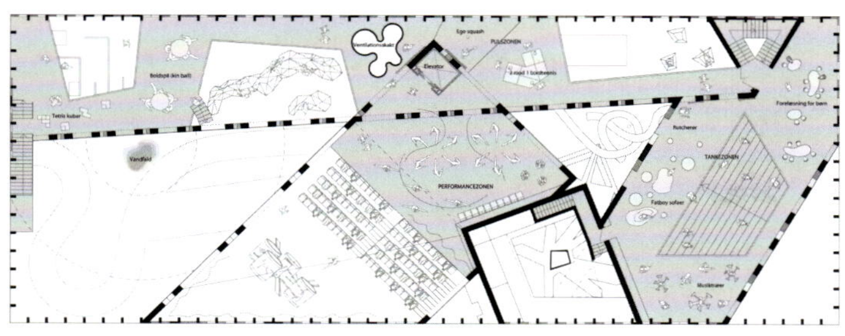

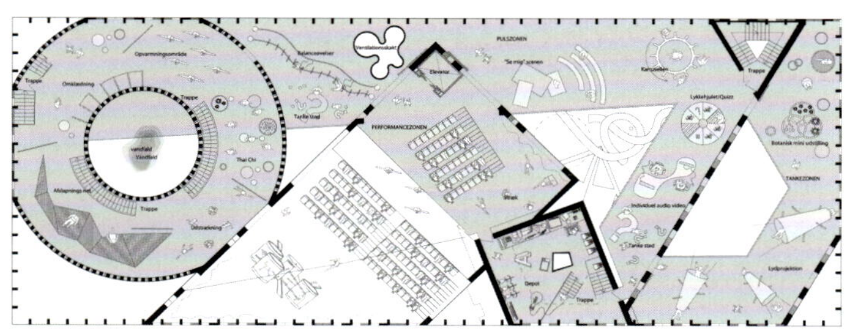

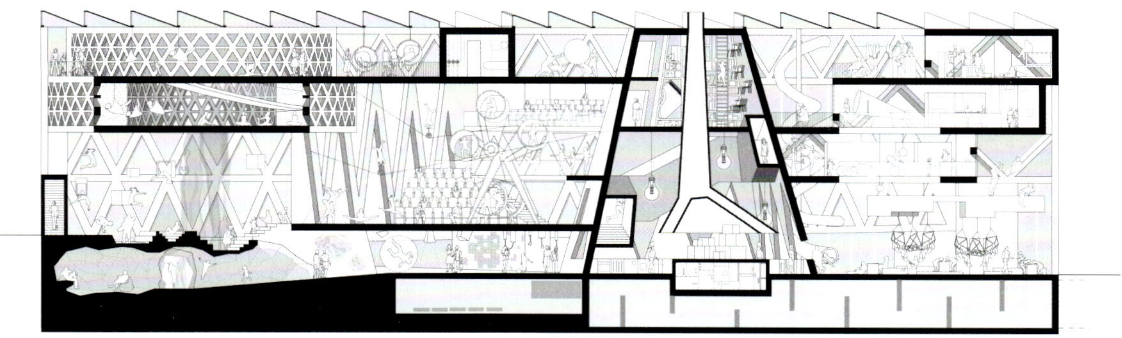

London South Bank University, Keyworth II Building
伦敦南岸大学，K2建筑

Grimshaw建筑事务所

项目地点	基沃思街，伦敦东南一区
项目面积	8500平方米
设计开始日期	2005年9月
施工开始时间	2007年7月
竣工时间	2009年10月
客户	伦敦南岸大学（LSBU）
格里姆肖团队：	
责任总指挥	英格丽•比尔
合作者	佩里•奥佩尔
项目建筑师	罗伯特•利勒
团队	亚当•杨
结构工程师	WSP
服务工程师	AECOM公司
工料测量师	特纳和汤森
主要承建商	Kier London公司
主要用料	玻璃、红陶、锌雨屏、编织钢丝网

Project Location	Keyworth Street, London, SE1
Project Area	8,500㎡
Design Start Date	09/2005
Construction Start Date	07/2007
Completion Date	10/2009
Client	London South Bank University (LSBU)
Grimshaw Team:	
Director in Charge	Ingrid Bille
Associate	Perry Hooper
Project Architect	Robert Lisle
Team	Adam Yang
Structural Engineer	WSP
Services Engineer	AECOM
Quantity Surveyor	Turner and Townsend
Main Contractor	Kier London
Principal Materials Used	Glass, terracotta, zinc rain screen, woven steel mesh

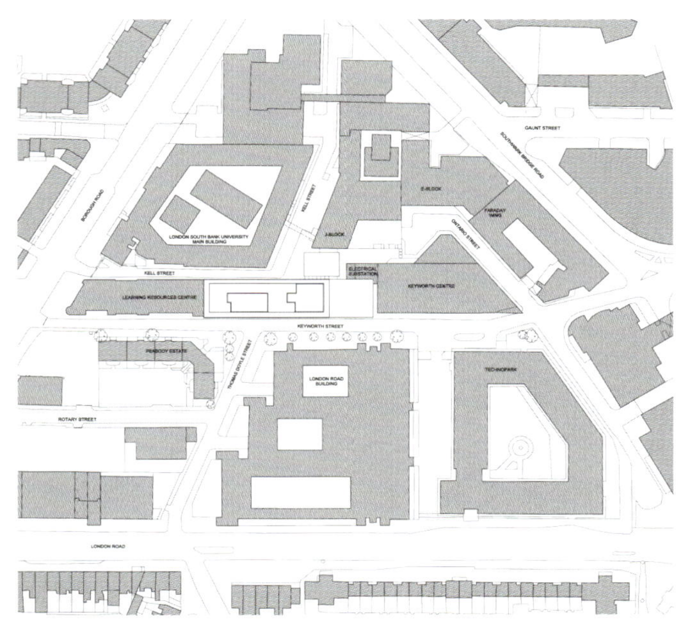

K2是为伦敦南岸大学设计的校园建筑，将成为费城校区的中心。新建筑为校园环境提供了灵活的办公以及学习空间，这样一来可以适应大学不断变化的需求与不断增长的学生人数。

新大楼位于Elephant和Castle校区的中间，将教学楼与健康保健职工楼相连。

大楼的室内设计实现了一种交换性，使得大楼间的设备共享成为可能，同时也保留了各自独立的学习空间。表面上，新建筑响应街道背景，与周围风格迥异的街区交织，来营造特有的街道氛围。

五层楼高的中庭作为重要参照点来定义大楼的入口。在内部形成了各功能大楼的连接，用于交流和平时的学习。旋梯是中庭设计的关键元素，同一系列光滑的连接桥一起为通往大楼的公共区域作出了清晰的指引。一至四楼的走廊空间则包括了上网以及进行其他休闲活动的空间，同时也通向可以俯瞰城市美景的位于中庭顶楼的会议空间。

本项目是突破性的低碳项目，其无论在排放量还是在可再生能源的创新性利用上都超越了政府所规定的标准。达到这一标准的关键是设计师在大楼总体设计方案的框架内制定了详细的综合化建筑设备策略。

大楼的冷热需求完全由地热系统提供，该系统开发利用了大楼地下土地的恒温条件。整个环境系统利用了大楼的热质并用40%实心不锈钢筋网来遮挡太阳。自然通风和光伏板进一步完善了该系统。

大楼所有的植物都位于大楼屋顶的橱窗围栏中，大楼和实验系统成为交互式的展示区，并由大学的工程以及建筑社区使用。

设计的质量以及大楼显著的造型也为伦敦南岸大学设立了更加鲜明的形象，同时也丰富了学生的整个学习经历。

K2 is a new academic building for London South Bank University that is set to become the heart of the Southwark campus. The new building provides a durable student environment with flexible office and learning space which

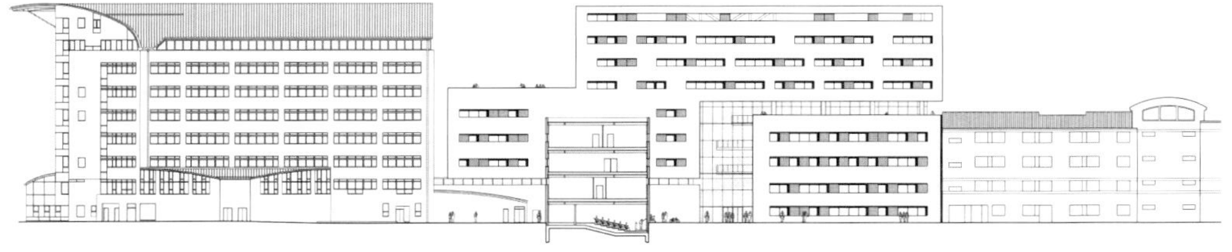

can be adapted over time to meet LSBU's changing demands and increasing student population.

Located centrally on the Elephant and Castle campus, K2 brings together the Faculty of Health & Social Care and the Department of Education.

The internal design of the building creates points of interchange and allows shared facilities between departments, while allowing them to retain their own unique learning spaces. Externally, the building responds to the street context, knitting together surrounding disparate blocks and providing a sense of street animation.

A five-storey atrium defines the entrance to the building as the key reference point for orientation. Internally it forms the link between the various departments with space for social interaction and casual study. The accommodation stair is the key sculptural feature within the atrium. Together with a series of glazed link bridges, it provides clear access to the more public parts of the building. Galleries at levels one to four contain Internet and informal breakout spaces, leading to a meeting room suite at the top of the atrium with spectacular views to the city beyond.

K2 is a groundbreaking low-carbon project that exceeds government criteria for both emissions and the creation of renewable energy on site. The key to this success is the careful integration of the building services strategy within the total building solution.

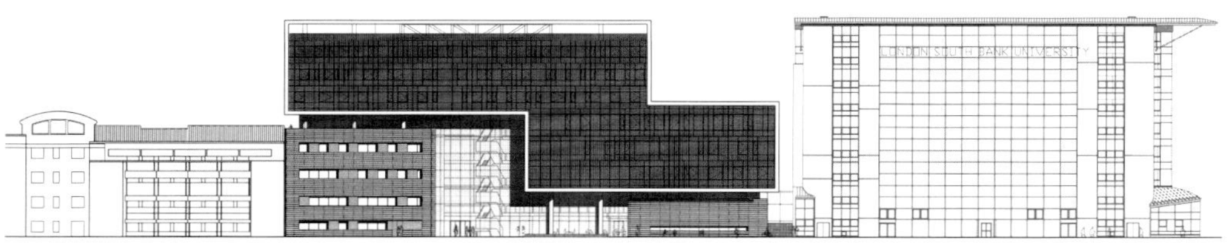

The heating and cooling demand is entirely supplied by a geothermal energy pile system that exploits the constant temperature of the ground beneath the building. Environmental systems utilise the building's thermal mass and benefit from an engineered façade using 40% solid stainless steel mesh for solar shading. These features are supplemented by natural ventilation and photovoltaic panels.

All of the building's plants are contained in a showcase enclosure at roof level, allowing Keyworth II and the experimental systems to become an interactive exhibit, studied and used by the university's engineering and architectural community.

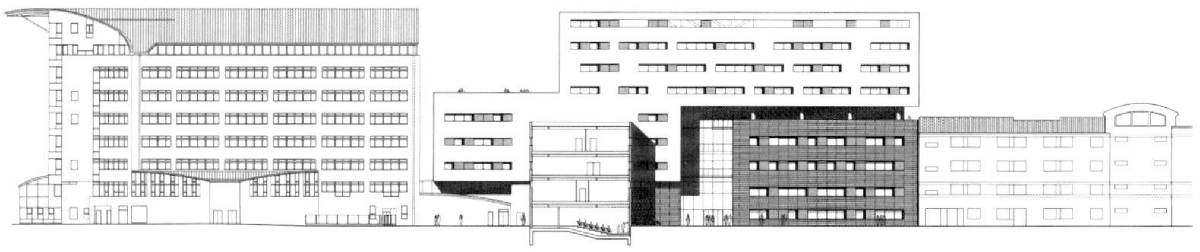

It is intended that the quality of the design and striking form of the building will help to create a strong identity for London South Bank University, assisting with enhancing the overall student experience.

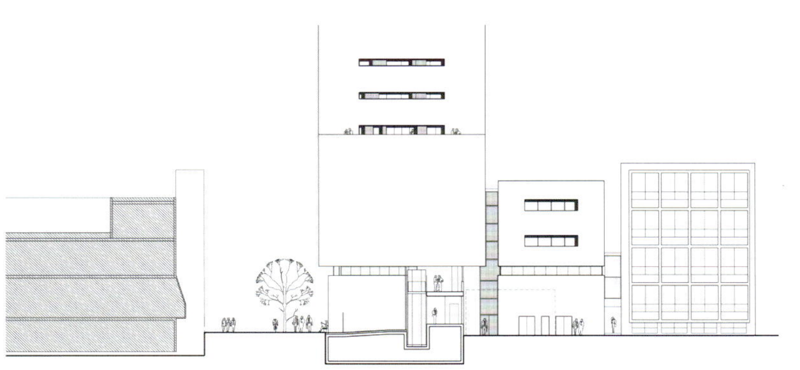

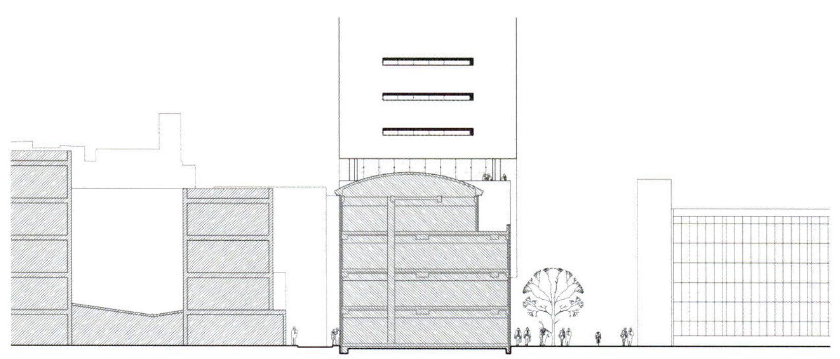

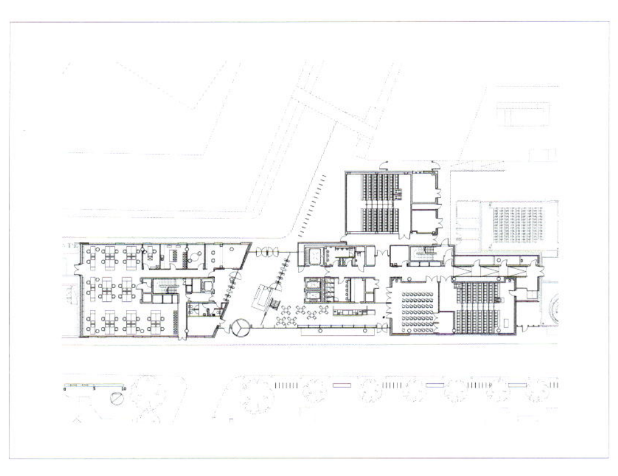

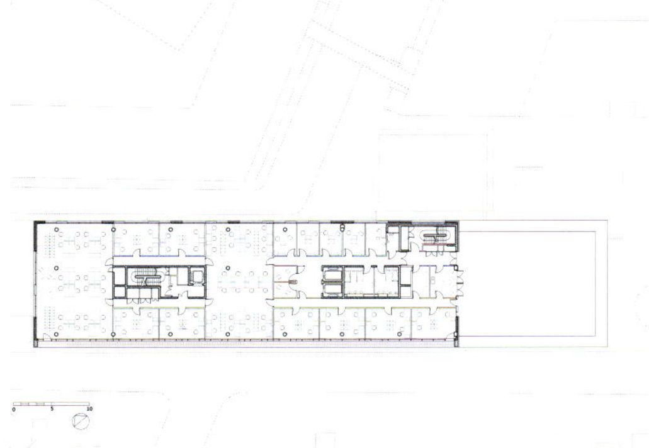

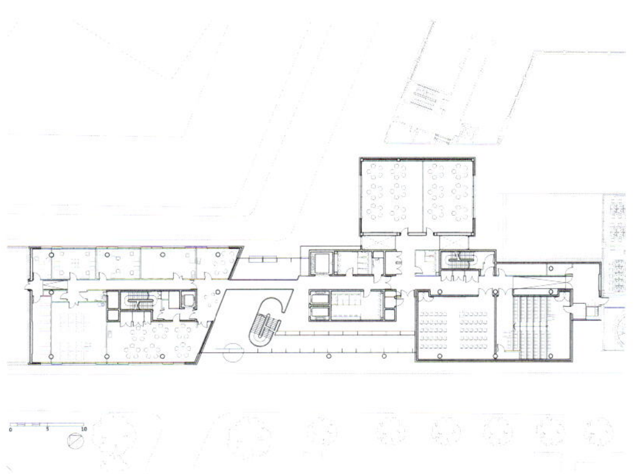

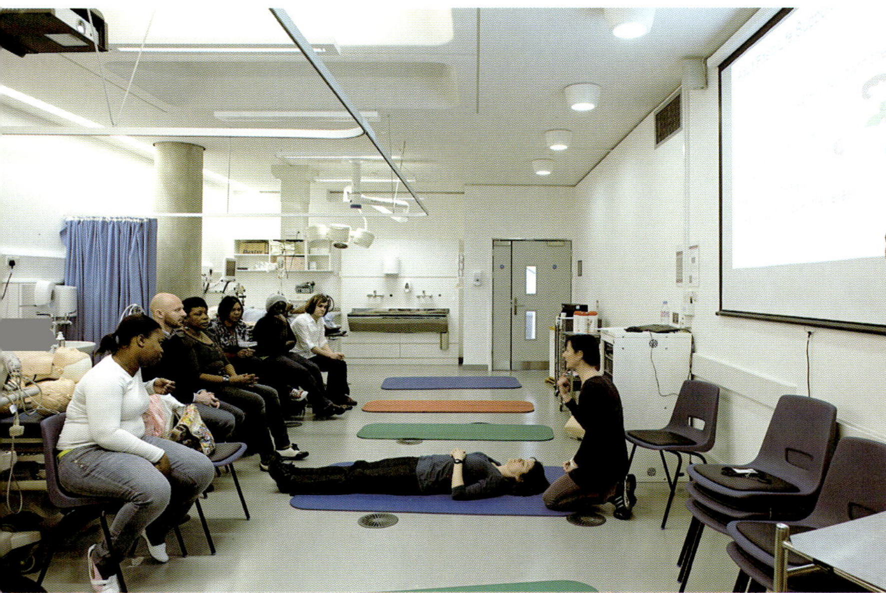

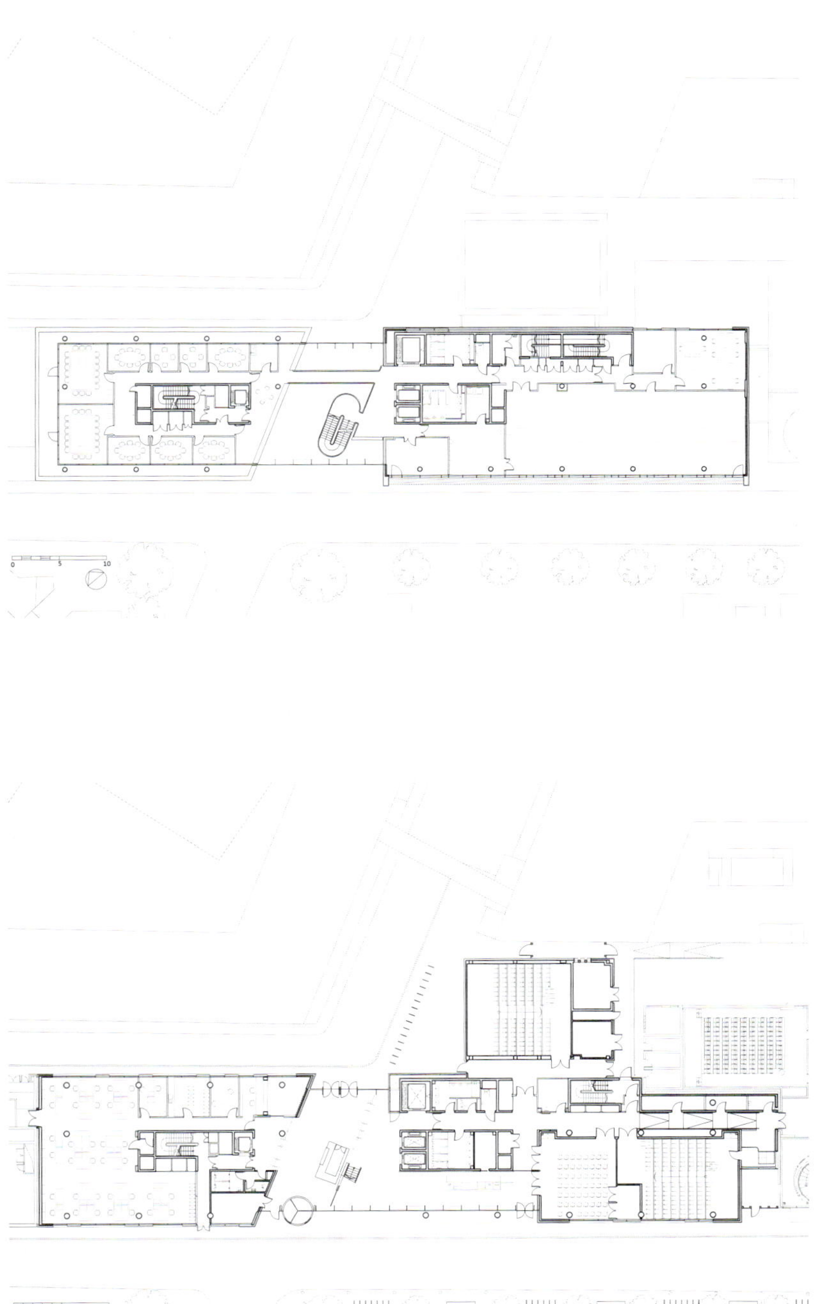

Cesar Chavez Library

凤凰城图书馆，西萨•查维兹分馆

线和空间建筑事务所/Line and Space

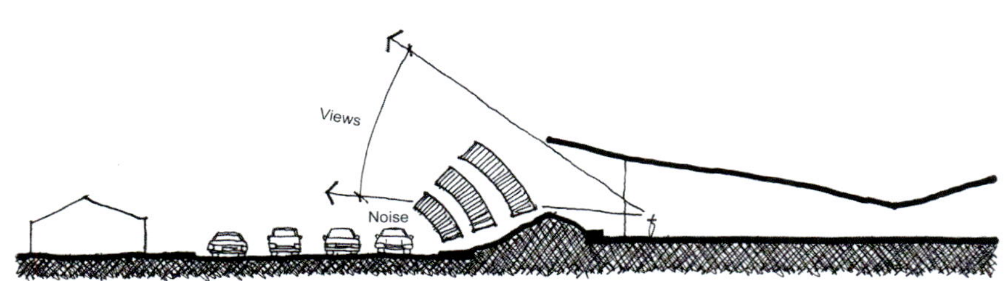

Site Section Diagram

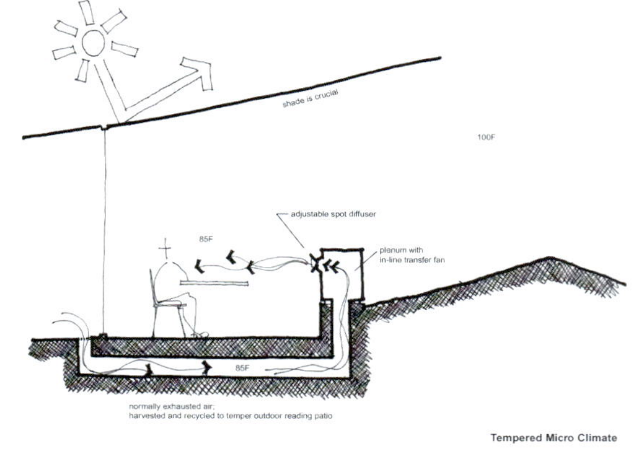

Tempered Micro Climate

凤凰城现已成为美国第六大城市，以特别设计的图书馆著称，本项目竣工于2007年1月，西萨·查韦兹分馆是凤凰城完工的四大地区性图书馆之一。

本项目毗邻公共公园的湖区，25 000平方米的图书馆坐落于Laveen村——凤凰城发展最快的区域，每月为40 000人提供服务。从周边的居住密度可见，该公园相当于整个社区的后院。同样地，图书馆被设计成了社区的"起居室"——一个供家庭成员和朋友交流聚会的场所，同时也是家庭个体成员独立地做各自事情的地方。图书馆内有专门为儿童、青少年和成人提供的舒适的场所，在其中人们可以享受阅读，尽情探索。

Phoenix, now the sixth largest American city, prides itself on providing exceptionally designed libraries to foster communities with information resources and works of the imagination. Completed in January 2007, the Cesar Chavez Library is one of four new regional libraries to be constructed for the city.

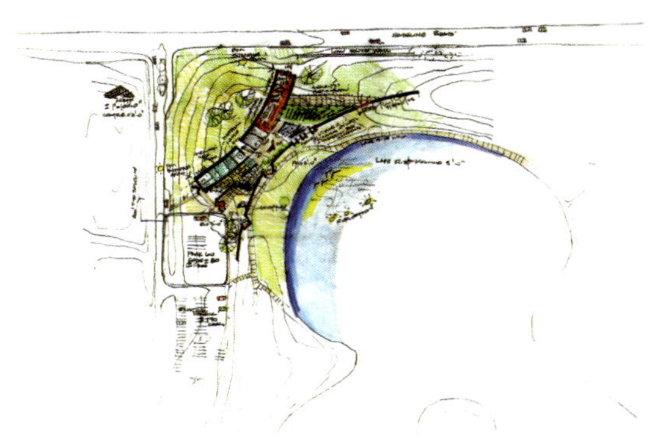

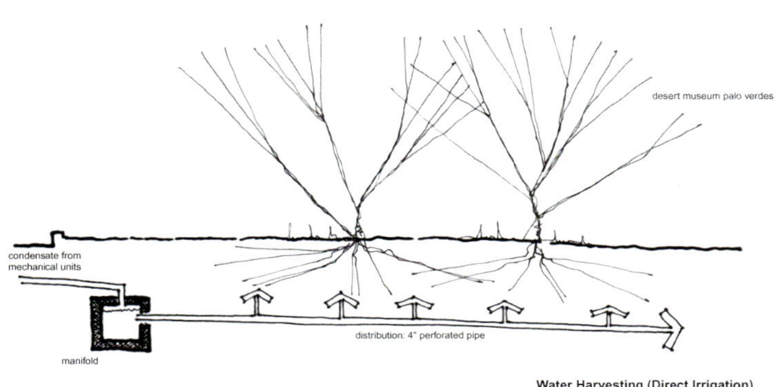

Water Harvesting (Direct Irrigation)

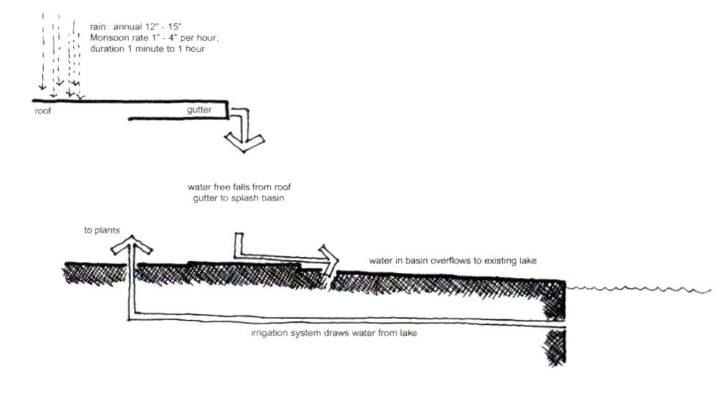

Water Harvesting (Roof Top Rainwater)

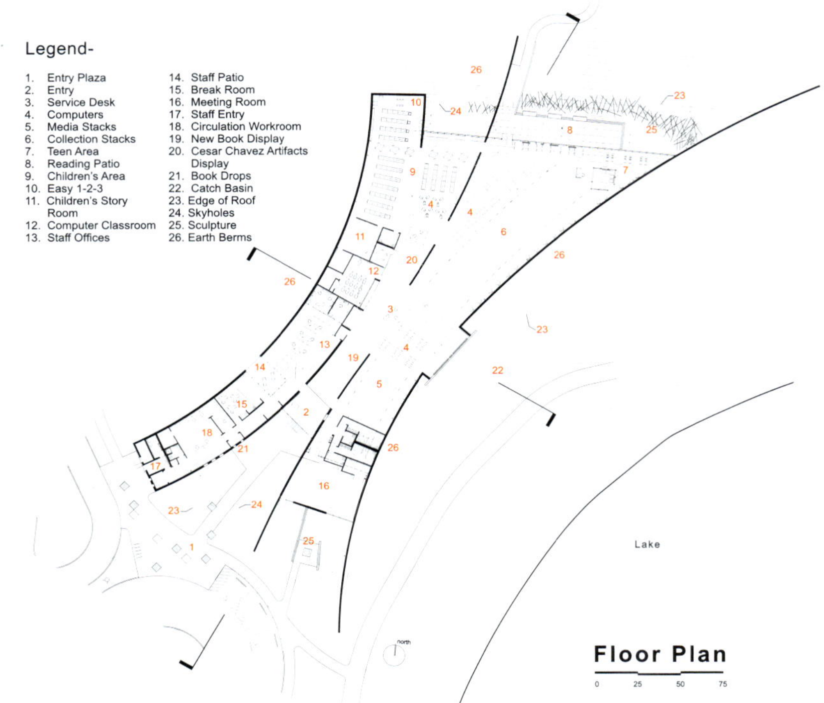

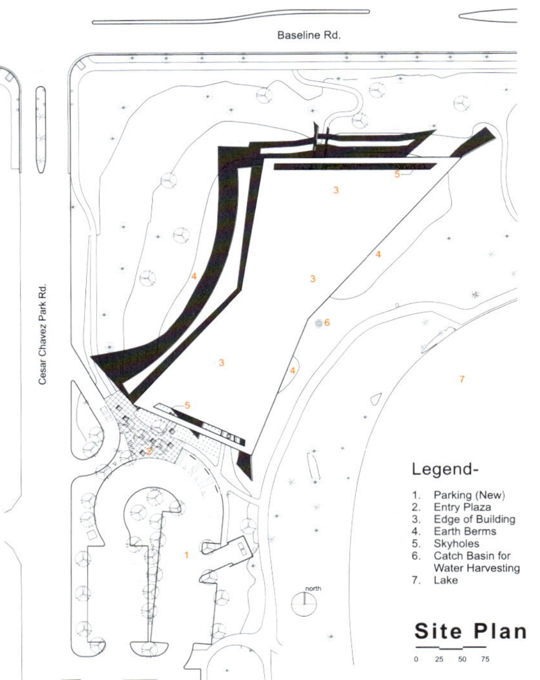

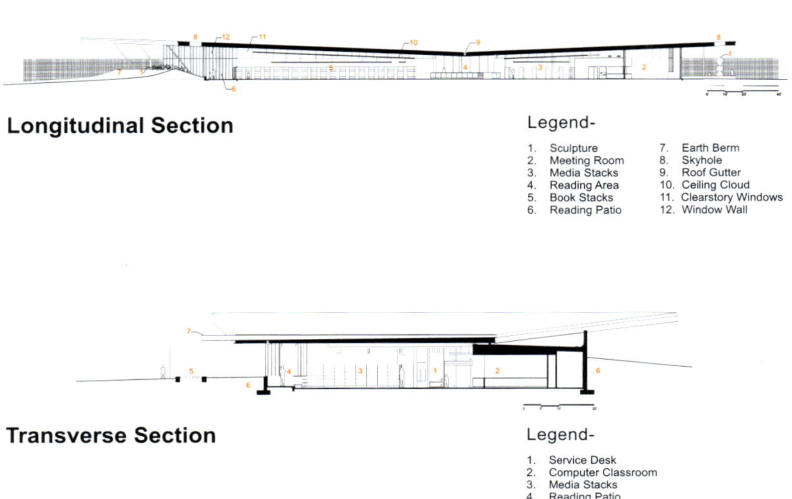

Located adjacent to an existing lake in a public park, the 25,000m² library is designed to serve a projected 40,000 visitors per month within one of the fastest growing areas of Phoenix, the Village of Laveen. Due to the density of nearby housing, the park is the backyard for the community, and in the same sense, the library is designed to be its living room — an interior place for interaction of families and friends, as well as space for individual family members to "do their own thing." There are comfortable special places for children, teens and adults to enjoy reading and other quiet explorations.

Mountain Rescue Bavaria
State Centre for Safety and Education
巴伐利亚国家山地安全教育中心

Herzog Partner

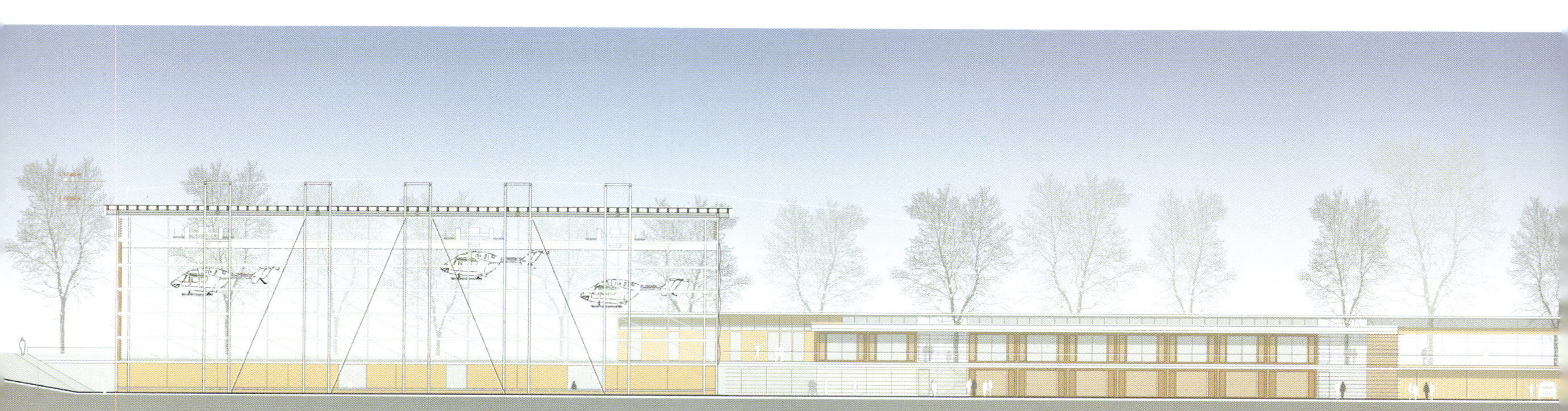

West

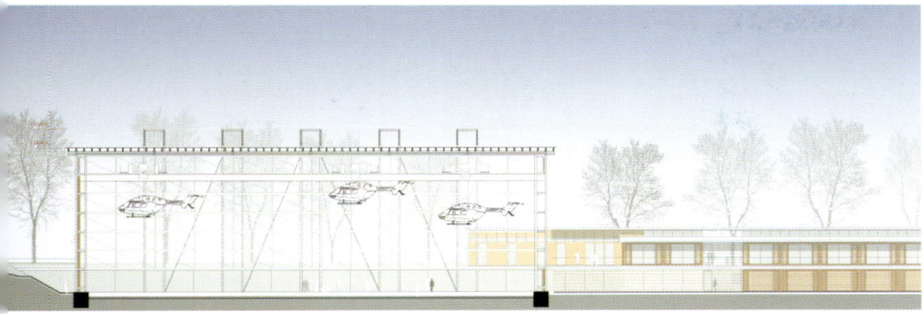

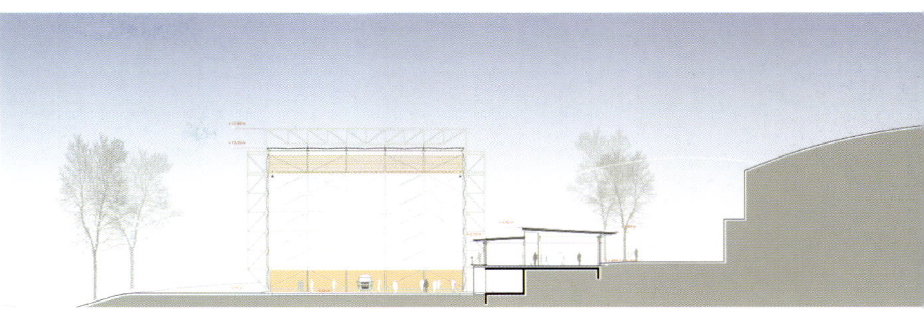

此次任务是设计一个全透明的大楼能够容纳模拟山地空中和地面营救的训练基地。相对于其他组织来说它也叫做班，例如水底营救、火旅和警察。大楼的另一边还必须容纳训练教室、技术设备和储藏室等房间。斜坡的特殊环境提供了在山坡上定位各种服务设施的可能性，并确定它们的界限，因此大厦建成后就可以看到整个大厦的结构。在建设一期只能实现北部的离大厦最近的服务脊椎，在二期建设时期计划延伸至南部的自由空间。

建筑坐落在一个小山顶上，鉴于其南侧与西侧均无遮挡，因此无论从邻近的街道，还是从远处的山上，都可以见到它。而建筑的自身的高度也使其从远处便可见到。建筑史无前例的使用功能从一开始就没有遵循传统的模式，因此我们将利用它的功能来创造建筑的身份。

该建筑物旨在建筑物内部使用现代技术，同时尊重巴伐利亚的文化遗产，例如实现富于创新同时公众可接受的综合方案。

大楼的建筑结构变得特别重要，另一个挑战是建筑的外观。相对于整体简单功能来说，大楼的整体规模是相当大的——这种组合不会经常出现。一开始我们便知道大型的、容易识别的大楼不仅要满足主要的防护目的，同时这个大型的与众不同的项目的外观也必须有创新。大楼突起的位置与风景相协调，它的比例、外观和部分结构要符合一定的美学标准。

大楼房间的温度规定不能太高，这与气温及湿度的强度有关。设计师这方面的设计非常重要，他们把大楼高度设计为20米，大楼高处常常遭受强风影响，以及部分范围被设计为可以承担超过平均标准的静态雪载荷。

建筑物主要结构特色是架空的三角形的钢铁大梁，它是由类似于真实的直升机机器人操控建造的。外观的次级横梁为先进的制造者运送标准隔膜。它只有0.3毫米厚，穿越外面的压缩门。这些拱门运用了有效的拉力原理，连门框都在非常精确的位置。因此，一个分量非

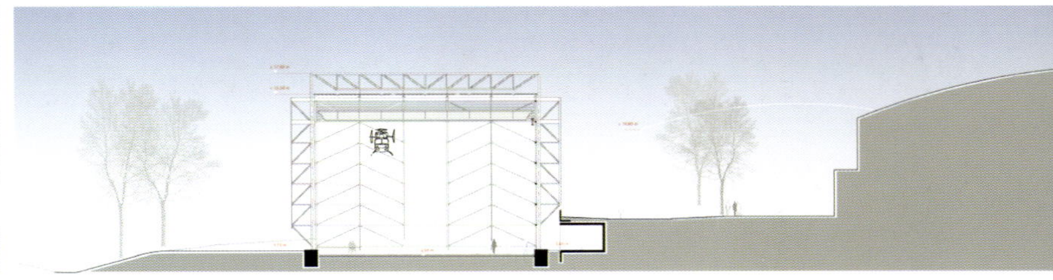

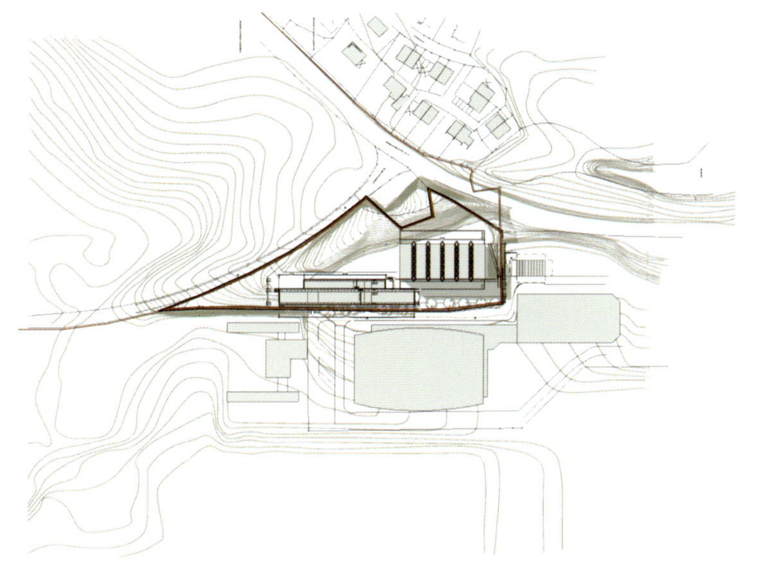

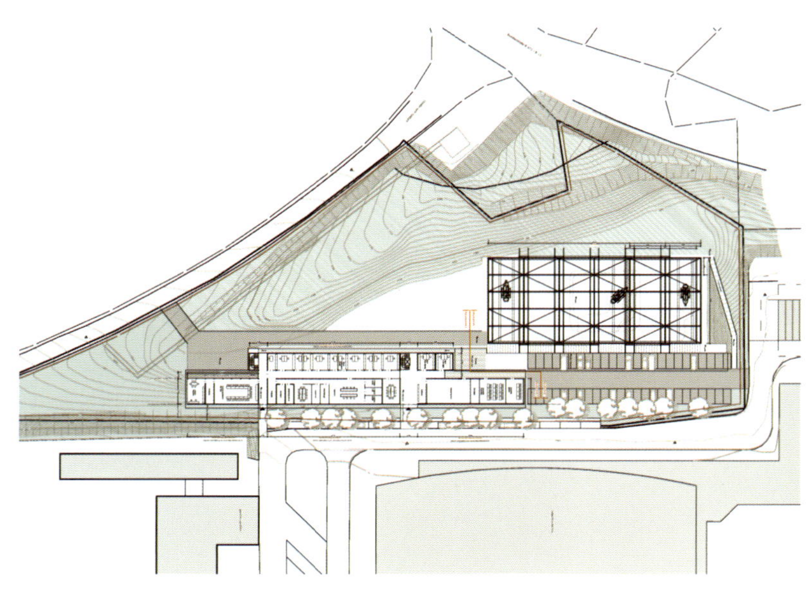

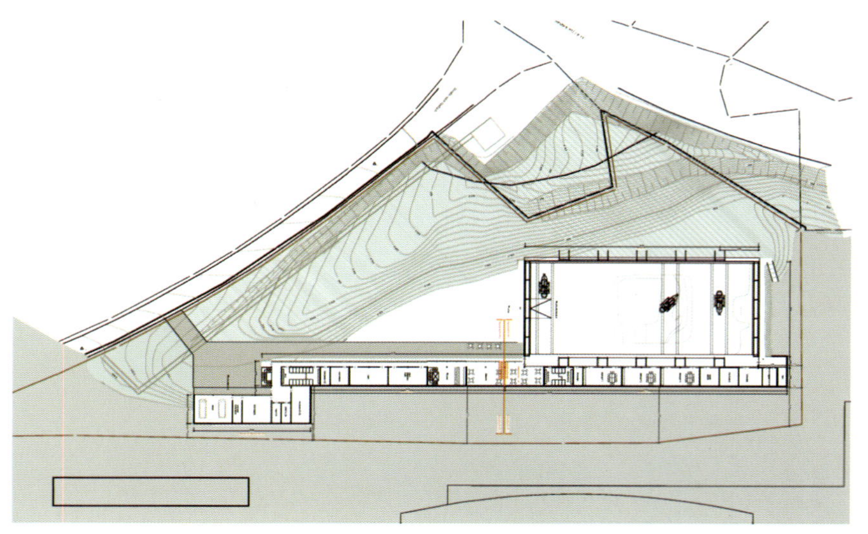

常轻但却大的、透明的、绝大部分为自动清洗的大厦外观建成了，它的建成仅仅充当天气防护者（风和降雨量）。

屋顶是由盖满纸板的木材横梁构成的。北面的外观为了安装攀登墙也是由木材建成的。因此外观的顶端可与屋顶相连，并可以打开，做到十字通风效果。

在底层强壮的实体墙中间有可折叠的，并且可以打开的出入门。

该项目实现了对环境的保护，而培训的目的也是大大地减少噪音和其他由于使用直升机而造成对的自然环境的损害。建筑物的维护在很大程度上不依赖于化石燃料，而是尽可能地使用免费的煤油。

此外，由于这些改进的训练条件，山区营救的成功率将会大大提高。

The task is to design a mainly transparent hall containing a simulation station for training the mountain and air rescue of the Mountain Rescue. It is also meant for squads of other organisations such as the water rescue, the fire brigade and the police. The hall also has to contain side rooms for services such as training classes, technical equipment and storage. The special circumstances of the inclined site offer the possibility of positioning these services on the hillside into the slope and thus defining it with a clear edge. So doing the completely different volume of the hall can show its form. In the first construction phase only the northern part of the service-spine close to the hall is realised. The extension to the south to the free space is planned for the second construction phase.

Due to its exposition to the south and west the building at the top of a little hill can be seen from the streets close by as well as from the big mountains further away. Its height alone makes it visible from a far distance and as the unprecedented use of the building does not follow traditional examples right from the start we took this use as the reason to create an architectural identity.

So the aim is to use modern technology, show it with the architecture and do this respecting the cultural heritage of Bavaria, i.e. to reach for a synthesis of both aspects which can be accepted as something new without just being a compromise.

The development of a respective bearing structure for the hall becomes specially important. Another challenging aspect is the construction of the skin. Opposed to the quite simple functions of the envelope the overall dimension of the hall is considerable—a combination that does not occur very often. From the beginning it is clear that the large, easily discernible building not only has to fulfil the primary purpose of protection in a simple way but this large unusual object should also show its innovation with its facade. Due to the prominent position of the building in the landscape, its proportion, form and sectioning have to meet certain aesthetical standards.

The standards for the room climate in the hall are not very high when it comes to the level and the tolerance of temperature and humidity. For the design the aspects are important that due to its height of almost 20 m the building is subject to high wind forces and that for the static snow loads far above the average standards the comparable dimensions had to be assumed.

A primary structural feature of the cube is the triangular

steel girders together with the overhead support robot construction for helicopters to be manipulated in space similar to a real situation. At the facade secondary beams are located carrying a modular membrane skin which has been developed together with the producer. Its only 0.3 mm thick ETFE skin is spanned to the outside with compression arches. Test-optimised tension elements of these arches hold the frames precisely in position. Thus a big, transparent, for the most part self-cleaning building skin with an extremely low weight, which only serves as a weather protection (wind and precipitation), is constructed.

The roof consists of timber beams covered with boards. The northern facade is timbered as well in order to install a climbing wall there. So are the upper parts of the facade touching the roof which can be opened and thus serve for cross ventilation.

At the base between robust concrete wall elements there are folding doors , which can be opened.

The claim for environment protection is generally already met by building a hall for training purposes as it means a large reduction of the noise and other emissions in free nature caused by the use of helicopters. A building largely independent of the use of fossil fuels for its maintenance makes kerosene-free training possible.

Furthermore the performance potential of the mountain rescue in our nature will increase considerably with these much improved training conditions.

Maosi Ecological Demonstration Primary School

毛寺生态实验小学

吴恩融，穆钧

项目团队	香港中文大学建筑学院
建筑师	吴恩融，穆钧
Project Team	Architecture College, The Chinese University of Hong Kong
Architect	Enrong Wu, Jun Mu

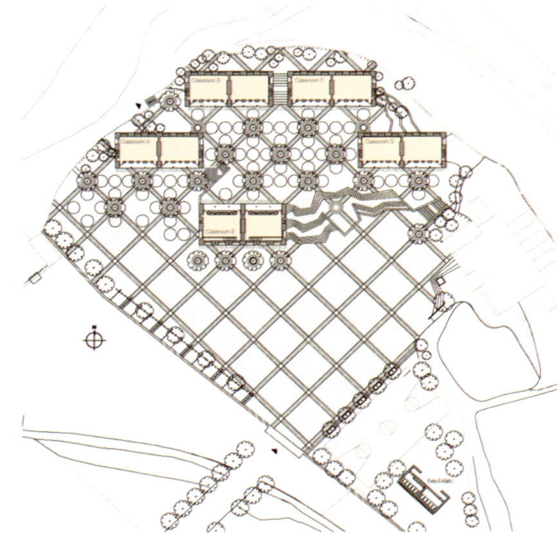

顺应所处的地形，学校所需的十间教室被分为五个单元，布置于两个不同标高的台地之上，使得每间教室均能获得尽可能多的日照和夏季自然通风。以绿化为主的院落环境有助于为孩子们创造一个舒适愉悦的校园环境。教室的造型源于当地传统木结构坡屋顶民居，不仅继承了传统木框架建筑优良的抗震性能，而且对于村民而言更容易建造施工。教室北侧嵌入台地，可以在保证南向日照的同时，有效地减少冬季教室内的热损失。宽厚的土坯墙、加入绝热层的传统屋面、双层玻璃等蓄热体或绝热体的处理方法可以极大地提升建筑抵御室外恶劣气候的能力，维护室内环境的舒适稳定。与此同时，根据位置的不同，部分窗洞采用切角处理，以最大限度地提升室内的自然采光效果。

小学的建设施工继承了当地传统的建造组织模式，施工人员全部由本村的村民组成。除平整土方所必须的挖掘机以外，所有施工工具均为当地农村常用的手工工具。同时，绝大部分建筑材料都是“就地取材”的自然元素，如土坯、茅草、芦苇等。由于这些材料所具有的“可再生性”，所有的边角废料均可通过简易处理，立即投入再利用。例如，土坯是由地基挖掘出

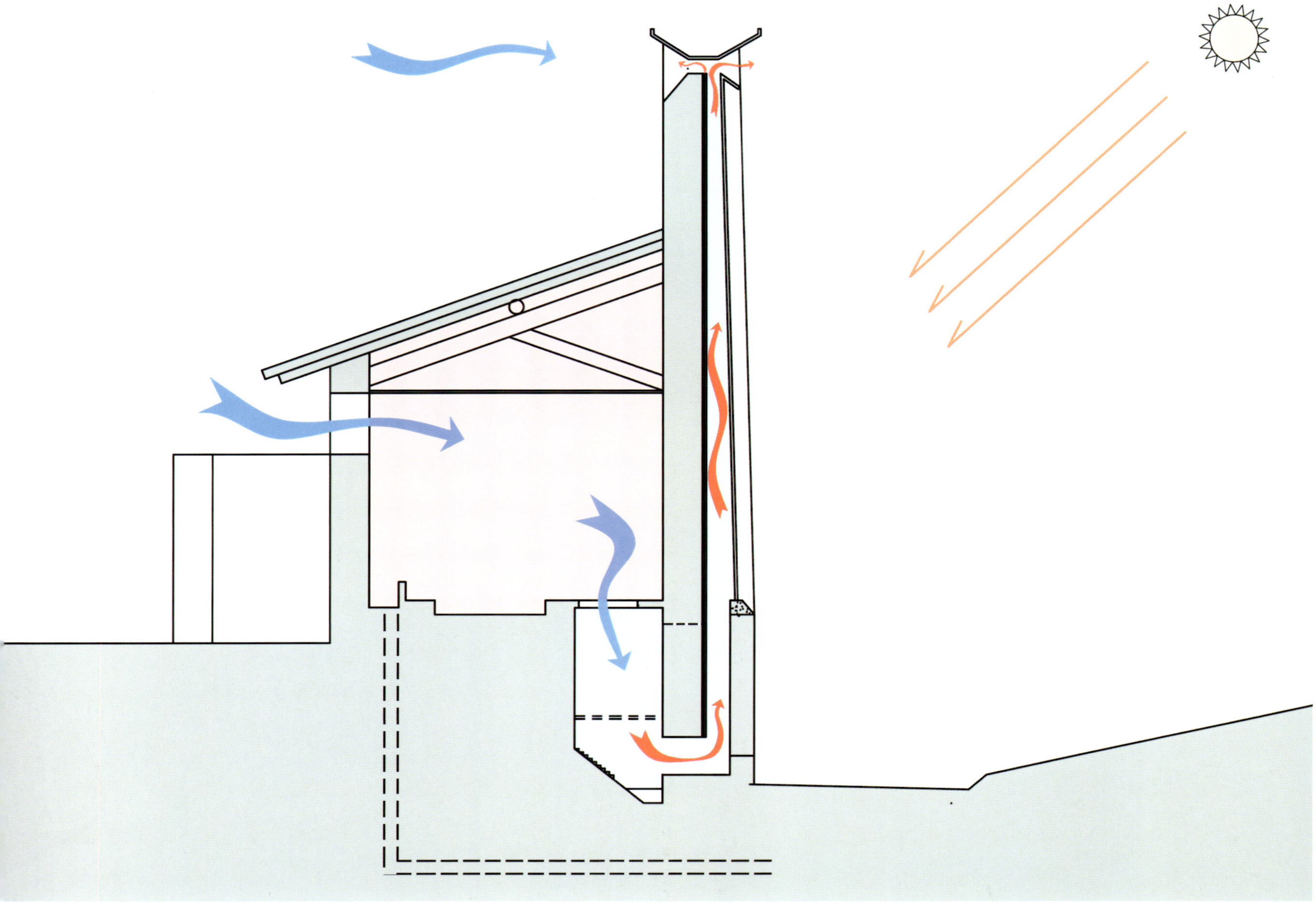

来的黄土压制而成，而土坯的碎块废料又可混合到麦草泥中作为粘接材料。再如，剩下的椽头与檩头被再利用到围墙和校园设施建造之中。以上措施不仅有助于最大限度地挖掘当地传统的建筑智慧，而且可以将由于施工而导致的能耗与对环境的破坏降到最低。

新校舍于2007年的夏天竣工。新教室的直接造价（包括材料、人工与设备）只有每平方米422港币，远低于当地由粘土砖和混凝土建造的常规学校建筑。而根据对教室在过去一年使用过程中的观测发现，与当地常规的学校建筑相比，新建教室的室内气温始终保持着相对稳定的状态，可谓冬暖夏凉。即便在今年初罕见的严冬，无需任何燃料采暖，教室仍可达到舒适且空气清新的室内环境。

从该项目建成的效果来看，总体而言，有三点值得强调。

首先，新学校为孩子们创造了一个舒适、宜人的学习环境。在室内舒适、能源消耗与环境保护方面，其具生态可持续效能远优于当地常规的建筑；

其次，由于施工建造大量地雇佣了当地的村民，作为慈善项目，除了学校本身，绝大部分的社会捐助得以惠及整个村落；

更重要的是，从这个学校项目中，村民们得以重新认识他们自己的传统。新学校的建造向他们诠释了一条适合于黄土高原地区发展现状的生态建筑之路。在有限的经济基础下，村民完全可以利用所熟知的传统技术和随地可得的自然材料，在改善自身生活条件的同时，最大限度地减少对环境的污染和破坏，实现人与建筑、自然的和谐共生。

该项目的总结报告已初步完成，以供未来的出版和推广。我们的工作还将继续下去，不仅仅是为了一座学校，而是为整个地区生态建筑的发展进行更加有意义的研究与示范。

最后，引用毛寺生态实验小学校长的一句话："从

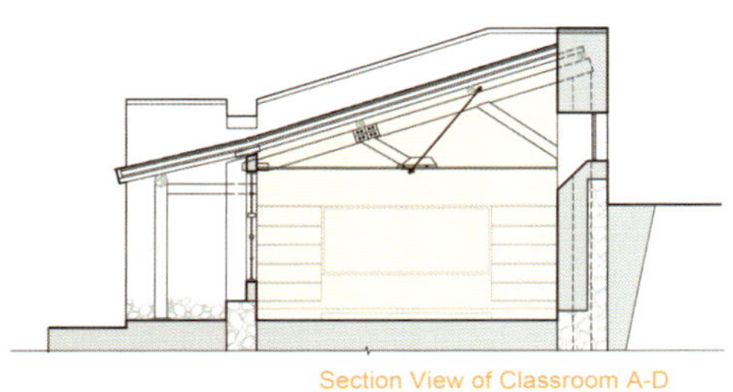

Section View of Classroom A-D

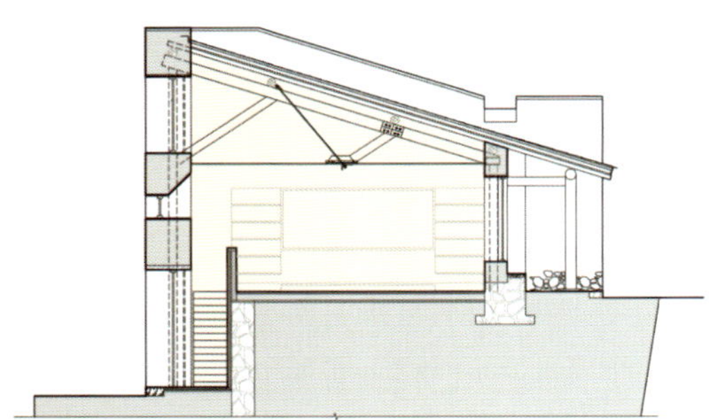

Section View of Classroom E

现在开始，学校不再需要烧煤来取暖了，省下来的钱可以为孩子们多买一些书了。”

“Do you think it is possible for you to build us a ‘good’ school?”

What is a “good” village school? The poor conditions of China’s Loess Plateau region (north-west China) demand serious re-thinking. The fragile ecological conditions, coupled with having some extreme climatic conditions pose severe environmental and sustainable challenges to designers. The poor economy has restricted interventions and solutions. Villagers needing to leave their village homes to find work in the city further drain the social lineage; they leave their children behind but still hope that they can be educated and one day make it in the society. They put their faith in their village schools. Like many donated schools that have been built in the region, a way they iconize them is as follow:

“The school is very modern with nice white tiles. We take pictures in front of them when they open. By the way, if they can also install air conditioners for us in the summer and heaters in the winter, it would be very good as otherwise it can be quite uncomfortable.”

We know for sure that something needs to be done.

The project commenced in 2002. We wished to build a "demonstration" school that can showcase appropriateness and can address the environmental, social and economical dimensions of sustainability.

The site

"The village head took the design team to a site that is south facing and surrounded by low hills. He told the team that it is the best site they have; and he hopes it is good enough for the school."

The project is located in the village of Maosi, in Gansu province, 6 hours drive from Xi'an. The tranquil and unspoiled setting reminds one that the village and its life have always been like that for hundreds if not thousands of years. It is only very recently that the onslaught of modernity and urbanisation is beginning to intrude. The village has around 2,500 villagers and 200 students. Their existing schools are either in caves or in simple single storey brick huts.

Methodology and design

"High science and low technology is a motto of the working."

The project emphasizes a scientific and transferable methodology: condition analyses, computer simulation experiments and field construction.

Condition analyses in economy & resource for building, climate, and vernacular architecture deduce that thermal design for this region is the most effective approach towards ecological architecture, and both design and construction should follow these principles: comfortable indoor ambience, cost-effectiveness, minimum embodied energy and construction ease. The investigation is further by with thermal simulation experiments. By filtering and optimizing locally available materials and techniques, it is found that the most basic techniques of thermal mass and insulation based on earth and natural materials can be very effective; they should be strategically employed.

Follow the topography, 10 proposed classrooms are planned into 5 units at two levels for maximum exposure to daylight and natural ventilation in summer. A tree-based landscape helps to create a desirable campus ambience for children. The classroom form is derived from local traditional houses with timber structures so as to be constructed easily by villagers. Thermal mass and insulation are employed in the forms of mud-brick walls, the insulated traditional roof, double-glazed windows, etc. The semi-buried form at the north side together with the direct-gain mode of passive solar system can further upgrade the thermal performance. Daylight to the indoor space is maximised by the angled opening of windows.

The construction inherited the local traditional means. It was implemented by the villagers themselves mostly with simple traditional tools. Most building materials, such as mud bricks, rubble, straw and reed, are sourced in or around the site with minimum embodied energy. Also due to the involvement of these natural materials, little waste was generated during construction. Off-cut were recycled back into construction, for example off-cut rafter reused for children's facilities,

spare mud bricks mixed with straw mud for plaster. The construction has almost no environmental impact.

Measurement and evaluation

The construction of the new school was completed in summer 2007. The direct construction cost for materials, equipments and manpower is around HKD$422/㎡, far cheaper than local conventional classrooms made of bricks and concrete. The new school has been used by children and teachers since September, 2007. According to field measurements last year, the indoor air temperature of new classrooms is always stable, cool in summer and warm in winter. Even in the last winter's infrequent cold temperatures, the indoor ambience could still reach an acceptable thermal comfort with fresh air, without needing coal for heating.

Deliverables

In conclusion, the followings are achieved:

Firstly, the new school creates a comfortable and desirable campus ambience for the children. Furthermore, compared to local conventional schools, during its whole life cycle it contributes a far better ecological performance in indoor comfort, energy consumption and environmental impact.

Secondly, as a charity project, most of the donation is shared within the village community due to employing the local villagers and using their resources.

The last and most important point, with the eco-school project the villagers can re-understand their own tradition. The school illustrates to the locals a feasible way towards an ecological architecture suited for the conditions of China's Loess Plateau region. In this way, by selectively employing their familiar techniques and materials the villagers can easily build themselves their most effective and affordable ecological buildings.

A book about this project will be published to share more widely our "demonstration" works and experience for the whole region.

Postscript

Last but not least, quoting the school master, "From now on, not one piece of coal needs to be burned to keep warm. All our money can be spent on books."

5 Hotel Residential

酒店 | 住宅

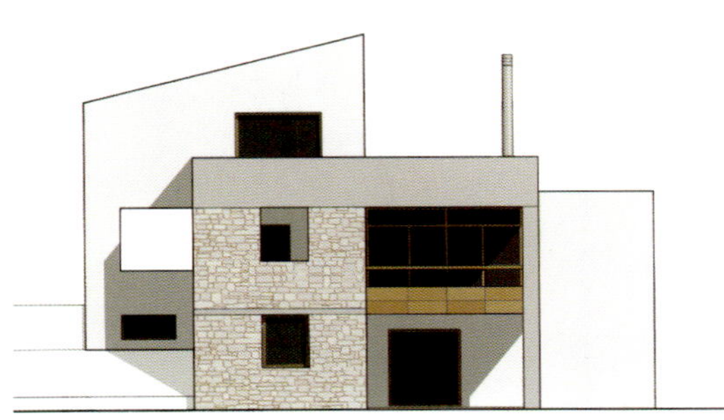
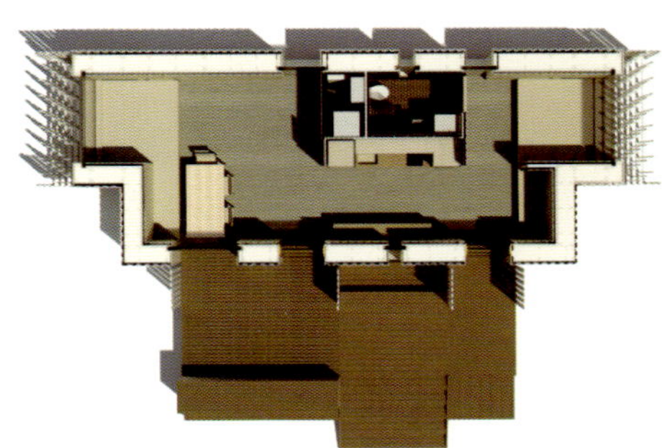

The North Slope Ski Hotel

北坡滑雪酒店

Michael Jantzen

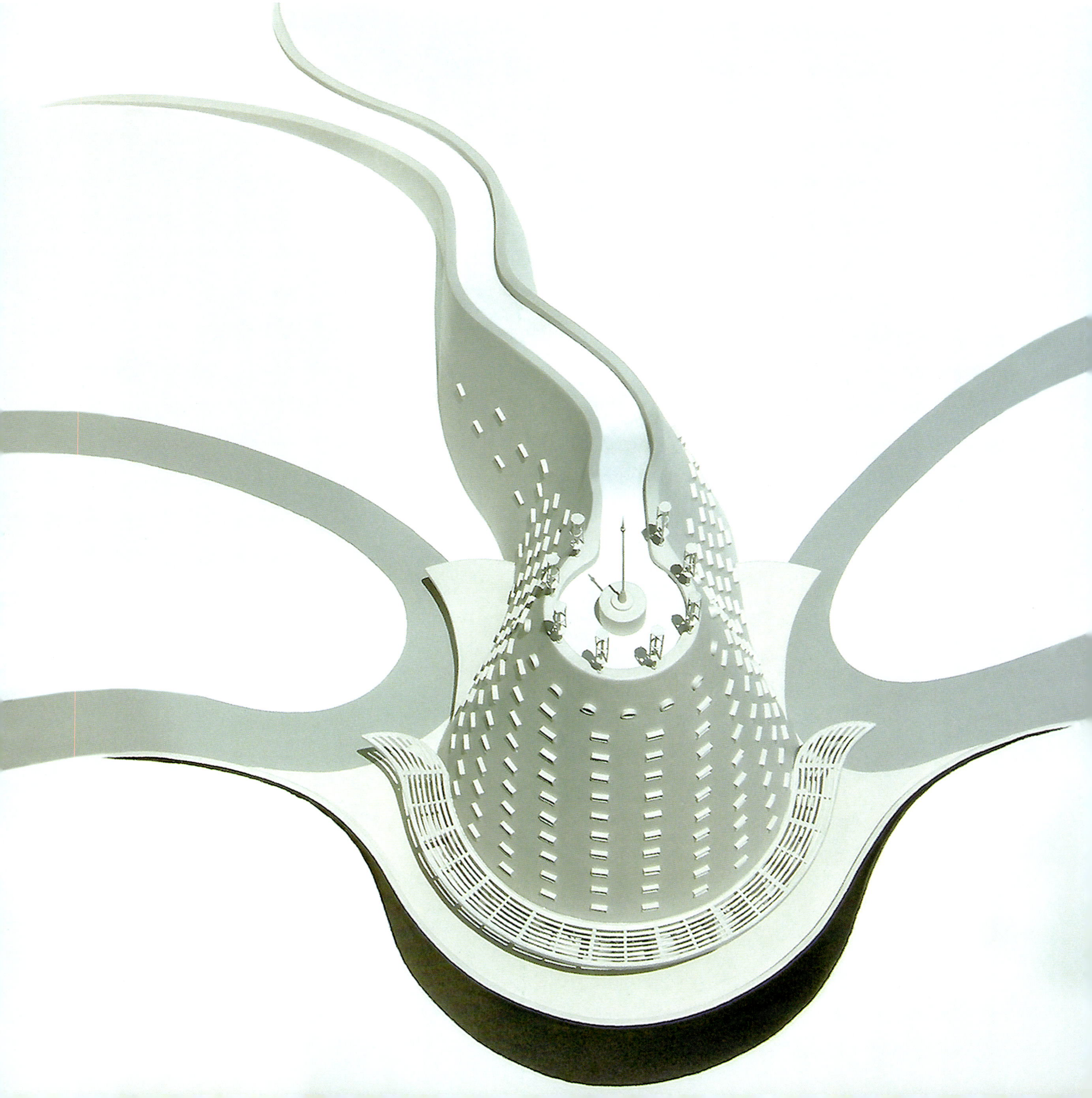

北坡滑雪酒店是为一家拥有95间客房的生态友好型酒店设计的概念性方案，其供电主要来自风和阳光。酒店顶部装有8个大型垂直轴风力发电机，其南面的部分地域被一组弯曲的大型光伏电池组所覆盖。这些风力发电机和太阳能电池被用来满足酒店的用电需求。市面上大部分适用于该设计且具有可持续性的建筑材料都运用到了整座酒店的建设中。在酒店的大厅一层，朝南的大型隔热窗户被安置在大量的热质材料的前端，这有助于被动式太阳能给酒店和深埋地下的管线进行加热。

夏天，由酒店南面结构所形成的巨大阴影将遮蔽这些窗户。这些大型的可控式窗户也为酒店的采光和通风提供了条件。此外，北坡滑雪酒店的每个房间都装有两扇大窗户，它们能控制光线和热量，并为酒店内部的房间提供新鲜空气。

除了这一点和许多其他的环保设计元素外，独特的造型是北坡滑雪酒店最大的特征。它特殊的外部形态构成了一个400英尺高的滑雪坡道，客人们可以乘坐电梯到达酒店顶层，并借用这个坡道进行滑雪运动。即使是夏天，人们也可以尽情享受冲下坡道时的快乐，因为坡道的表面早就经过了不需要雪也能使人们流畅前行的特殊处理。

滑雪坡道的另一个用处是将夏天的雨水和冬天融化的积雪引导进坡道表层下的收容器里，从而收集它们。这些水将被酒店循环利用。此外酒店还有许多不同寻常的设施，例如独具特色的生态水疗中心和健身房，在那里，当客人使用那些经过特殊设计的电力健身器材时，他们本身也参与到了酒店的供电工作中。此外，酒店的顶层还有各种各样独具一格的商店、餐厅和酒吧。

从概念上讲，北坡滑雪酒店的设计来自一幅山顶布满树林的雪山意象图。Michael Jantzen希望能够再次证明，即使是地球上最豪华的地方，也可以并且应该以生态友好的方式建造。

The North Slope Ski Hotel is a conceptual design proposal (in search of a client) for an eco-friendly ninety-five-room luxury hotel, which is powered primarily by the wind and by the sun. Eight large vertical axis wind turbines are mounted on the top of the hotel, and a very large array of flexible photovoltaic cells covers the lower south facing curved portion of the structure. These wind turbines and solar cells are used to supply most of the hotels electrical needs. The entire hotel is built from the most appropriate and sustainable building products available on the market. Large south facing insulated windows mounted in front of a large amount of thermal mass in the floor of the lobby, are used to help passively solar heat the hotel, along with deeply buried earth pipes.

The windows are shaded in the summer by a large overhang projecting from the south face of the structure. These large controllable windows are also used to provide natural light and ventilation for the hotel. In addition, each of the ninety five rooms in the North Slope Ski Hotel is fitted with two large windows. Each of these windows is designed to function like valves that can manipulate the appropriate amount of light, heat, and fresh air that enters each room.

In addition to these and many other eco-friendly elements designed into the hotel, its signature statement is its shape, which incorporates a four-hundred-foot ski slope. Guests can take an elevator to the top of the hotel, and ski down the side of the structure, into the surrounding landscape. They are able to also ski down the hotel's built-in ski slope during the summer, since it is fitted with a special skiable surface that will not require snow.

The ski slope is also designed to collect summer rainwater, and winter melting snow, as it is channeled into large storage containers buried at the base of the slope. This water is recycled and used in and around the hotel. There are many other unusual amenities in and around the hotel, like a special eco-spa, and a gym, in which guests can help to power the hotels electrical needs whenever they use the specially designed electrical generating exercise equipment. There are various unusual places to shop and to eat, and a specially designed bar, located at the top of the hotel.

Conceptually, the design for the North Slope Ski Hotel comes from a symbolic image of a large snow covered mountain, on which a small grove of trees grows at its peak. Michael Jantzen's hope with this design is to once again demonstrate even the most luxurious places on earth can, and should be built in an earth friendly way.

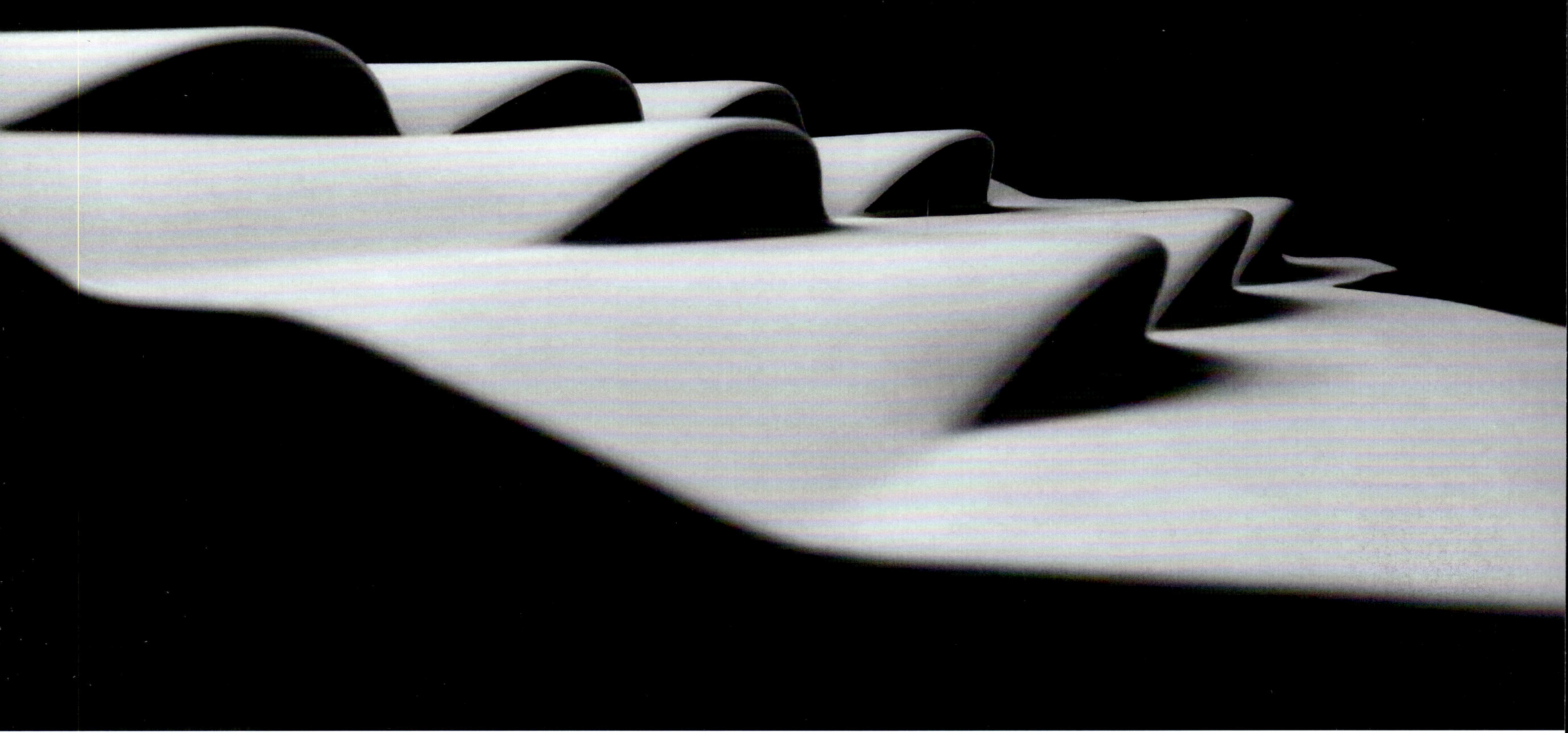

Perché Klima Hotel

Perché Klima酒店

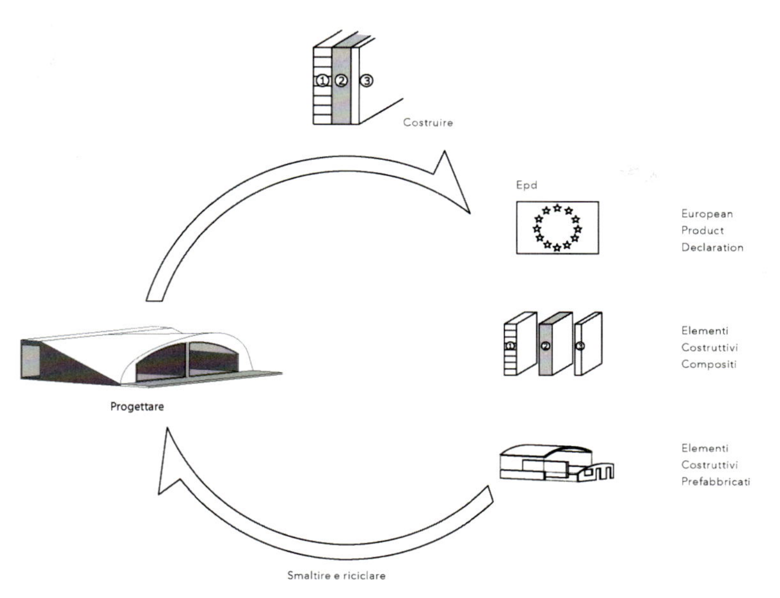

居住设施的认证是依据三大可持续性支柱：生态、经济和社会文化三个方面。

Klima Hotel的设计理念被引进酒店设施的建造，是技术性和战略性管理的需要。在标准的Klima Hotel的目录中三大可持续性支柱的概念分别是“自然”（生态）、“生活”（社会文化）和“透明度”（经济）。

在认证酒店的过程中，核数师起着连接机构和顾客的作用。该过程分为三个阶段：预认证、认证和再认证。

预认证是一个目标树立时期，通过标准的目录为未来经营管理和建筑设计奠定基础。认证是检验在施工时，如何重新确定预认证期间被确立的承诺和目标，一直到the Certificate Climate Hotel的发行。再认证是在开放设施两年后开始执行，验证酒店经营成功实施的可持续性战略，确保其在他们管理下的质量水平。

Klima Hotel Casa Klima想成为评估建筑可持续性的动力，这与它们的性能息息相关。此构造不仅对于能源分析是一个重要基准，而且对于有计划和可持续性管理的评估亦是如此。

The certification of accommodation facilities is based on the three pillars of sustainability: ecology, economy and socio-cultural aspects.

Klima Hotel is the instrument to be introduced within the hotel facilities as the necessary technical and strategic management. In the catalog of criteria Klima Hotel three pillars of sustainability are the concepts "Nature" (Ecology), "Life" (Socio-cultural) and "Transparency" (Economy).

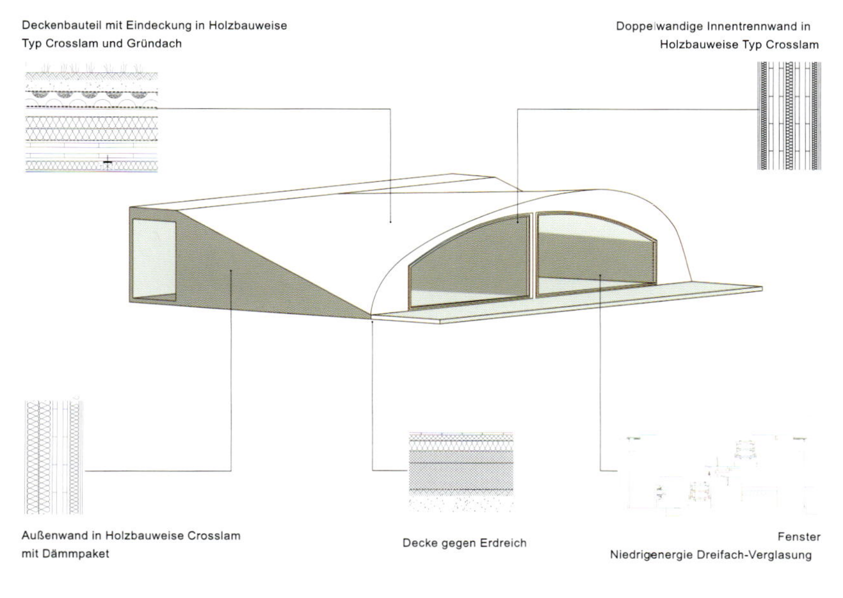

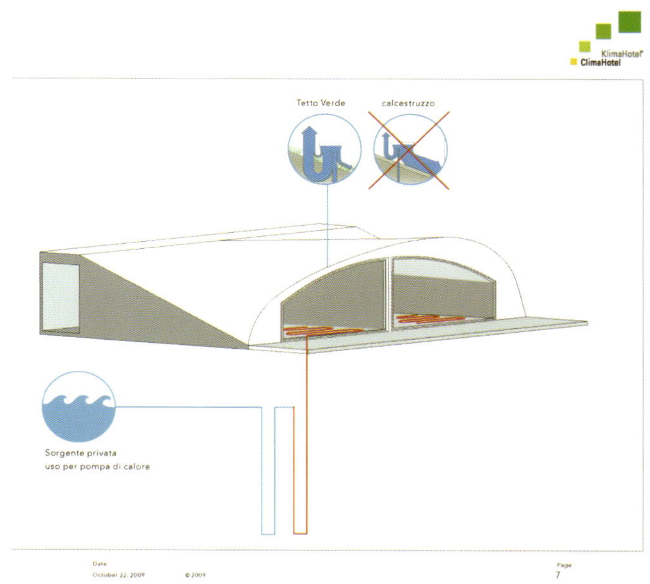

The certification process of the hotel is flanked by a competent auditor, acting as a connection between agency and clients. This process is divided into three key stages: the Pre-Certification, the Certification and Re-Certification.

The Pre-Certification is the phase during which the goals are established and the catalog of criteria are used as foundations for the future management and architectural design. Certification examines how the commitments and objectives defined during the pre-certification have been articulated in the construction phase and ends with the issuance of the Certificate Climate Hotel. The Re-certification takes place after two years from the date of opening of the facility and certifies that the hotel management has successfully carried out the objectives of sustainability ensuring the quality standards to which they had undertaken.

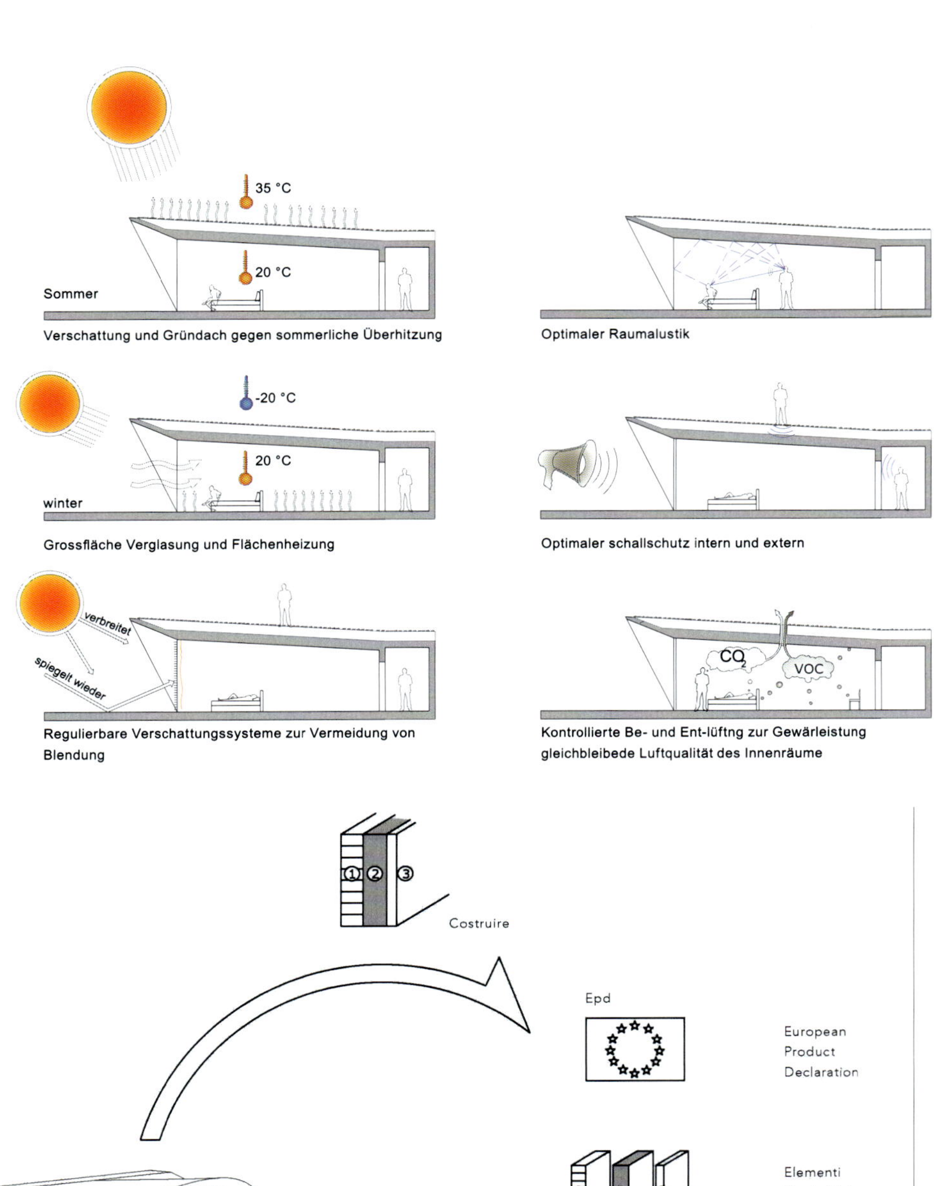

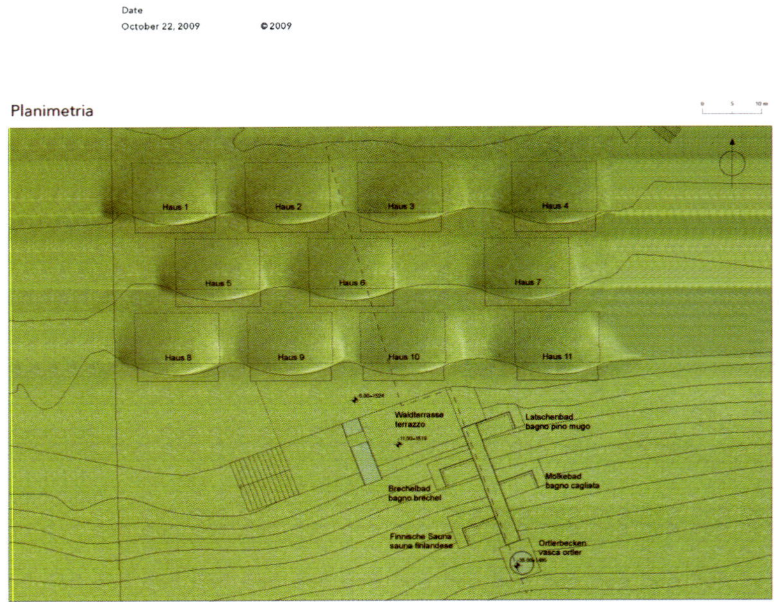

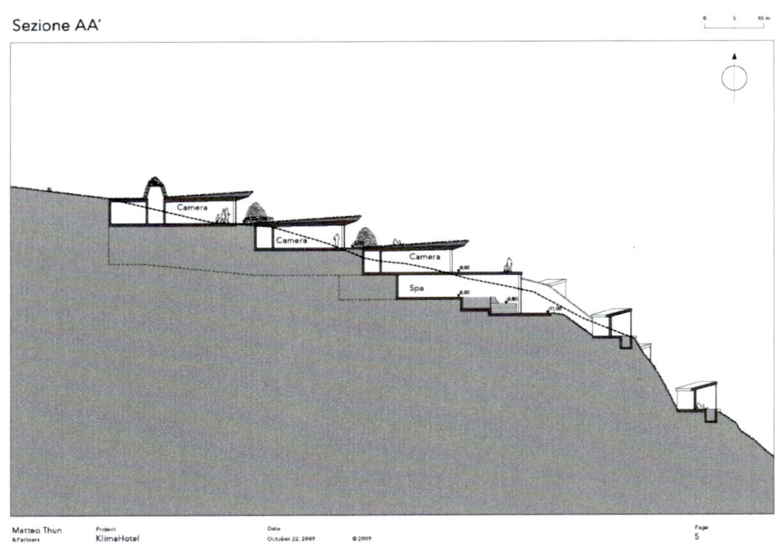

Blok K, Funen

荷兰阿姆斯特丹Blok K住宅

NL建筑师事务所

设计时间	1999~2006
竣工时间	2009
NL建筑师	Pieter Bannenberg, Walter van Dijk, Kamiel Klaasse, Mark Linnemann
合作者	Caro Baumann, Jennifer Petersen, Niels Petersen, Holger Schurk, Misa Shibukawa, Rolf Touzimsky
客户	IBC Vastgoed / Heijmans
结构工程师	Ingenieursbureau Zonneveld bv
机械工程师	Sweegers en de Bruin bv
建筑物理分析	Cauberg Huygen
承包商	IBC Woningbouw Amersfoort bv

Design Milestones	1999 -2006
Completion Date	2009
NL Architects	Pieter Bannenberg, Walter van Dijk, Kamiel Klaasse, Mark Linnemann
Collaborators	Caro Baumann, Jennifer Petersen, Niels Petersen, Holger Schurk, Misa Shibukawa, Rolf Touzimsky
Client	IBC Vastgoed / Heijmans
Structural Engineers	Ingenieursbureau Zonneveld bv
Mechanical Engineers	Sweegers en de Bruin bv
Building Physics	Cauberg Huygen
Contractor	IBC Woningbouw Amersfoort bv

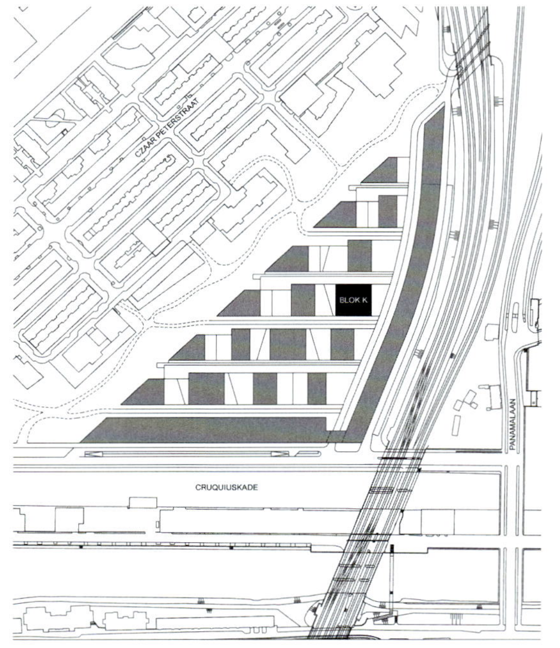

这十栋房屋是由Cie设计的包括500栋住宅和一个公园在内的总体规划的一部分，被称为"Het Funen,Hidden Delights"。这个三面体建筑位于历史中心和阿姆斯特丹东部的最近新开发的港口区之间，这里曾经是拖运式卡车的停车场。沿着东部和南部是一面包括三百个公寓楼和办公空间的"墙"，使其远离附近铁路的噪音干扰。这个半开放的街区内部被设计成松散的网络状，可以容纳16个位于公园中的小型街坊式住宅。这些"Hidden Delights"高低各不相同，从9至18米不等。实现了公共与私人空间之间的转换。整个地区规划包括公园在内都由城市的传统方法引入和操作，并由IBC Vastgoed公司进行开发与建造。公园的维护则由私人的事务所负责，而并不是由政府负责的，但是公园将仍然是对公众开放的。这里有三个几乎为方形的大楼，长30.5米，宽27.7米，可以容纳2.5层。最初排列成行的两层楼的建造是具有强制性的；第三层则应为50%的大楼和50%的屋顶平台/花园。在我们的大楼，体积是根据十栋房子平均分布的；每一栋的体积都为633立方米。房屋是根据背靠背的房屋类型学组织排列的，经由整个街区中央的走廊进入。这个"小型峡谷"为外立面节约了必要的储藏空间及在形式上需从公共区域进入的技术设备空间。

These ten houses are part of a masterplan for 500 dwellings and a park oy the Architect Cie, called "Het Funen, Hidden Delights". The triangular site is located between the historic center and the recently redeveloped harbor area in the east of Amsterdam, a former parking lot for towed cars. Along the east and the south side a "wall" containing over 300 apartments and office spaces shields the site from the noise of the adjacent railroad. Inside this semi-open block a loose grid is set up, containing 16 smaller housing blocks positioned in a park. These "Hidden Delights" vary in height from 9 to 18 meters. A shift from the public to the private has taken place. The urban plan including the park is initiated and commissioned traditionally operations directed by the City and developed and built by one single company, IBC

Vastgoed. The maintenance of the park will be handled by a private firm, not by the City, but the park will remain publicly accessible. There are three almost square blocks that measure 30.5 by 27.7 meters and should contain 2.5 stories. It is obligatory to build the first two stories in alignment; the third should be 50% building and 50% roof terrace/garden. In our block the volume is distributed evenly over the ten houses; each is allocated 633 m^3. The houses are organized according to a typology known as back to back housing. They will be accessed from an aisle in the middle of the block. This "mini-canyon" rids the facades of the obligatory storage spaces and technical facilities that formally have to be accessible from the public domain.

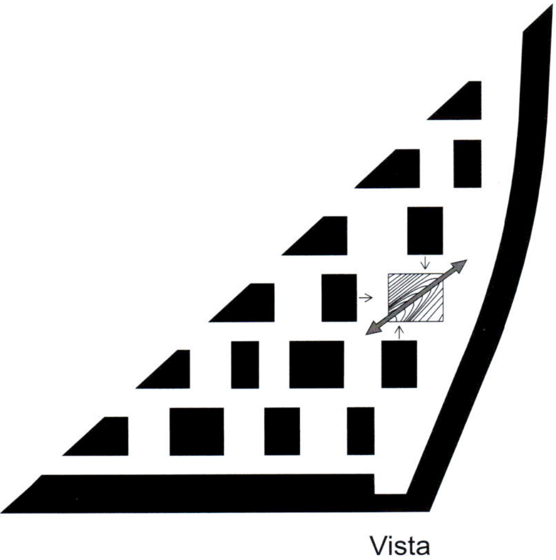

Vista

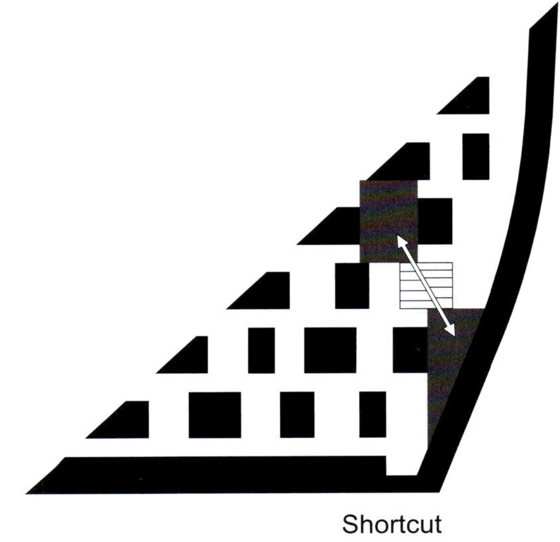

Shortcut

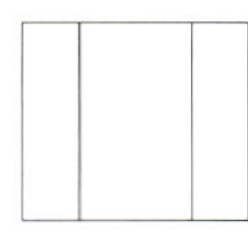

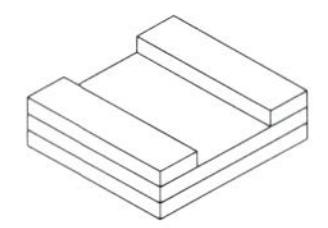

1 **2 + ½ Stories** Given envelope: The first two stories should be build in alignment (100 %), the third with a setback: 50% building + 50% roofgarden. Average building height = 7,5 meters. Total volume = 6336 m³.

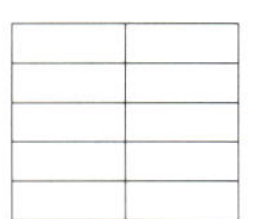

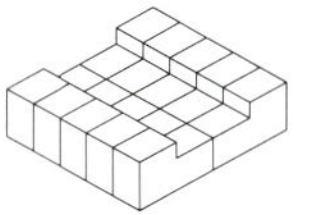

2 **Back to Back** Conventional building technique: 10 'ground related' identical houses.

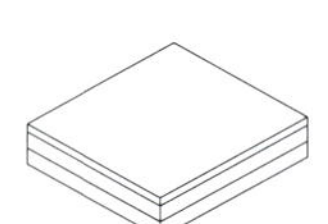

3 **Being John Malkovich** Re-interpretation of given envelope: 2½ stories (100% roofgarden): a block waiting to be touched.

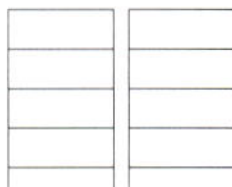

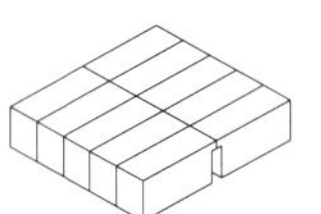

4 **Alley** The obligatory storage spaces, technical facilities and hallways are absorbed in and accessed from the 'center' o' the block: the facades open up to the light and the 'park'.

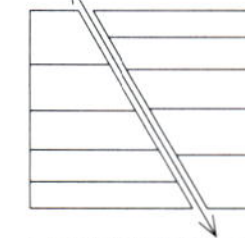

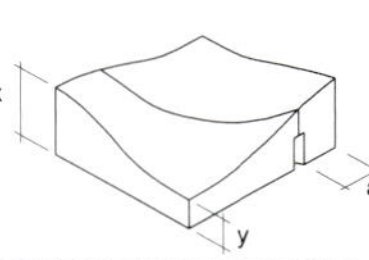

5 **Shortcut** By orienting the alley towards the two open spaces instead of two 'blind' walls an atttactive shortcut is created. As a consequence the block is deformed: northwest and southeast corners rise whereas northeast and southwest corners lower. The typology becomes elastic, a range from 1 ½ to 4 stories. Average building height remains 7,5 meters.

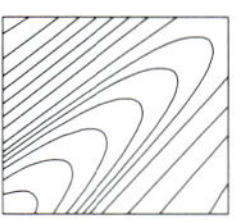

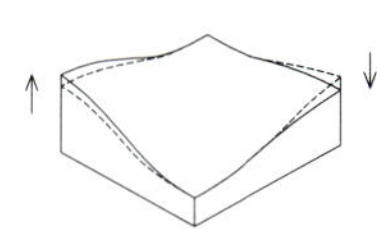

6 **Flex** Strategically positioning the volume towards the sun results in a lower south and higher north section. Amplitude of the building varies from 5 to 15 meters.

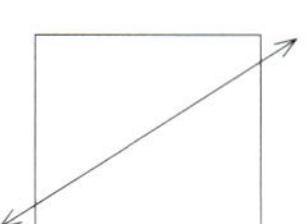

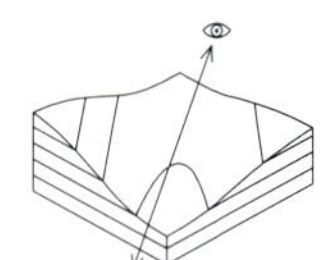

7 **Vista** The deformation as a consequence of the diagonal shortcut creates a 'void' in the otherwise dense master plan.

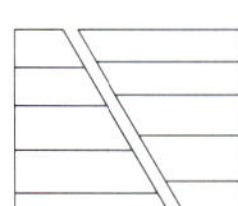

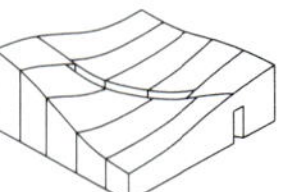

8 **Elastical Building Bay** The block is divided up in ten unique individual houses with equal volume (633 m³) but different floor areas. The houses on the north and south sides get daylight from two sides, whereas the six in the middle depend on only one facade (and the patio). The extra width is a positive side effect of the "equation".

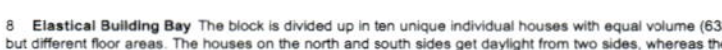

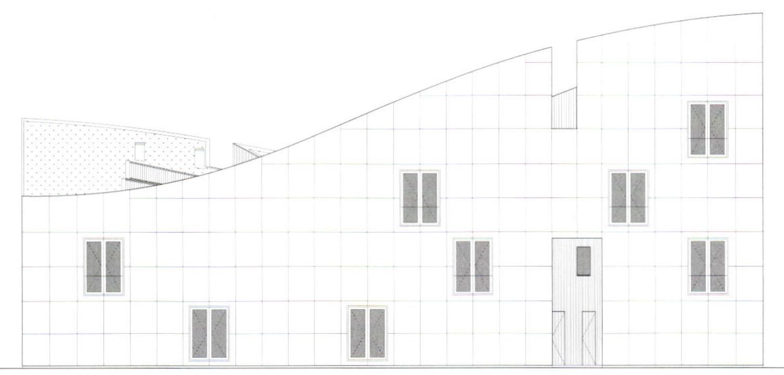

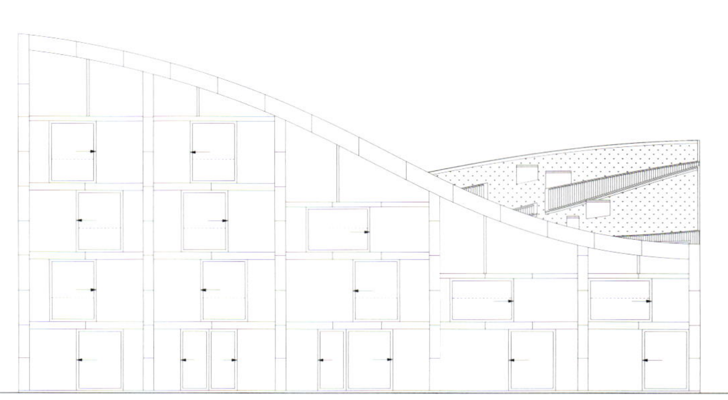

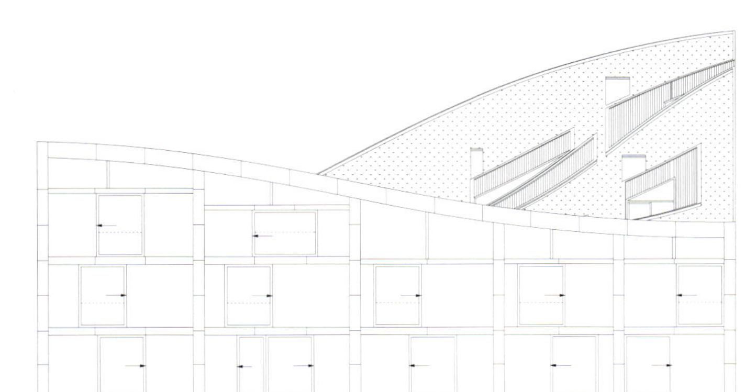

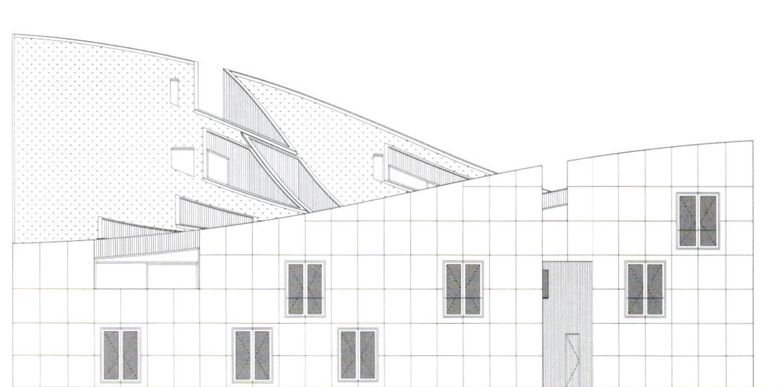

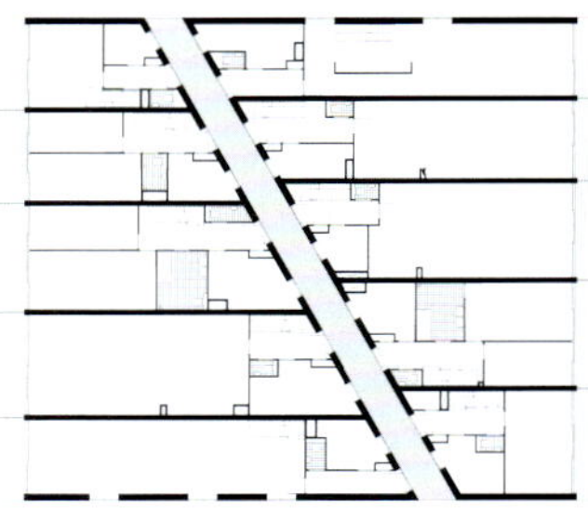
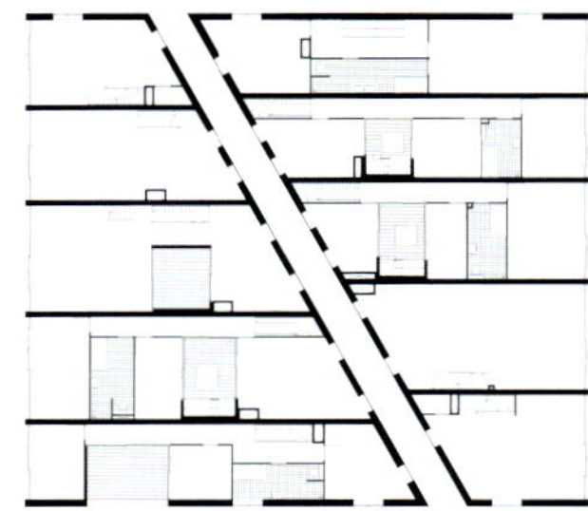
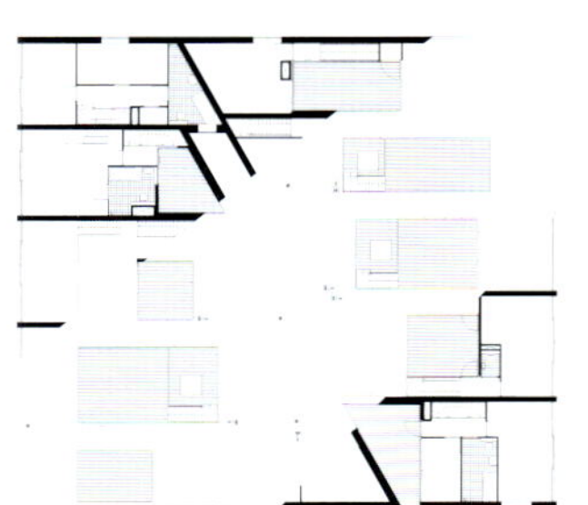

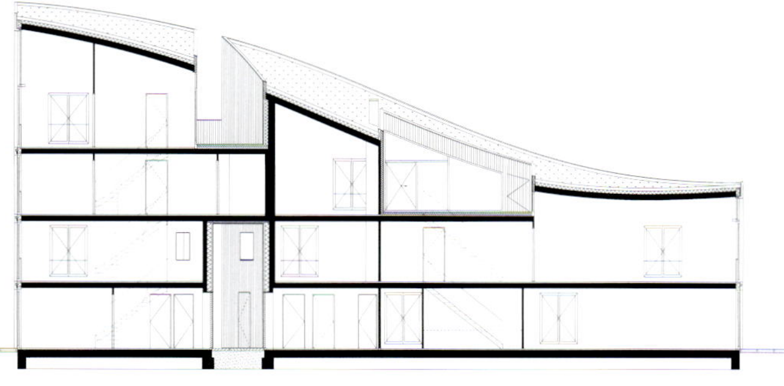
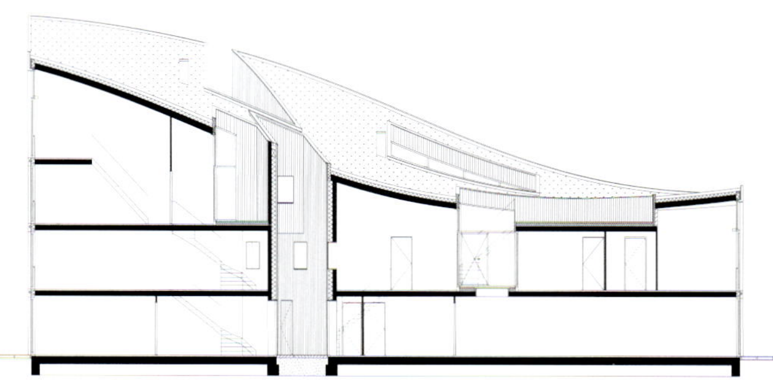
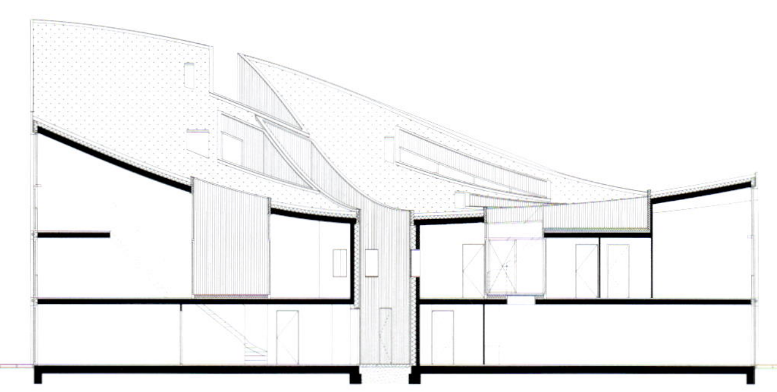
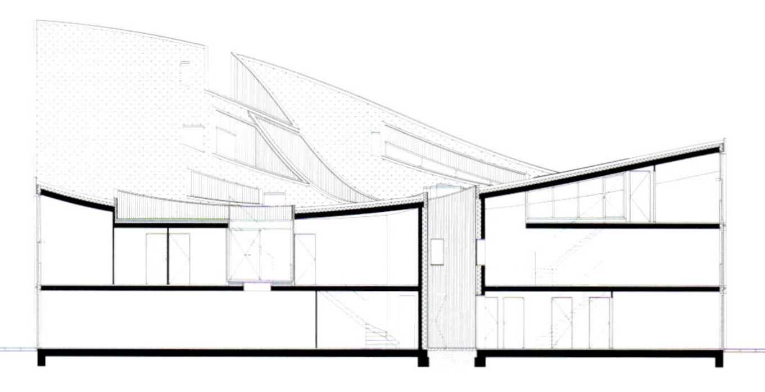
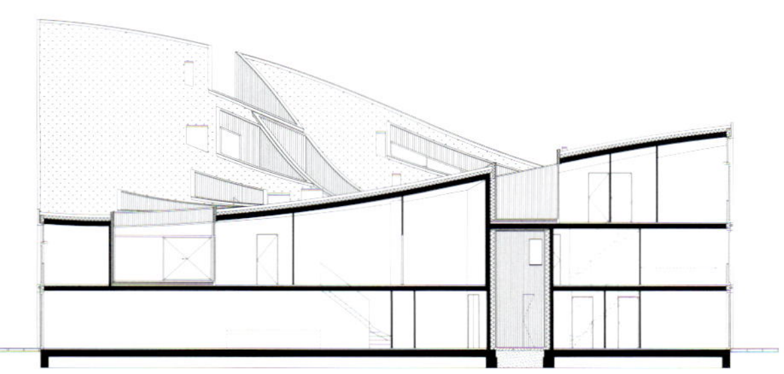
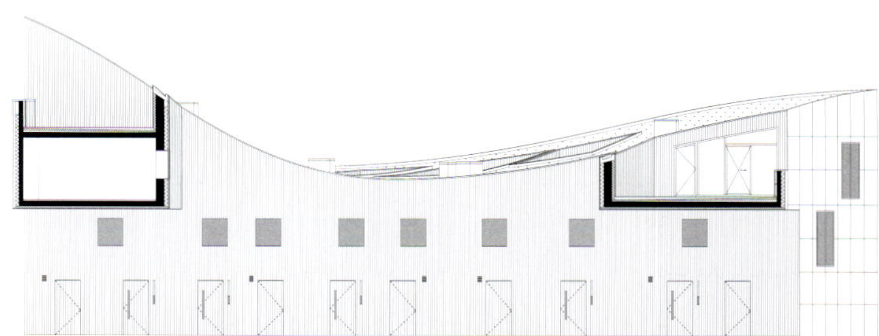

An Environment Friendly House in Thessaloniki

塞萨洛尼基生态住宅

Nicos Kalogirou

该房子坐落于一个非常险峻的东西走向的斜坡上，周围是茂密的松树林。依顾客的角度看，当前最重要的项目是建筑外壳的节能型设计，结合对房子内部的保护和对居住场所周围景观及绿化的规划，并且能够充分利用当地的自然物质和资源。

明确的需求和一个整体的独特功能形成大小不同的两个相交矩形体量。通过路口创造的固体体量是房子的主要核心部分，形成非对称交叉路口，以半开放空间为结束点。通达主要入口处的曲线提高了内部空间的私密性。建筑的线性发展明显地体现在其内部特征上，具体体现在具有连续性的朴素屋顶阳台和开放的起居室和中庭。一个抽象的整体展现了与传统的希腊建筑的关系，主要体现在类型的层次方面，自觉避免了所有关于形态和文体特征的参考。

建筑特征响应了所有节能方针的呼吁。位于北面的建筑物外壳外面的石头碰到外壳就会露出一些小孔通风，而入口处由一个封闭的走廊保护可以避开强风。

位于南面有非常大的玻璃口子——通过一种木质的耐用罩建造而成——冬天阳光可以进入这里。夏天可以阻止阳光的直接照射但允许暖空气的流通。在该建筑的西边，可以浏览城市的美丽风光，因此在这边有一个过渡空间，由一个有垂直的百叶窗的开放式画廊构成。所有这些因素对形态方面有着积极的作用，可以很容易地区分房子的表面特征。

该房子主要居住部分的特征是可以自然通风、高效可控的照明系统和内外空间的连续性。中央中庭的战略位置以及喷泉和一条小水池与上述元素共同构成一个开

放式的可以抵御强风的空间，是一个过渡性季节的居住场所。

建筑结构

建筑物大部分的外墙宽50~80厘米，有利于节约能源而且提供了有趣的开放概况。在其南边有一个画廊作为温室，它的内墙是实心砖构成的。

本项目特别重视材料的选择，在尽可能的情况下绝大部分都使用本地材料，且进行粗略的加工，保证施工中最低的能源损耗。使用了生态颜料除外它们没有污点也未抹灰浆。木质结构元素被广泛运用，原生态物质是专门木制流程的产品（仅在满月期间进行木材制造）以此保证最低湿度。镶板和窗户用木制的，屋顶和地板没有任何污点。双固体墙用10厘米厚的岩棉片隔离开，此举有利于环境保护。在双层玻璃窗户内充满了氩气。

在希腊很少用到地板供暖和隔板。该建筑没有常规的锅炉而是有一个热水泵来处理一个深1.20米的外观网络工作管而不是处理一个冰冷塔。该外观网络工作管中的温度一般比较稳定，一年之中温度保持在10~20摄氏度。据估计房子的供暖费用是常规费用的五分之一，而且不会造成任何的环境污染。

该建筑有一个水资源管理系统。地下的蓄水池用来收集屋顶上的雨水。住宅内的排水最初汇聚到粪水池，在粪水池中固体部分被分离开。液体部分被引入一个生态锅中，这个生态锅是一个盆形状，防水型，大约长4米，宽10米，高2.5米，具有不同的物质层——主要有厚砂、铁和涂报纸的石灰等。地表上种植芦苇，它的根部为地下层提供了氧气。最后水会聚集在水槽中，有时也会通过比较小的连续池系统进行回收。因此，水和氧气有着直接的联系，于是便有了寒蝉效应和自来水。所有回收的水储备用于园林灌溉和外观建筑的清洗保养。

The house is set on a very steep inclined site which develops on an east-west axis, densely planted with pine trees. The critical items in the clients' brief are an energy conservation design of the building's shell, the combination of introvert protection of the private realm of the house and the opening up of the living area to the view and the greenery, while exploiting all local natural materials and resources.

The demand for clarity and distinct presence of each function within a well defined whole, results in the formation of two intersecting rectangular volumes of different sizes. The solid volume created through the intersection is the main core of the house while the aisles of the asymmetrical cross formed, ended up in semi-open spaces. The curve that leads the way to the main entrance enhances the feeling of a well enclosed interior space. The linear development of the building is also evident in its interior disposition with the austere spatial continuity of the roofed balcony, the open plan living room and the atrium. The overall abstract composition presents an inner link to the traditional Greek domestic architecture, mostly evident a typological level, consciously avoiding all clear references to morphological and stylistic features.

The architectural disposition responds to all energy saving and orientation facts. The closed curved stone outer shell of the building to the north, referring to a shell, allows only small openings for ventilation, while the entrance is well protected from the wind by a closed portico.

To the south large glass openings allow—through a system

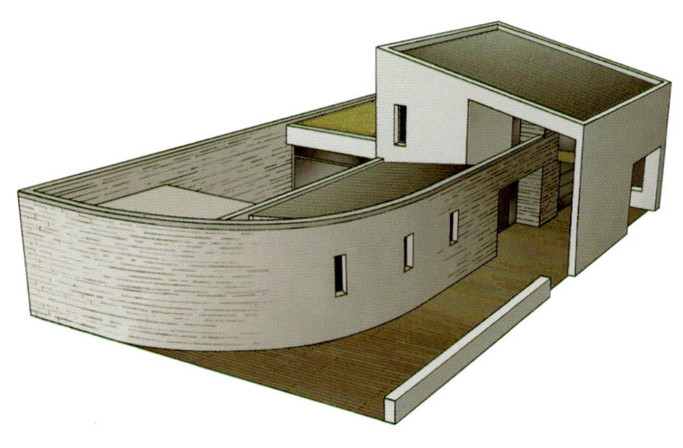

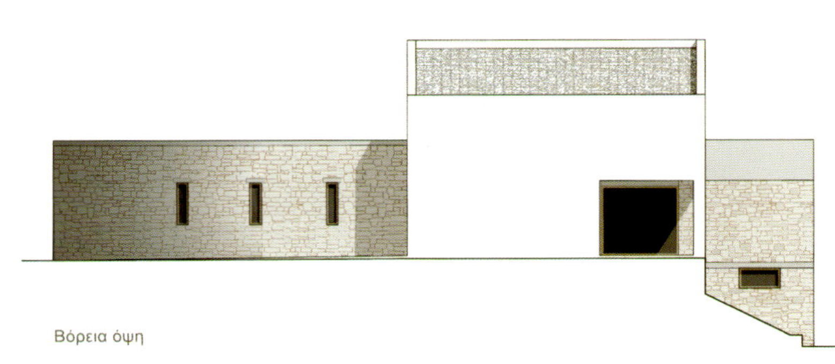

Βόρεια όψη

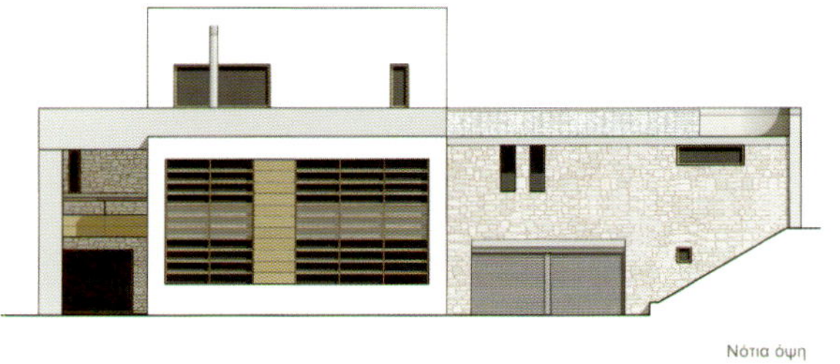

Νότια όψη

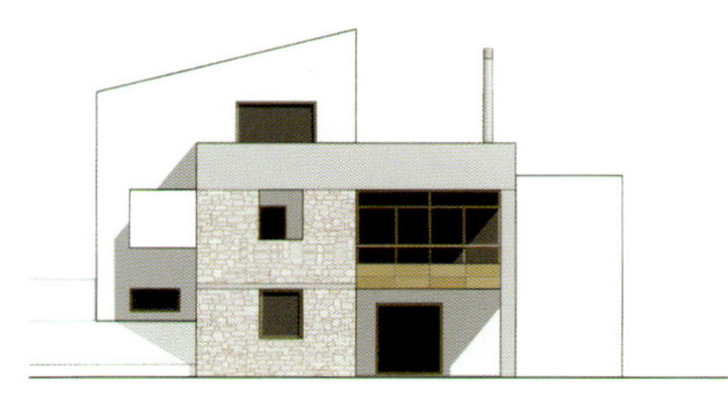

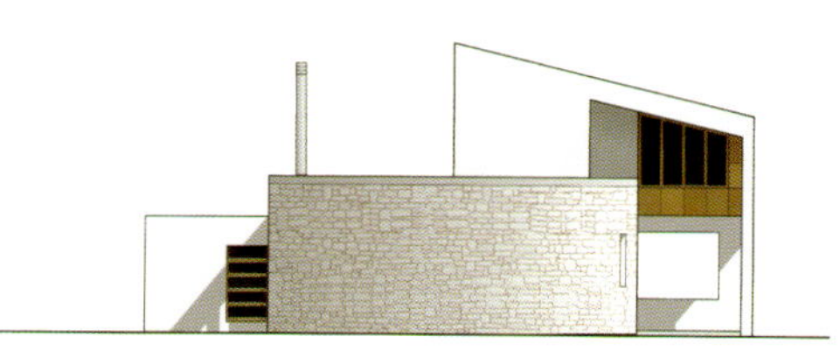

of wooden permanent shades — the sunlight to enter during the winter. During the summer protection from direct sunlight is provided while permitting the warm air to circulate. To the west, where there's usually the best view of the city, there is a transitional space with the form of an open gallery protected by vertical shutters. All of these elements have also a very positive contribution on a morphological level while actually distinctly differentiate the facades of the house.

The disposition of the main living areas of the house allows their natural ventilation, a well controlled light scheme as well as the continuity between inner and outer space. The strategic location of the central atrium, with a fountain and a small water pool works together with all the above stated elements providing an open air room well protected from the wind, a living space ideal for the transitional seasons.

Construction

Most of the exterior walls of the building have a width of 50-80 cm resulting in an energy saving behavior while providing interesting opening profiles. To the south there is a closed gallery functioning as a greenhouse and having an inner wall made from hand made solid brick.

Special care is taken of the materials which are mostly local, in the most natural condition possible, roughly crafted in order to incorporate the minimum quantity of energy during the construction. They have been left unstained or unplastered with very few exceptions where ecological colours have been used. Wooden construction elements are used extensively and the raw material is the product of a highly specialized timbering process (i.e. wood timbered only during full moon periods) so that it would contain the lowest quantity of humidity. With that timber, without any stain or vernice roofs and floors, paneling and windows are constructed. From the products of the construction process of the wooden elements a gutex is prepared to isolate the roofs and floors. The double solid walls are insulated by the use of 10 cm thick sheets of rockwool, a material chosen for its most friendly to the environment substance. The double glazing of the windows have been filled with argon gas.

There is a combination of floor heating and wall panels that has been very rarely used in Greece. There is no conventional boiler but a heat water pump which disposes instead of a chilling tower an exterior network of tubes in a depth of 1.20 m, where the temperature remains stable, all around the year, between 10-20 °C. It is estimated that the heating expenses for the house is 1/5 of the conventional cost while there is no pollution for the environment whatsoever.

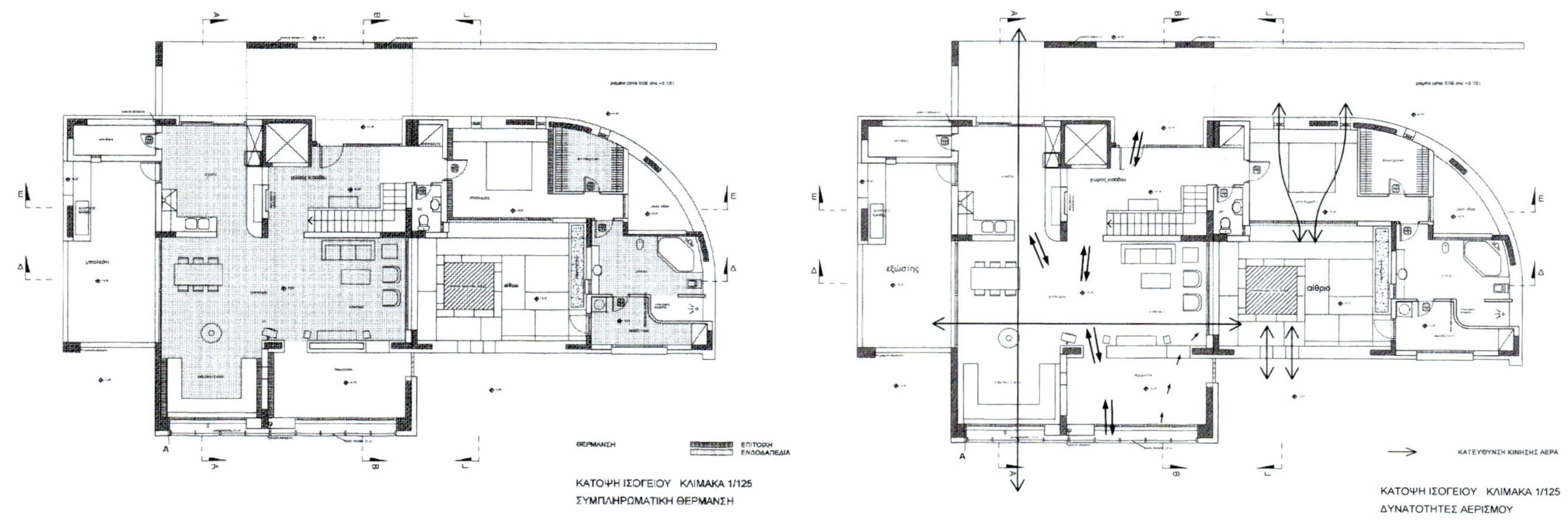

There is a system of water management. Rain water from the roofs is collected in underground tanks. The domestic discharges are gathered initially in the septic tank where solid parts are being separated. The liquid discharges then are led to an ecological pot, a basin 4x10x2.5 m, totally water proof and filled with layers of different materials — i.e. a variety of aggregates, thick sand, rust, lime coated newspapers etc. On the surface reeds are being planted whose roots provide oxygen to the underlayers. Finally the waters are gathered in a cistern while they are recycled occasionally through a system of small successive pools. So the water has a direct contact with oxygen, providing a chilling effect and the pleasant sound of running water. All water reserves are recycled in the watering of the garden and for exterior housekeeping.

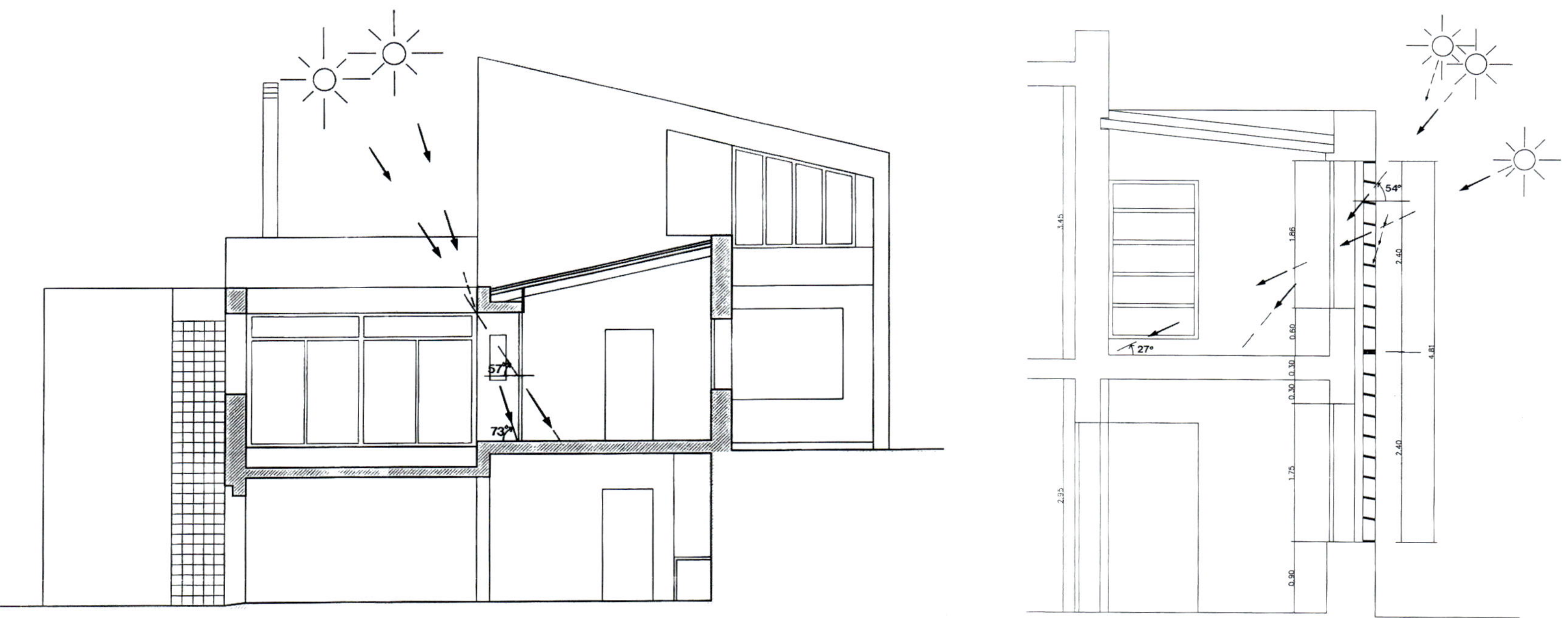

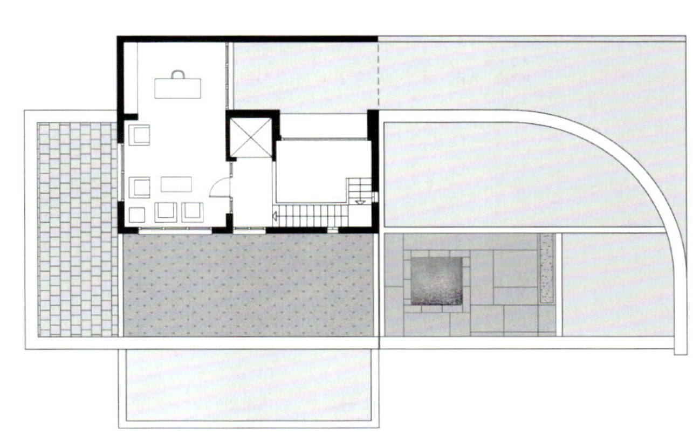

Los Altos Hills Family Residence
美国加州洛斯阿尔托斯山私家住宅

CCS Architecture

项目地点	美国，加州，洛斯阿尔托斯山，米兰达路14355号
建筑类型	新型住宅，（6000平方英尺，加上全部地下部分）
竣工时间	2008年6月
建筑团队	首席设计师：卡斯•科尔德•史密斯
	室内设计总监：芭芭•拉特平– Vickroy
	项目建筑师：德拉–吉尔•拉蒙塔涅
	设计师：Lynne Riesselman
顾问	土木工程: Triad Holmes联合公司
	结构工程：RC咨询工程师
	景观设计：布拉德•伯克
	能源工程: EnergySoft
总承包商	Consilium Construction、卡察夫•葛雷、帕特•卡瓦纳
摄影师	马修•米尔曼

Project Location	14355 Miranda Way, Los Altos Hills, CA, USA
Building Type	New Residence, 6,000 sf, plus full basement
Completion Date	June 2008
Architecture Team	Design Principal: Cass Calder Smith
	Interior Design Director: Barbara Turpin-Vickroy
	Project Architect: Dera-Jill Lamontagne
	Designer: Lynne Riesselman
Consultants	Civil Engineer: Triad Holmes Associates
	Structural Engineer: RC Consulting Engineers
	Landscape: Brad Burke
	Energy Designer: EnergySoft
General Contractor	Consilium Construction, Moshe Gray, Pat Kavanagh
Photographer	Matthew Millman

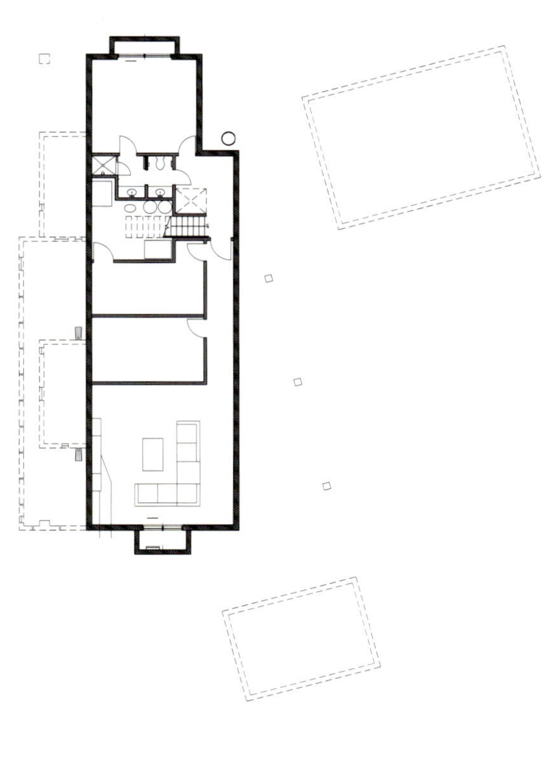

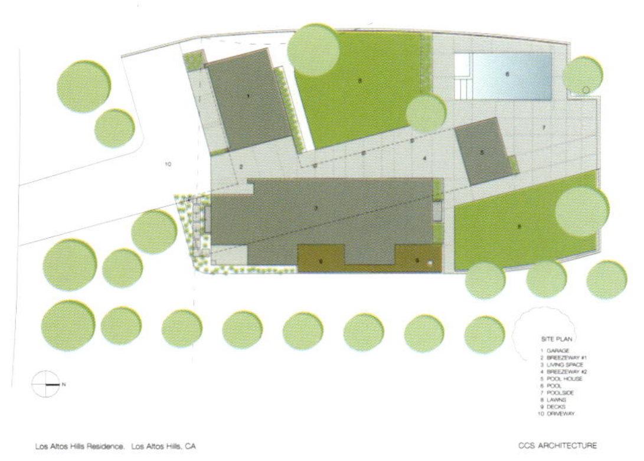

CCS建筑事务所为一个生活在加利福尼亚州的五口人的家庭设计了这套6 000平方英尺的现代公寓。同在硅谷工作的两位主人希望他们的住房既休闲又温暖，在高效节能的同时，还必须顺应当地气候。

为了发挥这座公寓在通道、景观及室外空间方面的优势，设计师充分利用了每一寸土地。他们发现，将庭院和游泳池建在公寓的西面是最合适不过的——那有一片长满树木的平缓草坡，它不仅能用作天然隔断，还给该区域增添了私密性和观赏性。除此之外，露天活动区也是朝西而建的，往东可以看到硅谷。室内开敞的平面布局最大限度地实现了客厅与室外的连接。主客厅和起居室毗邻而建，并留有两条去往室外的通道。大型滑动门使得西面的泳池、东面的休息桌与家庭活动室、厨房、餐厅连接在一起。公寓的东面种着杏树和加州胡椒树，它们为公寓创造出一个私密空间，使公寓很好地与周边街道分离开，并且也与洛斯阿尔托斯山过去的农业传统建立起历史联系。

别墅的大体形式由两个相扣的体量构成。这个以L形水泥石膏板为基础的建筑，经由一条走廊被分成两部分，包含了主要的起居空间及车库。连廊不仅打开了通向室外生活区域的景观视线，同时也是别墅的主入口通道。二楼卧室立面为垂直干挂的雪松板，此部分为一长长的条状形体，与一楼的轴线呈旋转的关系，从而在为一楼提供遮挡屋面的同时，也创造了二楼西侧的露台空间。在二楼的体量向下折转，并与地面交接的地方出现了第二个连廊，并成为了室内泳池空间。这个连廊提供了从树的高度望向硅谷的景框。通过连廊将别墅划分为三个部分，不仅最大限度地实现了室内与室外生活空间相互间的景观交流及朝向硅谷的景观视线，而且为别墅增添了当代围场的氛围。

这里的气候大部分时间都非常温和，但夏天会异常炎热。为了最大化地使用室外生活区域，二楼深远的悬挑为与室内紧邻的空间创造了遮蔽条件。炎热的天气下，它将成为一个舒适的遮荫处，因而可以常年使用。游泳室的三面都设有推拉门，这样可以在提供附加的遮荫功能的同时，也最大限度的加强与池边活动的参与性。在庭院的西侧边缘上有一片长长的弯曲的挡墙，确保了别墅的室外空间的私密性，室外空间包括一个可用的草坪，彩色混凝土铺地的院子，由高高的景墙围成的水池。

业主想要一个现代的别墅，但同时他们也要一个温馨、随意，且与周边环境相协调的家。为了营造温馨感，设计在白色橡木和胡桃木的地面，与白色和特色色彩的墙面之间取得平衡。别墅的外立面采用了亚光的拉毛水泥和雪松木边条，且利用竖向的构图方式来创造肌理。尽管其主要结构非常简约，两种对比鲜明的材料，各个体量之间的渗透，由旋转的形式所产生的独特的几何图形创造了建筑的复杂性。其复杂性所呈现出的尺度感使其与周边的大农场风格的别墅保持一致，也使得设计方案容易被接受。

屋顶上大型的光伏电池组为公寓提供了电力，太阳能板则用来供应热水。屋顶由节能的结构绝热板（SIP）制成。电动分体式热泵实现了房屋的供暖和制冷。大部分窗门自带的遮阳棚能在夏季有效减少阳光直晒和热量。泳池的热池盖能够通过被动式太阳能直接加热水体。冬天的时候窗户可以保暖，夏天则能够散热。墙壁材料是棉，比起玻璃纤维它拥有更好的绝缘性。

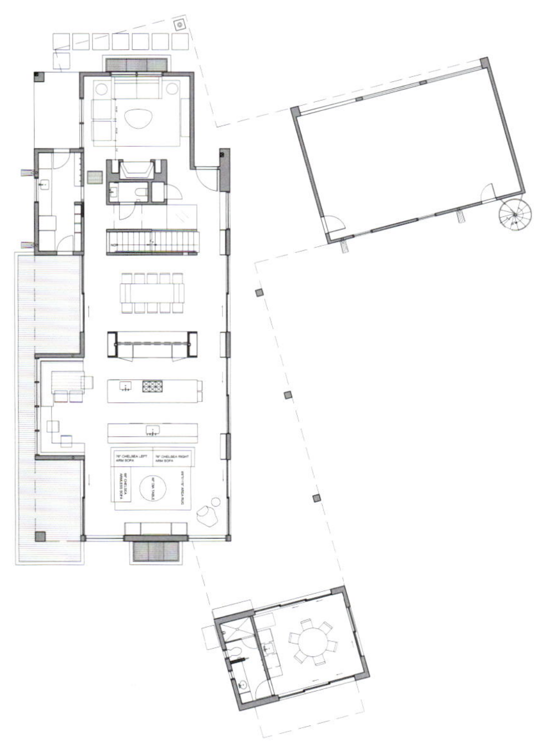

CCS Architecture designs this 6,000 square foot modern home for a family of five in Los Altos Hills, California. The clients, both working parents in the Silicon Valley, want their dwelling to be as casual and family friendly as possible, energy efficient, and responsive to the local climate conditions.

In order to take advantage of the site in regard to access, views, and outdoor space, architects carefully position the house on its one-acre site. They determine that the best part of the site to locate the main yard and pool is to the west. This space is bordered by a sloping grassy meadow that rises up to a line of trees above that give it more definition and a nice sense of place. The house is then designed to open up to this outdoor living area to the west and take advantage of the views to Silicon Valley to the east. Inside, the open floor plan maximizes connections between the living and outdoor space. The main living areas are connected to one another and to outdoor spaces on both sides of the house. Large sliding doors open the family room, kitchen and dining room to the pool in the west and to decks in the east. The east side of the property is planted with apricot orchards and California pepper trees, creating privacy from the adjacent street, and making a historical connection to the agricultural past of Los Altos Hills.

The general form of the house is created by two interlocking volumes. An L-shaped cement plaster base building, split by a breezeway, houses the main living spaces and the garage. The breezeway provides views into the main landscaped outdoor living area and access to the main entrance to the house. The second floor bedroom wing, clad in vertical cedar slats, is a long bar that is rotated in relation to the axis of first floor, creating a sheltered space below and a roof deck to the west. Where the volume of the second floor turns down to meet the ground, it creates a second breezeway and becomes the pool house. This breezeway frames a tree-level view of the Silicon Valley. By breaking the house into three parts separated by breezeways, views to and from the outdoor living spaces and views of the Silicon Valley are maximized, and the house has the feeling of a contemporary compound.

Weather in the area is moderate for most of the year, but hot during the summer. To maximize the use of the outdoor living areas, spaces closest to the house are within a deep overhang created by the second floor above. This creates a nice sheltered place and shade when it's hot so it can be enjoyed year round. The pool house has sliding doors on 3 sides to provide additional shade and maximize connections to poolside activity. A long curved retaining wall borders the west side of the yard. The home's outdoor space is protected and private, with usable lawns, colored concrete patios, and a pool screened with tall landscaping.

The clients want a modern home, but they want theirs to be warm and casual and to integrate well into the neighborhood. To keep it warm, we use a balance of white oak, walnut wood finishes, and white and accent color painted walls. On the exterior of the outside, there is muted painted stucco and cedar siding, with texture created by its vertical slatted configuration. The two contrasting materials, the permeability of the volumes, and the unique geometries created by the rotated forms create complexity even though the main forms of the house are quite simple. This complexity provides a sense of scale that is in keeping with the other

ranch style homes that are typical of the neighborhood and helps the project win easy design approval.

A large photovoltaic array on the roof generates some of the home's electricity, and additional solar panels provide hot water. The roof is made of energy efficient, structurally insulated panels (SIPs). An electric, split-system heat pump provides heating and cooling. Sunshades on most windows and doors combine with deep roof overhangs to temper the light and decrease heat gain in the summer. The pool has a thermal pool cover which warms it directly through passive solar action. High performance windows help keep heat in during winter and heat out during summer. Walls are insulated with cotton rather than fiberglass.

TUSCANY
SCARPA

Mobile Floating Architecture
移动漂浮式建筑

瑞士RAFF建筑设计公司

概念

传统房屋的概念和移动式水上建筑是不能混为一谈的。创造新类型的时候无疑要考虑到其可用性、空间以及相关技术。形式化、概念化需求与同时营造家一般的氛围恰恰是相反的。设计师必须慎重考虑在水上生活的问题。这也是设计师选择将移动性与独立性作为移动式漂浮建筑的设计重点的原因。顾名思义，"最后的度假胜地"就是希望将本项目打造成独一无二的度假区，力图打破常规的理念，展示另一种生活方式。

组成部分

本项目分为两层，为了给上层甲板留出2.5米足够的高度，床铺、技术设备以及舱门都被建造在下层甲板中。下层甲板是作为上层甲板的额外空间而存在的：床和沙发被嵌入底板内，有需要时才打开使用。外测量为5米 x 15米。在前方有个带顶平台，在那里有通往屋顶的楼梯。厨房和起居室共同组成了一个开放空间。浴室将卧室和其他房间分开。卧室可以用滑动面板隔开，用作独立书房。总而言之，这里一共有六张床（包括两张甲板床）。

劳济茨的湖泊区

本方案的设计灵感来自于滨水区水平的波浪形状。此外，这也是居民居住方式的一种全新定位。其周边的景观也可以理解为起居室的扩展空间。无阻隔的自然风光同时提高了设计的空间质量。

结构

天花板和地板两个表面使整个房间呈现自然的框架：构成一幅图画的上下两边。由于各层的高度和形状可以调整，所以周边的景色与视角也在不停地变换。例如，地板弯曲到某一点然后消失于水中。另一边，屋顶垂下来与地面相接，这样就营造了一个"相吻时刻"。屋面在某些部位伸展开来，从而为楼梯提供了空间。这一切随着慢慢的扩张，在风与水面倒影之间形成了强烈的自然体验。

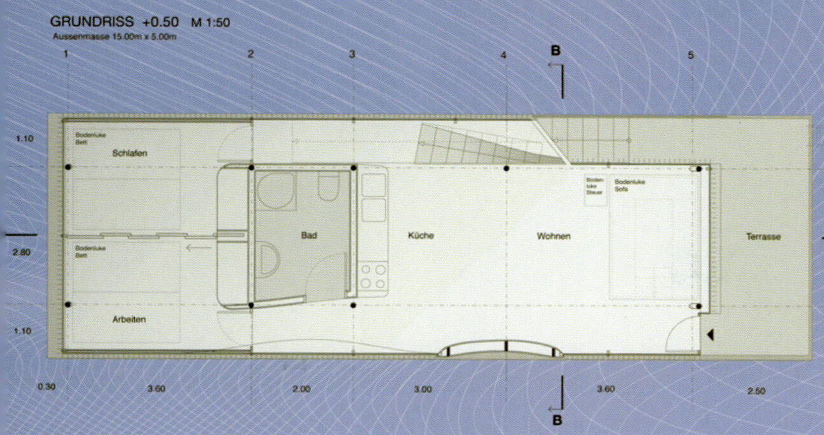

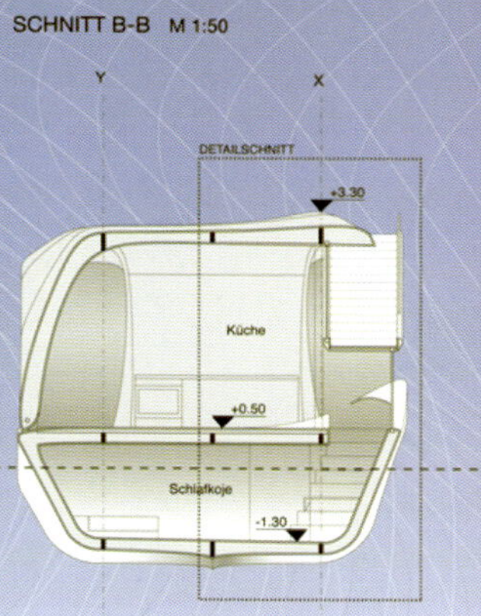

GRUNDRISS +0.50 M 1:50

Aussenmasse 15.00m x 5.00m

Schlafen
Bad
Küche
Wohnen
Terrasse
Arbeiten

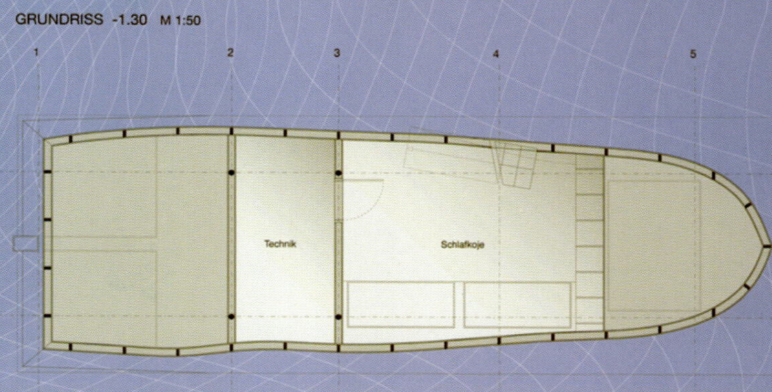

GRUNDRISS -1.30 M 1:50

Technik
Schlafkoje

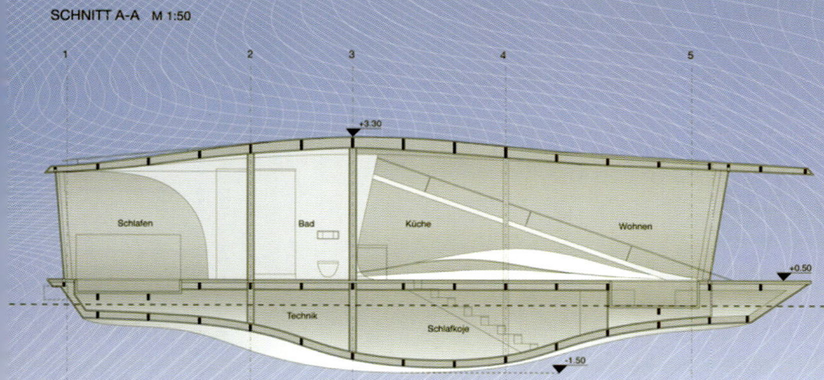

SCHNITT A-A M 1:50

Schlafen
Bad
Küche
Wohnen
Technik
Schlafkoje
Wasserlinie +0.00

SCHNITT B-B M 1:50

DETAILSCHNITT
Küche
Schlafkoje

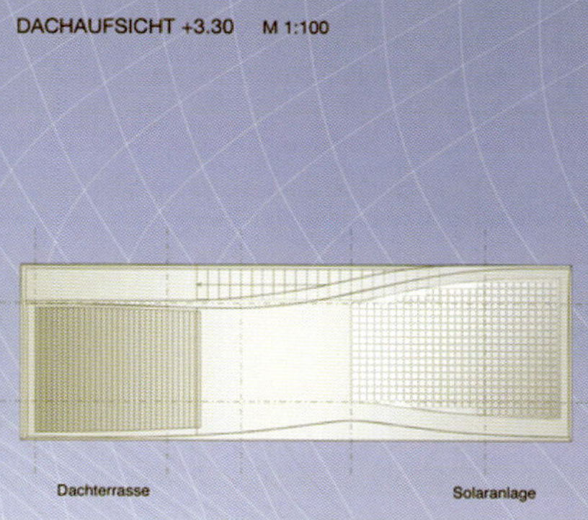

DACHAUFSICHT +3.30 M 1:100

Dachterrasse
Solaranlage

公共/私人空间

总体来说，它有两种用途。当它在停泊处着陆的时候，要充分考虑到高度的隐私问题，因为周围有邻居。然而，当在没有标记的地方就可以自由自在，不受约束。因此必要的开放性和隐私保护都是需要的，这不仅仅是由于天气的原因。如果需要的话，还可以将窗帘房与其他的区域区分开来。

材料/建造阶段/静态

房顶和地板是根据常规的造船结构方式建造的，能够承载重物。由纤维增强塑料制成的加强筋组成。根据3.6米的最大跨度，它的静态高度和支柱立面大大减少。它的支撑点通过核心部分起作用。房顶支撑在地板的某一点上，超过其平台2.5米。顶棚的形状必须适应静态的需求。屋顶的形态在荷载效应中必须计算其几何正确性。为了能够批量生产，我们必须考虑CNC–machines的运用。专业化的数字化生产方式和传统的造船技术将会成为劳济茨湖泊区未来的主导。

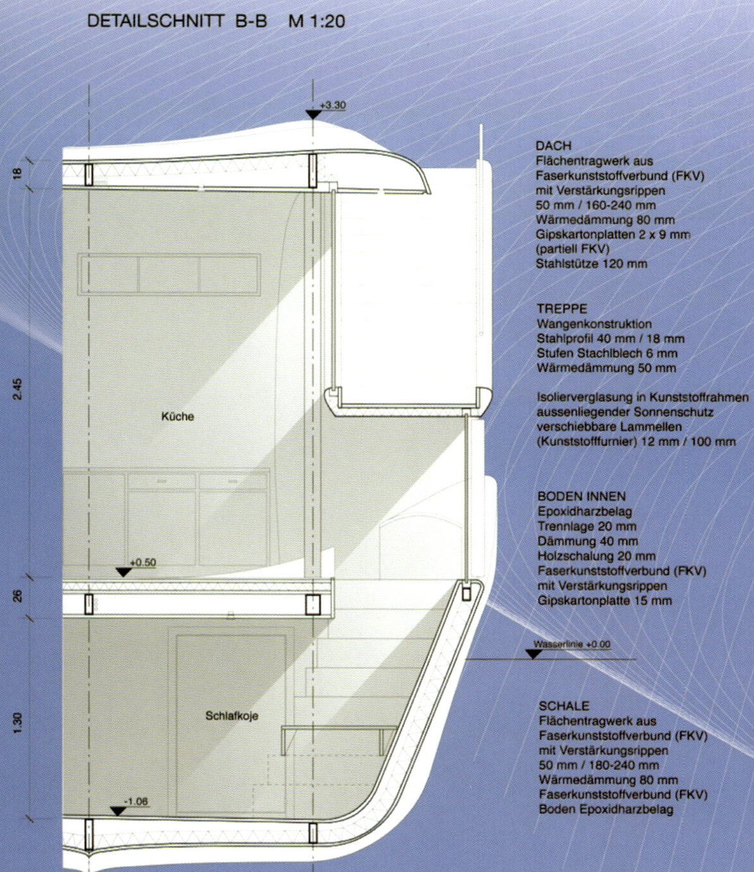

DETAILSCHNITT B-B M 1:20

Küche
Schlafkoje
Wasserlinie +0.00

DACH
Flächentragwerk aus
Faserkunststoffverbund (FKV)
mit Verstärkungsrippen
50 mm / 160-240 mm
Wärmedämmung 80 mm
Gipskartonplatten 2 x 9 mm
(partiell FKV)
Stahlstütze 120 mm

TREPPE
Wangenkonstruktion
Stahlprofil 40 mm / 18 mm
Stufen Stachlblech 6 mm
Wärmedämmung 50 mm

Isolierverglasung in Kunststoffrahmen
aussenliegender Sonnenschutz
verschiebbare Lammellen
(Kunststofffurnier) 12 mm / 100 mm

BODEN INNEN
Epoxidharzbelag
Trennlage 20 mm
Dämmung 40 mm
Holzschalung 20 mm
Faserkunststoffverbund (FKV)
mit Verstärkungsrippen
Gipskartonplatte 15 mm

SCHALE
Flächentragwerk aus
Faserkunststoffverbund (FKV)
mit Verstärkungsrippen
50 mm / 180-240 mm
Wärmedämmung 80 mm
Faserkunststoffverbund (FKV)
Boden Epoxidharzbelag

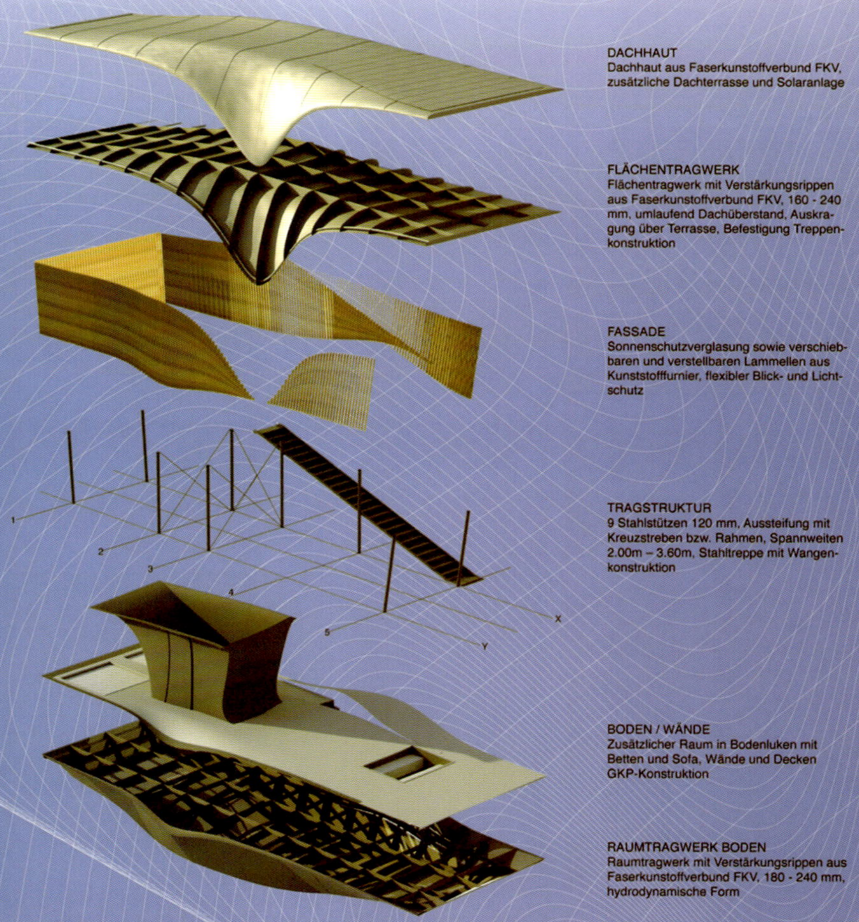

DACHHAUT
Dachhaut aus Faserkunststoffverbund FKV, zusätzliche Dachterrasse und Solaranlage

FLÄCHENTRAGWERK
Flächentragwerk mit Verstärkungsrippen aus Faserkunststoffverbund FKV, 160 - 240 mm, umlaufend Dachüberstand, Auskragung über Terrasse, Befestigung Treppenkonstruktion

FASSADE
Sonnenschutzverglasung sowie verschiebbaren und verstellbaren Lammellen aus Kunststofffurnier, flexibler Blick- und Lichtschutz

TRAGSTRUKTUR
9 Stahlstützen 120 mm, Aussteifung mit Kreuzstreben bzw. Rahmen, Spannweiten 2.00m – 3.60m, Stahltreppe mit Wangenkonstruktion

BODEN / WÄNDE
Zusätzlicher Raum in Bodenluken mit Betten und Sofa, Wände und Decken GKP-Konstruktion

RAUMTRAGWERK BODEN
Raumtragwerk mit Verstärkungsrippen aus Faserkunststoffverbund FKV, 180 - 240 mm, hydrodynamische Form

Concept

The concept of a conventional house and mobile floating architecture cannot be easily merged. There is no doubt that a new typology has to be invented regarding usability, space and technology. The formal and conceptual requirements of being in motion stand in opposition with the desire to simultaneously feel at home. Living on water has to be rethought. This is why we propose a mobile floating architecture focusing on movement and autarchy. As its name suggests, the project "the last resort" follows the strategy of a last resort and tries to challenge known and familiar concepts and show alternative ways of living.

Organisation

The programme is organised on two levels. In order to be able to have an adequate height of 2.5m in the upper deck, sleeping bunks, technical equipment and hatches are built into the lower deck. They serve as "extension rooms" of the upper deck: beds and couches are embedded into the floor and can be opened as the need arises. The external measurements are 5m x 15m. In the front area, there is a covered terrace; there, you can find a staircase leading to the roof. Together, the kitchen and the living room form a generous open room. The bathroom consists of a core separating both bedrooms from the common room. The bedroom can be separated by sliding panels and used separately as a study. All in all, there are six beds (incl. 2 bunk beds).

Lusatian lakeland

The horizontal, undulated shape of the waterfront is inspiration for the design of the project. Furthermore, it serves as means for orientation for the residents. The landscape can be understood as an extension of the living room. Therefore, an unobstructed view of the landscape enhances the spatial quality of the design.

Form

Two surfaces (floor and ceiling) frame the room experience "nature"; they form the upper and lower margin of the picture. Since the height and shape of the levels can be varied, the perspective and view of the landscape keep changing. The floor, for example, bends downwards at one point and disappears into the water. At another point, the roof curves down to the floor, thus creating a "kissing moment". At some other spot, it unfurls in order to make

space for the staircase. All this generates a play on forms that, together with the swell, the wind and the water reflection results in an intense nature experience.

Public/Private space

Generally, there are two types of use. When landing at a moorage, a high level of privacy is necessary due to possible adjacent neighbours. However, there is a practically boundless freedom when landing on unmarked sites. A flexible view and light protection are therefore necessary, not only for climatic reasons. It allows for the screening of certain rooms from others, if required.

Materials/Construction phase/Static state

The roof and the floor are built with conventional methods of boat construction as load-bearing structure. They have reinforcing ribs made of fibre-reinforced plastic (FRP). With a maximum span of 3.6m, the static height as well as the section of stanchion can be considerably reduced. The bracing is effected through the core. The roof lies on the floor on certain spots and protrudes 2.5m over the terrace area. The shape of the ceiling adjusts itself to the static requirements. The roof's forming during load effect has to be calculated for the accuracy of the geometry. For serial production, the use of CNC-machines has to be considered. The specialisation of digital production methods, together with traditional craft technology can be the future direction for the Lusatian dockyards.

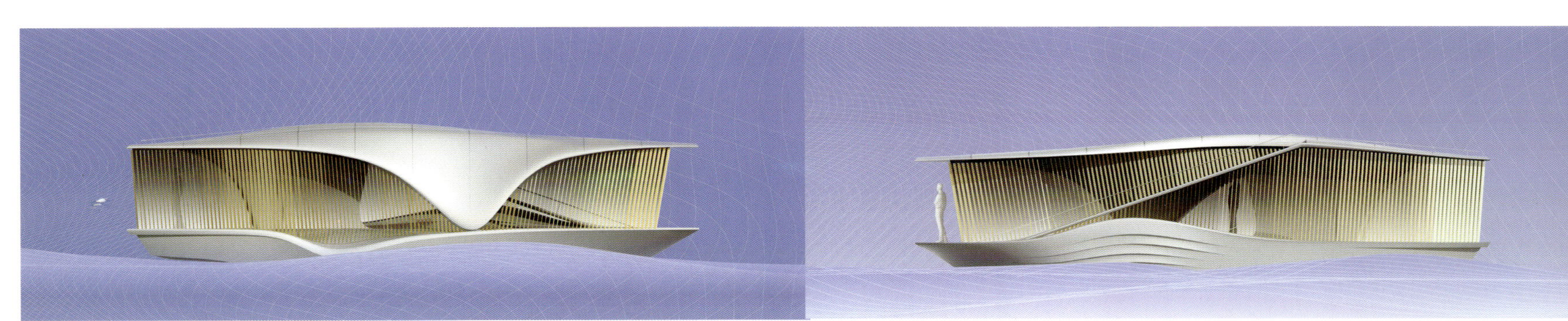

Linked Hybrid
北京当代MOMA 联接复合体

斯蒂文•霍尔建筑事务所/Steven Holl Architects

项目地点	中国，北京
设计时间	2003年~2009年
项目规划	689套公寓面积达135 000平方米，包括电影院、幼儿园、画廊、商店、健身馆、咖啡馆,以及1 000车位的地下停车场
总建筑面积	221 462平方米
设计建筑师	Steven Holl,李虎
负责合伙人	李虎
项目建筑师	Hideki Hirahara
助理建筑师	陈彦玲
技术顾问	Chris McVoy, Tim Bade
项目设计师	Garrick Ambrose,陈绮薇, Rodolfo Dias,董功, Peter Englaender,Guido Guscianna, Young Jang, Edward Lalonde, JongSeo Lee, Richard Liu, James MacGillivray, Matthew Uselman
合作建筑师	北京首都工程建筑设计有限公司
结构工程师	Guy Nordenson and Associates, 中国建筑科学研究院
机电工程师	Transsolar, 北京首都工程建筑设计有限公司, Cosentini
照明顾问	L'Observatoire International
幕墙顾问	Front Inc.,西飞集团公司，沈阳远大铝业工程有限公司, 北京江河幕墙装饰工程有限公司
景观建筑师	斯蒂文•霍尔建筑师事务所, 易道(北京), 北京创新景观园林设计有限公司
室内设计	斯蒂文•霍尔建筑师事务所, 中国装饰有限公司

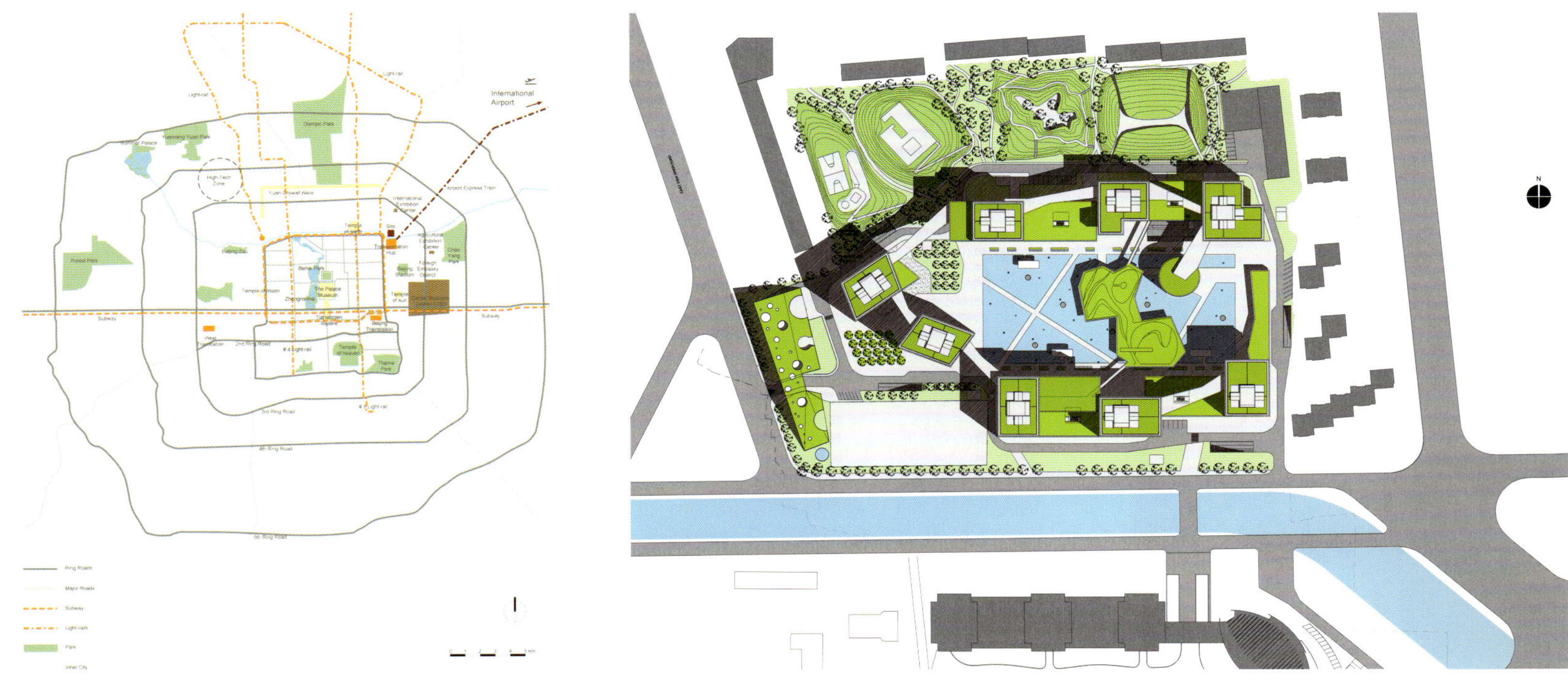

Project Location	Beijing, China
Design Milestones	2003 – 2009
Project Program	The 135,000m^2 apartment building has 689 units, including cinema, kindergarten, gallery, shop, gym, café, and underground parking for 1,000 cars.
Gross Floor Area	221, 462m^2
Design Architect	Steven Holl, Hu Li
Partner in Charge	Hu Li
Project Architect	Hideki Hirahara
Assistant Project Architect	Yanling Chen
Technical Advisor	Chris McVoy, Tim Bade
Project Designer	Garrick Ambrose, Yiwei Chen, Rodolfo Dias, Gong Dong, Peter Englaender, Guido Guscianna, Young Jang, Edward Lalonde, JongSeo Lee, Richard Liu, James MacGillivray, Matthew Uselman
Associate Architects	Beijing Capital Engineering Architecture Design Co. LTD
Structural Engineer	Guy Nordenson and Associates, China Academy of Building Research
Mechanical Engineer	Transsolar, Beijing Capital Engineering Architecture Design Co. LTD, Cosentini Associates
Lighting Consultant	L'Observatoire International
Curtain Wall Consultant	Front Inc., Xi'an Aircraft Industry Decoration Engineering Co. Ltd., Shenyang Yuanda Aluminum Industry Engineering Co., Ltd., Beijing Jianghe Curtain Wall Co., Ltd
Landscape Architect	Steven Holl Architects, EDAW Beijing, Beijing Top-Sense Landscape Design Limited Co.
Interior Designer	Steven Holl Architects, China National Decoration Co., LTD

660 GEOTHERMAL WELLS / 100 m DEEP
5000 KW COOLING / HEATING CAPACITY

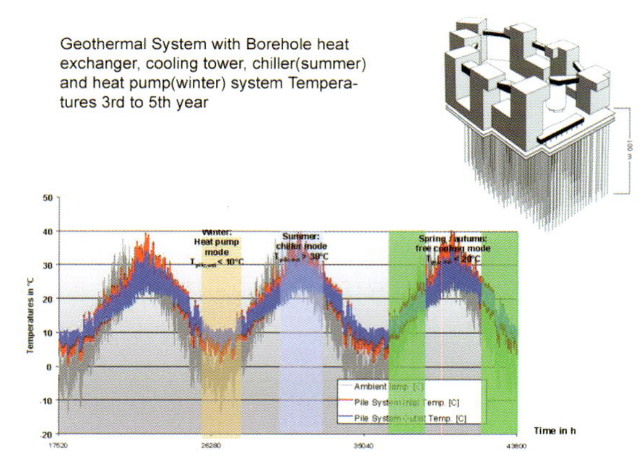

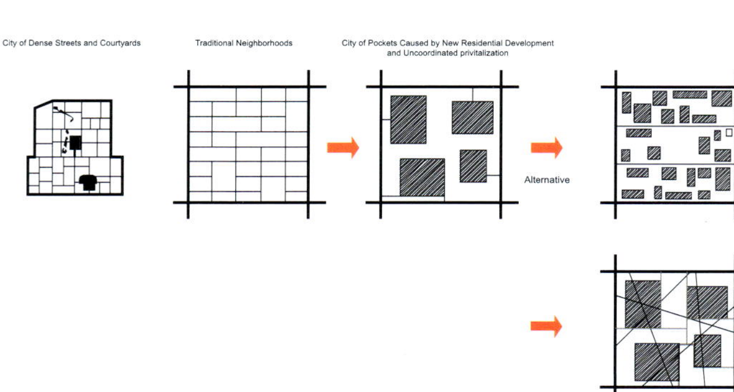

在北京，220 000平方米的Linked Hybrid复合建筑，适应中国目前私有化城市发展为目的，创造了二十一世纪穿透的城市空间。该建筑物每一个角度都向公众开放。电影城市空间穿越多方面的空间板层，许多旅行者游览此项目，都把Linked Hybrid称为“城市中的开放城市”。项目创立理念力求商业、住宅、娱乐教育各方面相互协作与激励，是一个三维公共城市空间。

项目底层是一些开放式通道，让人们（包括居民和参观者）步行穿越。这些通道内包括各种小面积的商店，它们经常吸引人们围绕在大型中央反射池塘周边。低层建筑物的中间层，公共的屋顶花园提供一个安静的绿色空间，而在八层最顶端的住宅内有一个私人的屋顶花园与楼顶相连。底层所有公共功能区域包括饭店、宾馆、蒙台梭利学校、幼儿园和电影院，都与周围的绿色空间连接，为项目增添亮彩。另一区域高层的旅客，用电梯取代“跳切”。从12楼到18楼是一个多功能区域，有天桥与游泳池、健身房、咖啡厅、画廊连接，另外它还与八栋住宅楼和宾馆相连，可以纵览城市景观。标志性的环状设计理念取代了单调的长形设计理念。我们希望高层的环状设计与底层的环状设计不断地生成无规则的联系。这些功能就好像是一个“小社会”，为居住者和参观者提供特殊体验的城市生活。

专注于通道空间设计，建筑物把活动、时间和序列均考虑进去。景观变化的要点设计是让通道向上倾斜，慢慢地向右旋转。其余围绕着的建筑物也都从单独或越来越多私有化岛屿概念转变为同样的设计理念。我们希望呈现出一个崭新的“Z”字形的城市脉搏，追求公共空间城市居住的个性化。

地暖设备（100米深处，655个换热器）给Linked Hybrid复合建筑物在夏天提供冷气，在冬天提供暖气，同时让Linked Hybrid复合建筑成为大型绿色住宅项目之一。此项目中心宽敞的城市空间因有着水百合和青草的水循环系统而更具吸引力，实验电影院和宾馆均漂浮在此之上。冬天水池结冰，这里就变成了一个溜冰场。实验电影院不仅仅是一个人们聚集的场所，更是一个聚集视觉的地方。设计实验电影院的建筑师运用浅色反射原理，放映室在其正面，影片在里面放映。建筑物的1楼可以纵览社区全景，并向社区开放。上面的中国佛教建筑装饰物颜色鲜艳。桥的下侧和一部分悬臂处有彩色的薄膜，夜晚的灯光投射到桥底部和悬挑部分的彩色膜上，熠熠发光。而窗框的侧柱则根据《周易》，采用了古代寺庙中的色彩进行随机涂色。

整个项目的水资源均采用水循环系统。脏水由管道进入水箱，经过紫外线过滤后进入大的反射池塘，然后再利用，用于浇灌周围景观。将施工中开挖的土方用于基地北边的五个带有娱乐设施的景观丘的建设。“儿童时代的娱乐场”与幼儿园连接在一起，有一个入口通过。“青年时代的娱乐场”主要是篮球场、滚轴和溜冰场。“中年时代的娱乐场”，我们可以找到对所有人都开放的咖啡馆和茶室，太极拳场地和两个网球场。“老年时代的娱乐场”主要被品酒吧所占据。“无限制的娱乐场”有圆形的无边无际的银河空间供人们沉思。

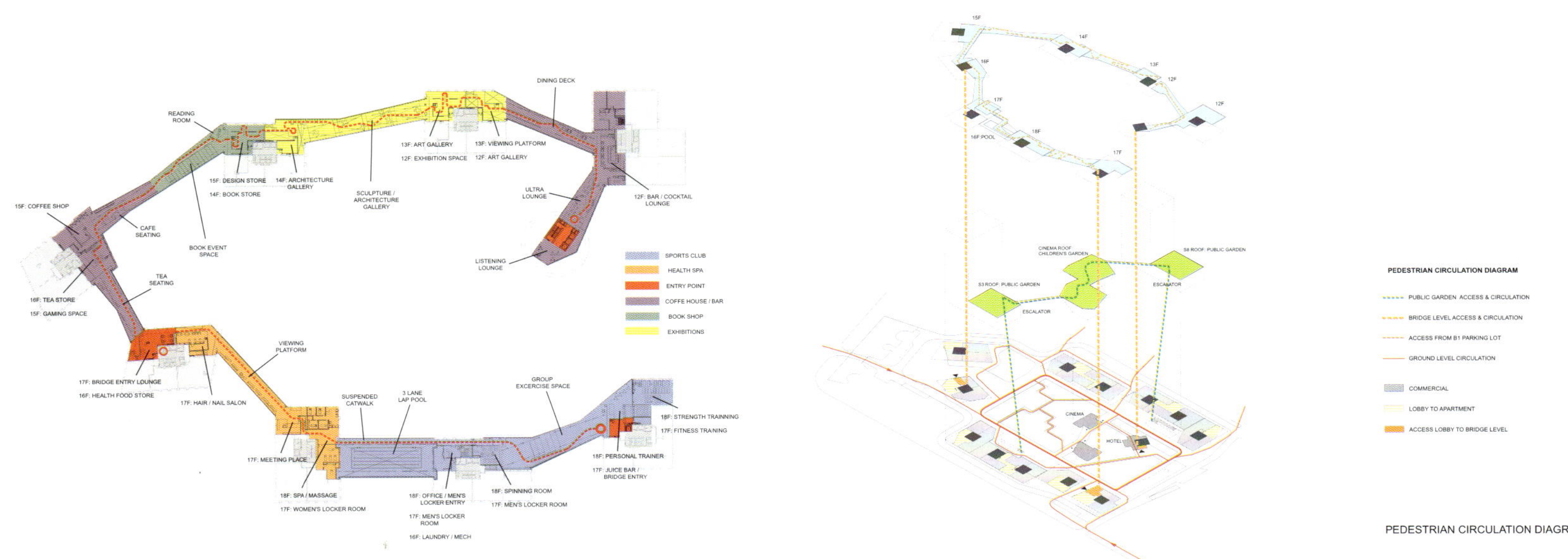

PEDESTRIAN CIRCULATION DIAGRAM

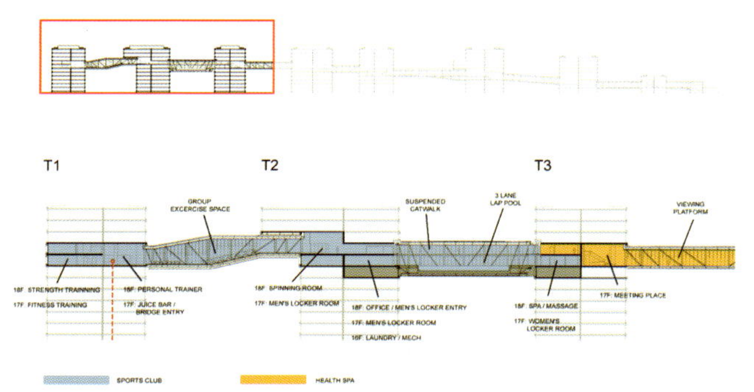

T1
T2
T3
GROUP EXCERCISE SPACE
SUSPENDED CATWALK
3 LANE LAP POOL
VIEWING PLATFORM
SPORTS CLUB
HEALTH SPA

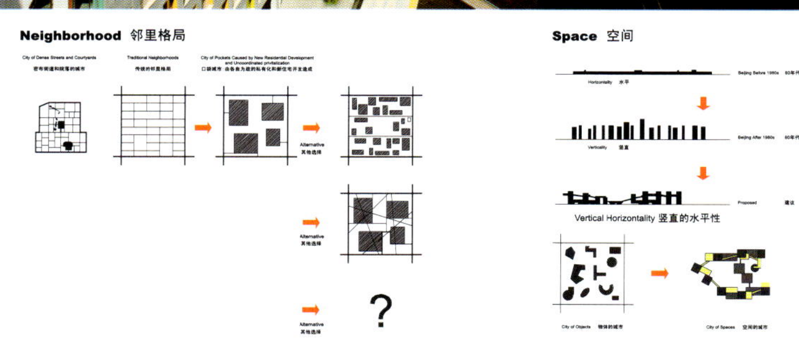

Neighborhood 邻里格局
Space 空间
Vertical Horizontality 竖直的水平性
?

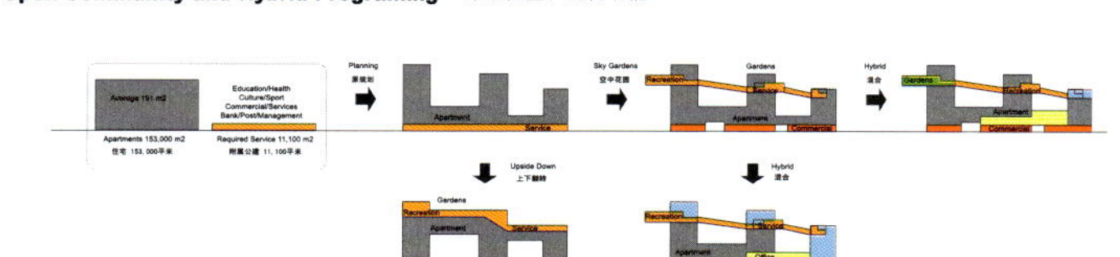

Open Community and Hybrid Programing 开放社区和混合功能

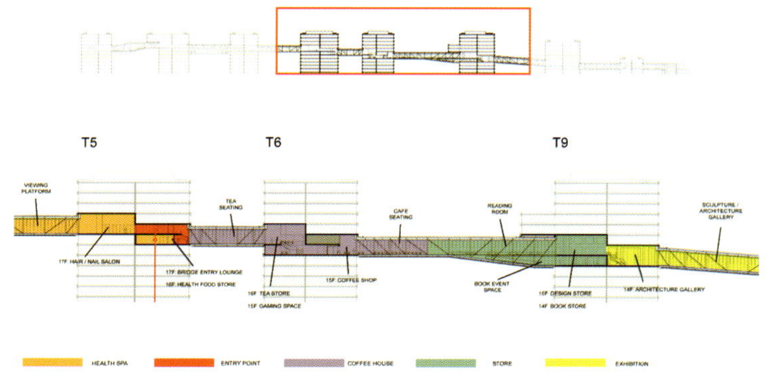

T5
T6
T9
HEALTH SPA
ENTRY POINT
COFFEE HOUSE
STORE
EXHIBITION

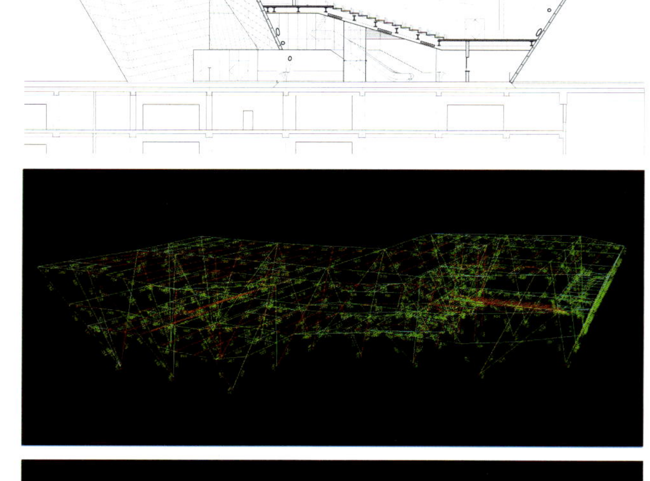

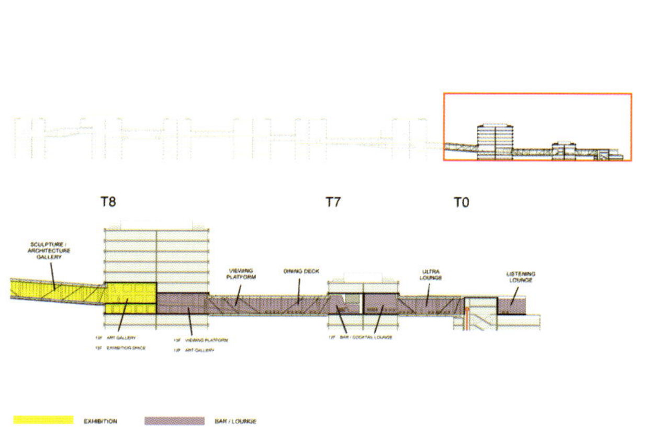

T8
T7
T0
EXHIBITION
BAR / LOUNGE

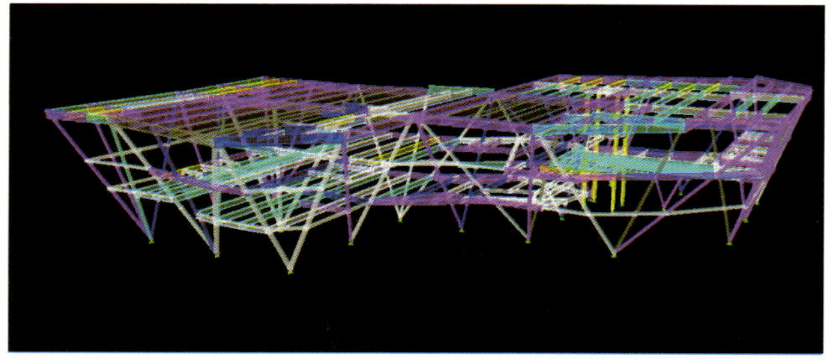

The 220,000 m² Linked Hybrid complex in Beijing, aims to counter the current privatized urban developments in China by creating a twenty-first century porous urban space, and open to the public from every side. A filmic urban experience of space; around, over and through multifaceted spatial layers, as well as the many passages through the project, make the Linked Hybrid an "open city within a city". The project promotes interactive relations and encourages encounters in the public spaces that vary from commercial, residential, and educational to recreational; it is a three-dimensional public urban space.

The ground level offers a number of open passages for all people (residents and visitors) to walk through. These passages include "micro-urbanisms" of small-scale shops which also activate the urban space surrounding the large central reflecting pond. On the intermediate level of the lower buildings, public roof gardens offer tranquil green spaces, and at the top of the eight residential towers private roof gardens are connected to the penthouses. All public functions on the ground level including a restaurant, hotel, Montessori school, kindergarten, and cinema, have connections with the green spaces surrounding and penetrating the project. Elevators displace like a "jump cut" to another series of passages on higher levels. From the 12th to the 18th floor a multi-functional series of skybridges with a swimming pool, a fitness room, a café, a gallery, connects the eight residential towers and the hotel tower, and offers views over the unfolding city. Programmatically this loop aspires to be semi-lattice-like rather than simplistically

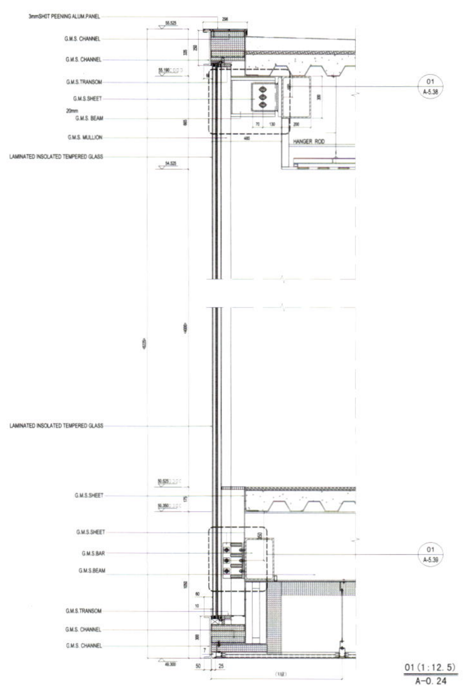

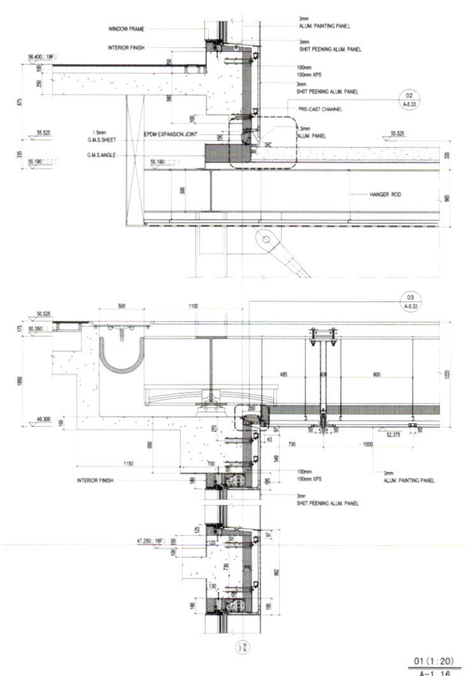

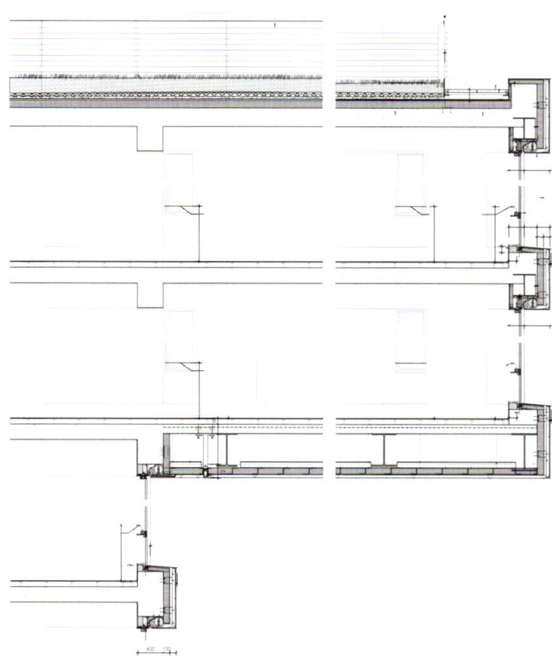

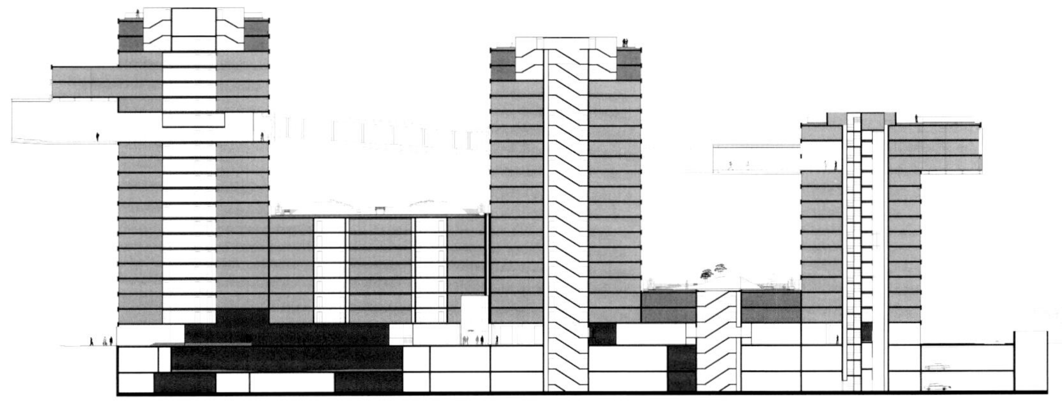

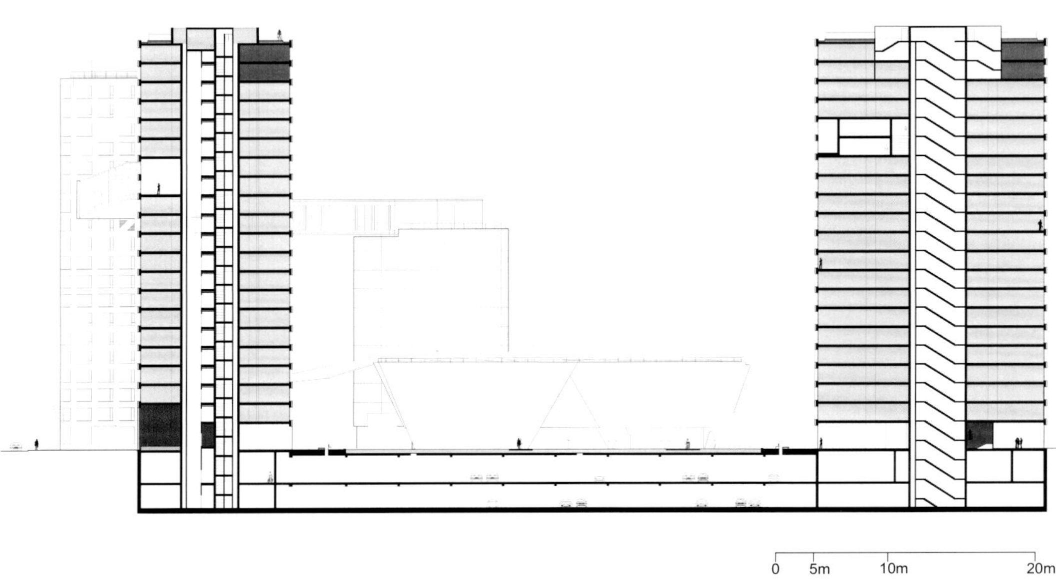

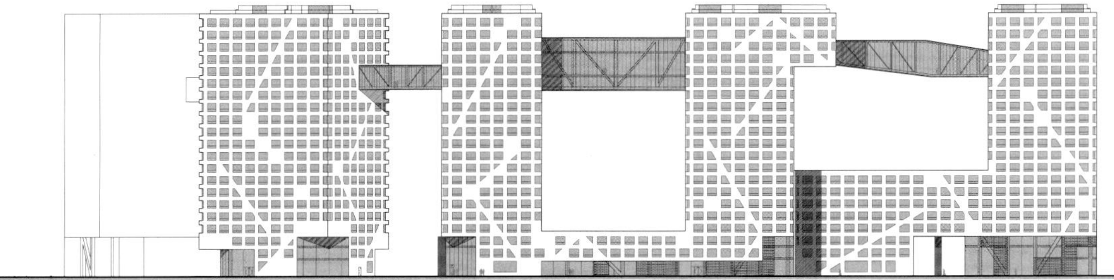

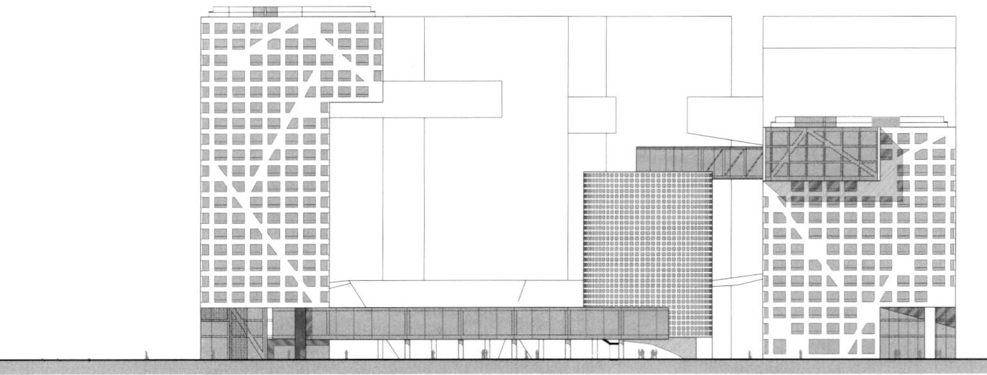

linear. We hope the public sky-loop and the base-loop will constantly generate random relationships, functioning as social condensers in a special experience of city life to both residents and visitors.

Focused on the experience of passage of the body through space, the towers are organized to take movement, timing and sequence into consideration. The point of view changes with a slight ramp up, a slow right turn. The encircled towers express a collective aspiration; rather than towers as isolated objects or private islands in an increasingly privatized city. Our hope is for new "Z" dimension urban sectors that aspire to individuation in urban living while shaping public space.

Geo-thermal wells (655 at 100 meters deep) provide Linked Hybrid with cooling in summer and heating in winter, and make Linked Hybrid one of the largest green residential projects. The large urban space in the center of the project is activated by a greywater recycling pond with water lilies and grasses in which the cinematheque and the hotel appear to float. In the winter the pool freezes to become an ice-skating rink. The cinematheque is not only a gathering venue but also a visual focus to the area. The cinematheque architecture floats on its reflection in the shallow pond, and projections on its facades indicate films playing within. The first floor of the building, with views over the landscape, is left open to the community. The polychrome of Chinese Buddhist architecture inspires a chromatic dimension. The undersides of the bridges and cantilevered portions are colored membranes that glow with projected nightlight and the window jambs have been colored by chance operations based on the "Book of Changes" with colors found in ancient temples.

The water in the whole project is recycled. This greywater is piped into tanks with ultraviolet filters, and then put back into the large reflecting pond and used to water the landscapes. Re-using the earth excavated from the new construction, five landscaped mounds to the north contain recreational functions. The "Mound of Childhood" integrated with the kindergarten, has an entrance portal through it. The "Mound of Adolescence" holds a basketball court, a roller blade and skate board area. In the "Mound of Middle Age" we find a coffee and tea house (open to all), a Tai Chi platform, and two tennis courts. The "Mound of Old Age" is occupied with a wine tasting bar and the "Mound of Infinity" is carved into a meditation space with circular openings referring to infinite galaxies.

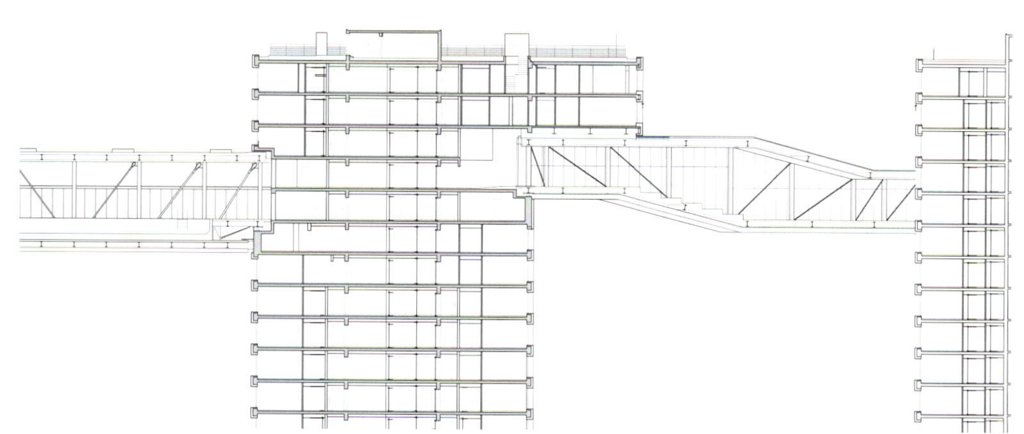

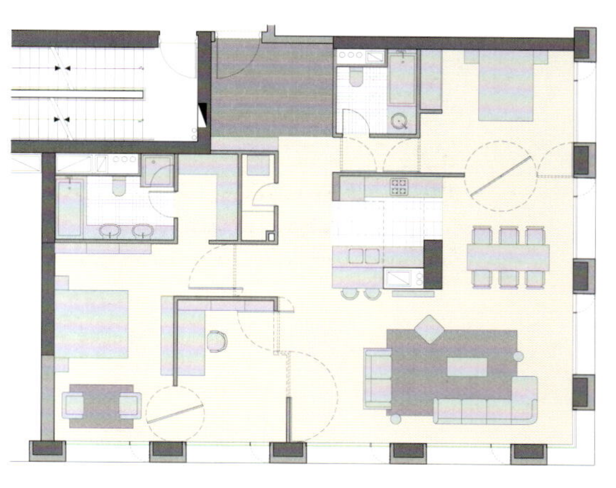

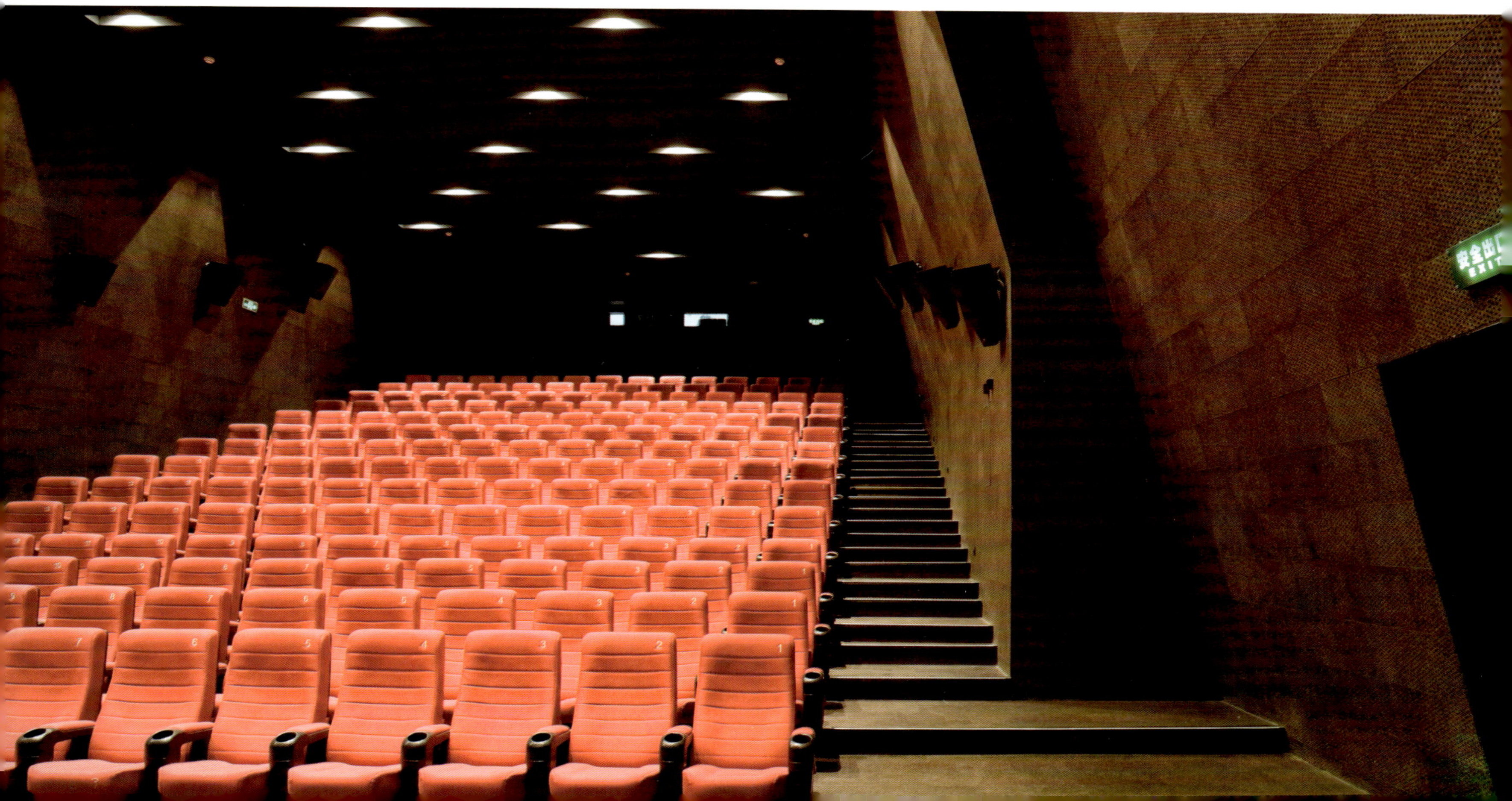

Vanke Center

万科中心

斯蒂文·霍尔建筑事务所/Steven Holl Architects

项目地点	中国，深圳
总建筑面积	80 200平方米
设计时间	2006~2009
设计建筑师	Steven Holl，李虎
合作建筑师	中建国际(深圳)设计顾问有限公司
结构工程师	中国建筑科学研究院，中建国际(深圳)设计顾问有限公司

Project Location	Shenzhen, China
Gross Floor Area	80,000m^2
Design Milestones	2006 – 2009
Design Architect	Steven Holl, Hu Li
Associate Architects	CCDI
Structural Engineer	CABR, CCDI

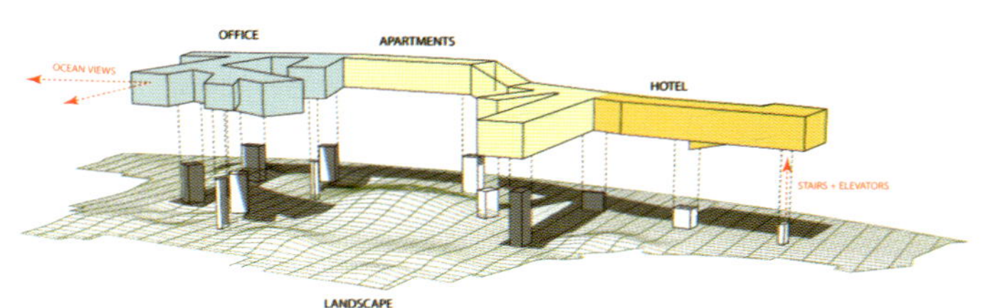

都市主义:斯蒂文•霍尔+李虎

——斯蒂文·霍尔建筑师事务所在中国的四个作品展览开幕式

此巡回展览是针对2003年至2009年间中国四项显示雄心的建筑设计过程的回顾与展示：南京艺术与建筑博物馆，北京当代MOMA联接复合体，深圳万科中心，以及成都来福士广场。

正当中国经历着人类有史以来最大规模的城市化进程时，这些作品对如何在今日语境下，创造城市与集体使用空间一而非标新立异建筑一开启了宏观层面的探索。

此次展览的目的是为呈现从初期概念阶段到真实现状的完整设计过程，透过大型多媒体影片和建筑模型来表现；影片中不但将会展现平时鲜为人知的工作室内部运作情况，并且记载了建筑模型、图纸和项目尾期动画的集体制作过程。所有展览作品皆为斯蒂文·霍尔建筑师事务所纽约和北京工作室协力合作的成果，两地时差提供了24小时连续工作的周期，而最终实现了景观、城市与建筑融合为一体的卓绝设计。

此次展览将于2009年12月7日至2010年2月12日，在深圳大梅沙新建成的“水平摩天楼—万科中心”中举行。

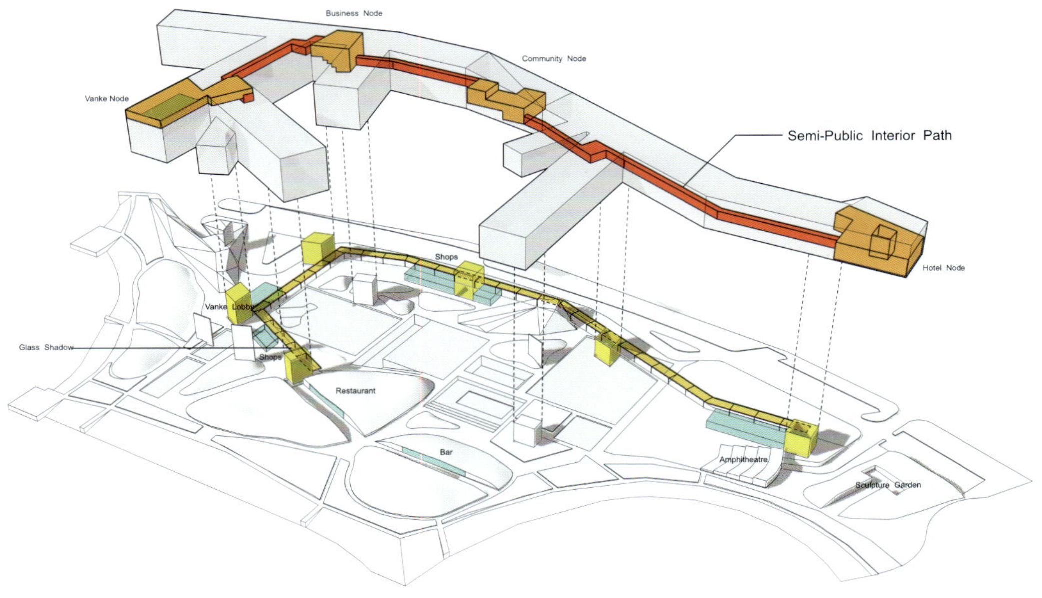

Path Diagram

STRUCTURAL GLAZING
3850 MM FLOOR TO FLOOR SPACES

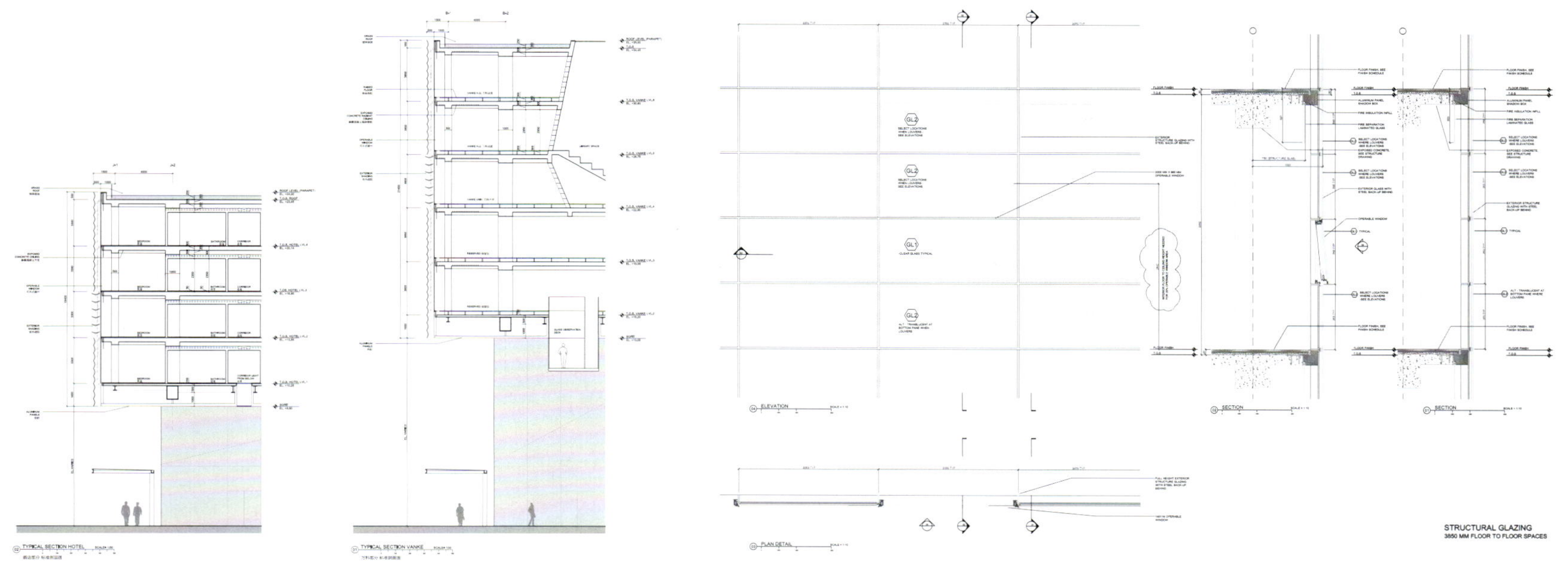

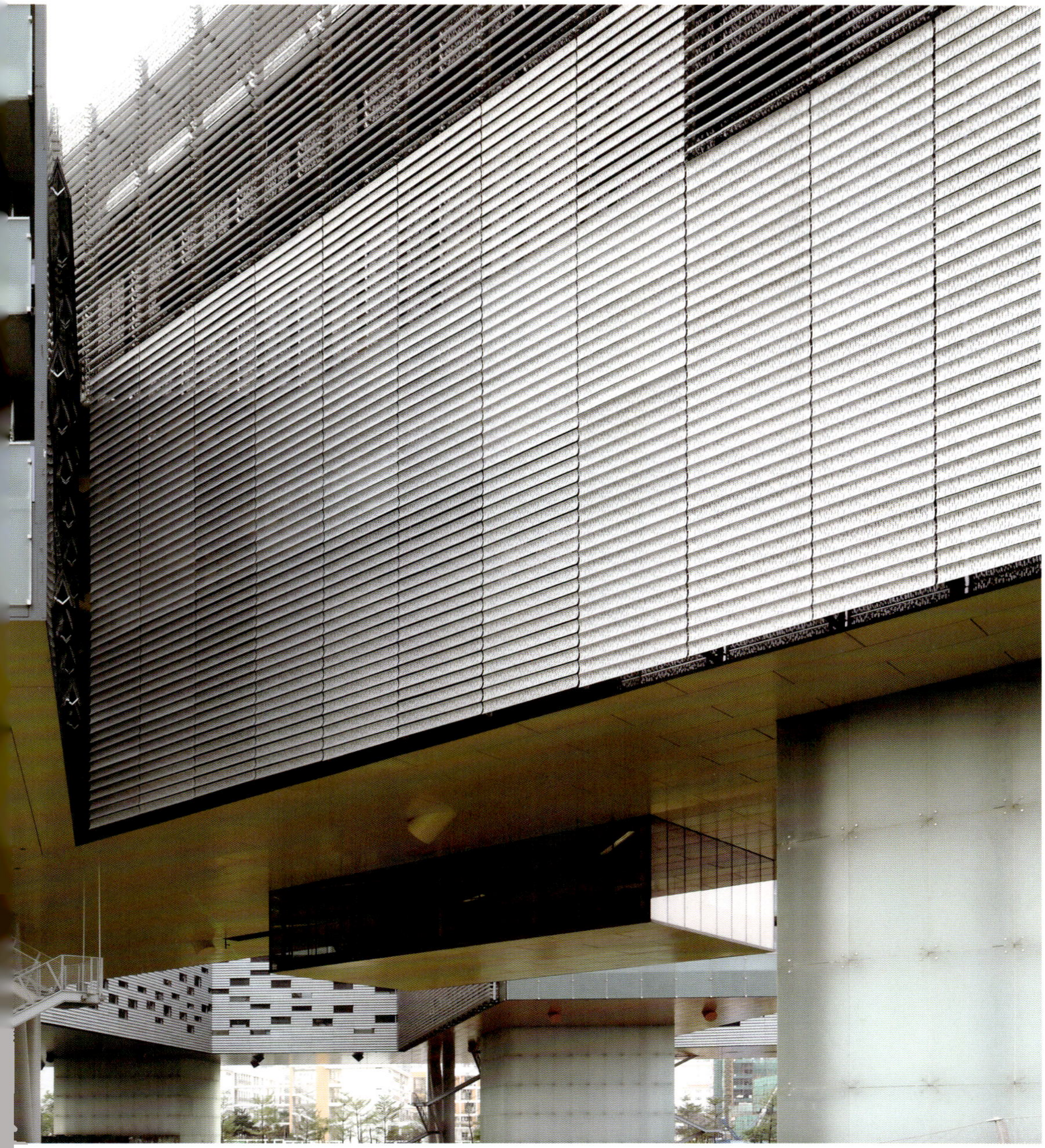

New York City, November 30, 2009 – On Monday, December 7, Steven Holl Architects opens the exhibition Urbanisms: Steven Holl + Hu Li :4 Projects in China by Steven Holl Architects in the Horizontal Skyscraper/Vanke Center, in Shenzhen, China.

The exhibition tracks the process of designing four ambitious projects in China from 2003 to 2009: Nanjing Museum of Art and Architecture, Beijing Linked Hybrid, Shenzhen Horizontal Skyscraper, and Chengdu Sliced Porosity Block.

As China experiences one of the world's largest urbanizations in history, these works explore the creation of collective urban space as opposed to object buildings.

The exhibition illustrates the design process from initial conception to current status, documenting the collaborative process of model making, drawing, and animation. The works presented are the product of a cooperative effort between Steven Holl Architects' offices in New York and Beijing, where the difference in time zones often facilitates a continuous 24-hour cycle of production, the result of which are unprecedented works that are a fusion of landscape, urbanism, and architecture.

The exhibition will be on view in the newly finished Vanke headquarter offices in the Horizontal Skyscraper/Vanke Center in Dameisha, Shenzhen.

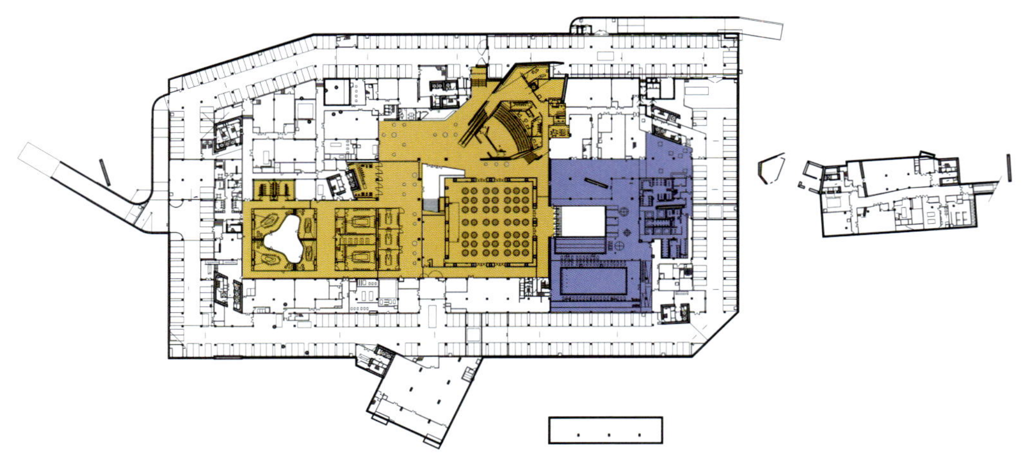

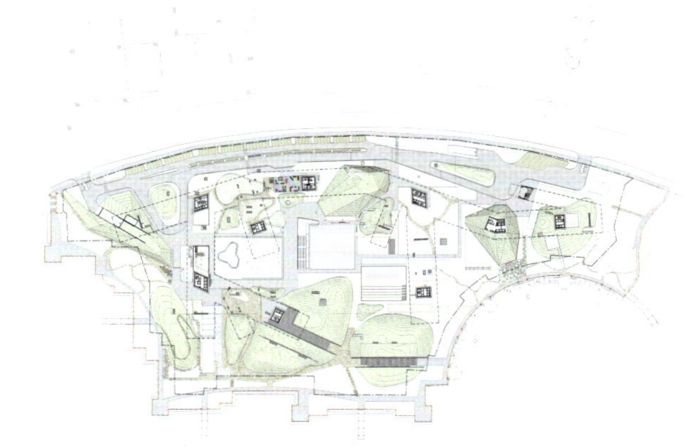

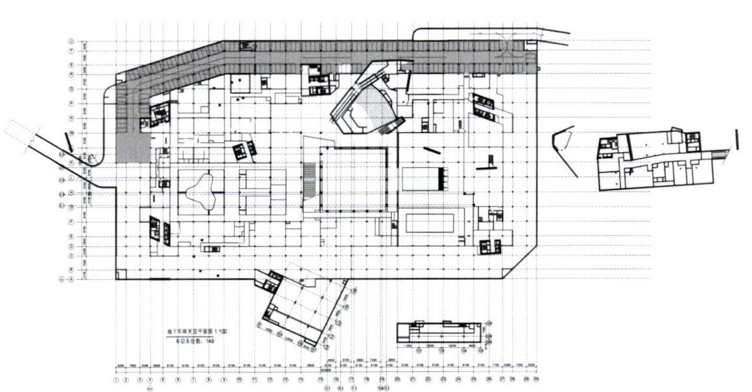

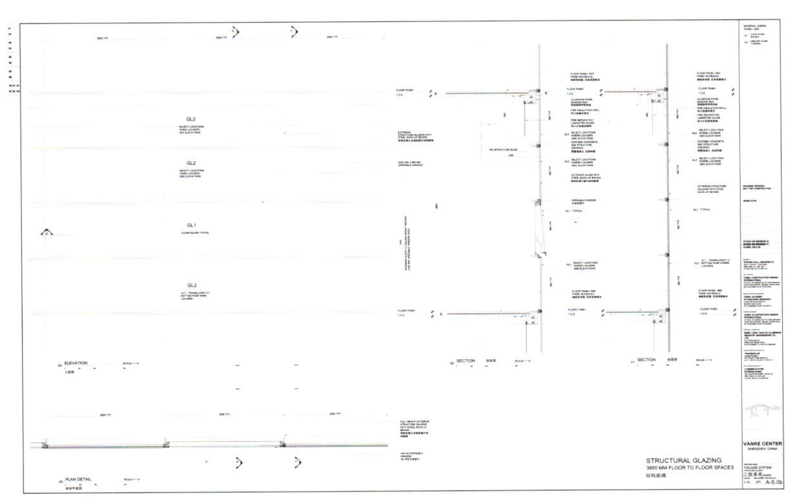
STRUCTURAL GLAZING
VANKE CENTER

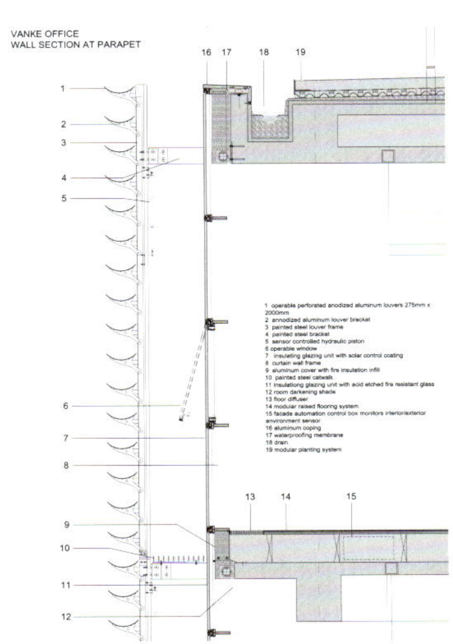
VANKE OFFICE
WALL SECTION AT PARAPET
1 operable perforated anodized aluminum louvers 275mm x 2000mm
2 annodized aluminum louver bracket
3 painted steel louver frame
4 painted steel bracket
5 sensor controlled hydraulic piston
6 operable window
7 insulating glazing unit with solar control coating
8 curtain wall frame
9 aluminum cover with fire insulation infill
10 painted steel catwalk
11 insulationg glazing unit with acid etched fire resistant glass
12 room darkening shade
13 floor diffuser
14 modular raised flooring system
15 facade automation control box monitors interior/exterior environment sensor
16 aluminum coping
17 waterproofing membrane
18 drain
19 modular planting system

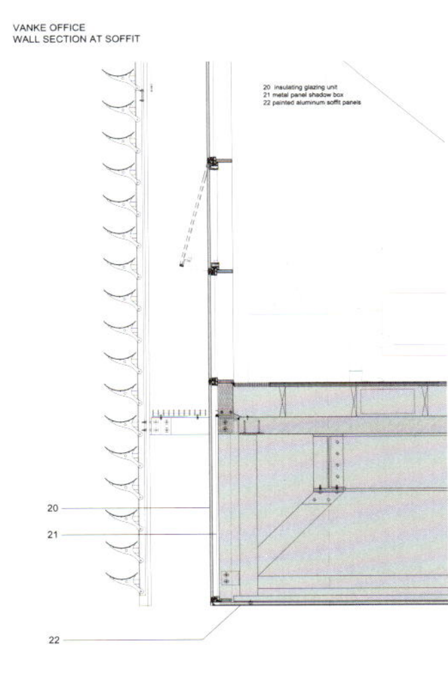
VANKE OFFICE
WALL SECTION AT SOFFIT
20 insulating glazing unit
21 metal panel shadow box
22 painted aluminum soffit panels

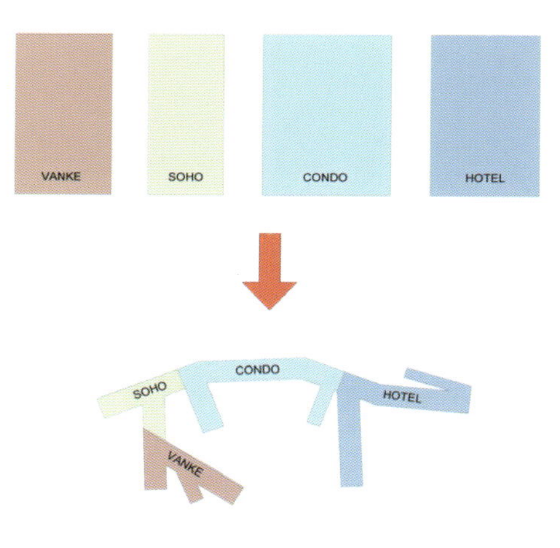

ONE BUILDING

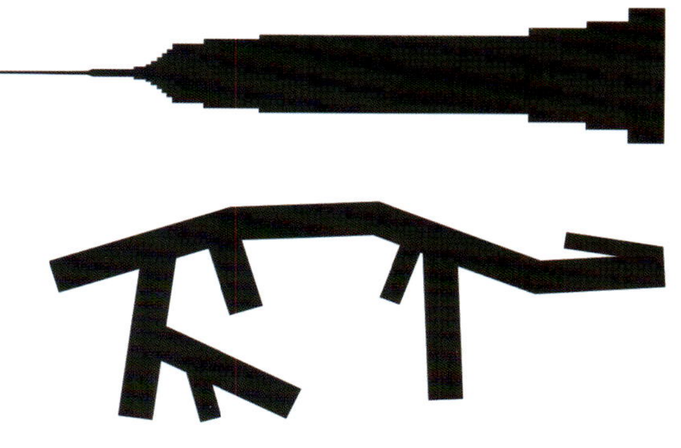

SIZE COMPARISON

Prisma

Prisma住宅项目

NL建筑师事务所

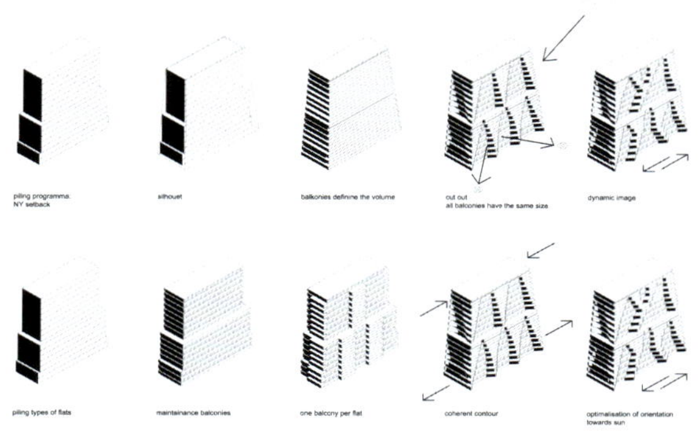

balconies

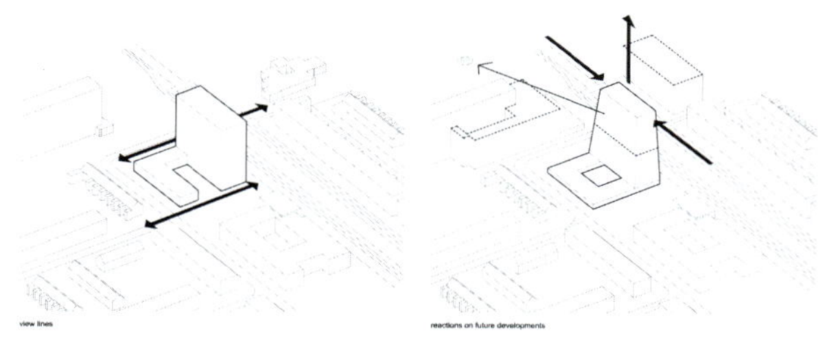

shape

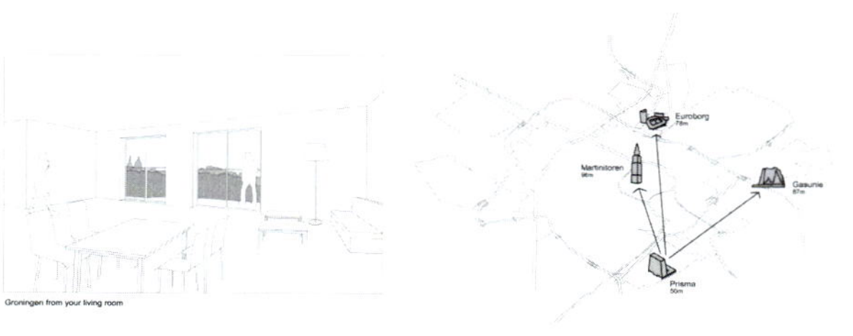

vista

项目地点	荷兰，格罗宁根，Siersteenlaan 482
地块面积	2600平方米
卫星覆盖区	1600平方米
建筑总面积	8650平方米—育儿室面积，1000平方米；住房面积，7 650平方米
楼高	49.5米（16层）
建筑面积率	3.3
建筑预算	9000000欧元
建筑师	NL Architects, Pieter Bannenberg, Walter van Dijk, Kamiel Klaasse
项目指导者	Sören Grünert (Design Phase), Gert Jan Machiels (Execution Phase)
建筑工程师	ABT Velp, Rob Nijsse, Erwin ten Brincke
服务工程师	Van der Weele Groningen, Frits Weijers, Radek Nosek
消防顾问	DGMR Arnhem
客户	Stichting De Huismeesters, Ed Moonen / Roelof Jong
承包商	Schutte Bouw en Ontwikkeling, Zwolle

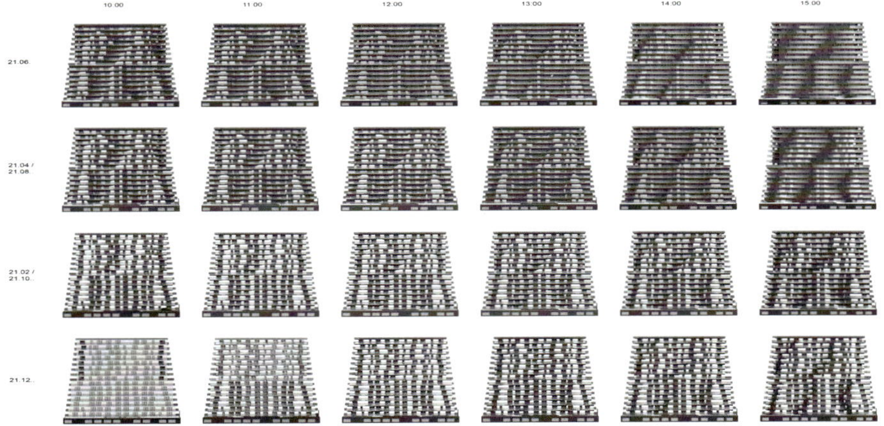

Project Location	Siersteenlaan 482, Groningen, The Netherlands
Plot Size	2,600 m^2
Footprint	1,600 m^2
Total GFA	8,650 m^2 — nursery 1,000 m^2, housing 7,650 m^2
Height	49.5 m (16 floors)
Floor Area Ratio	3.3
Project Budget	9,000,000 euro
Architect	NL Architects, Pieter Bannenberg, Walter van Dijk, Kamiel Klaasse
Project Leader	Sören Grünert (Design Phase), Gert Jan Machiels (Execution Phase)
Structural Engineer	ABT Velp, Rob Nijsse, Erwin ten Brincke
Building Services Engineer	Van der Weele Groningen, Frits Weijers, Radek Nosek
Fire Consultancy	DGMR Arnhem
Client	Stichting De Huismeesters, Ed Moonen / Roelof Jong
Contractor	Schutte Bouw en Ontwikkeling, Zwolle

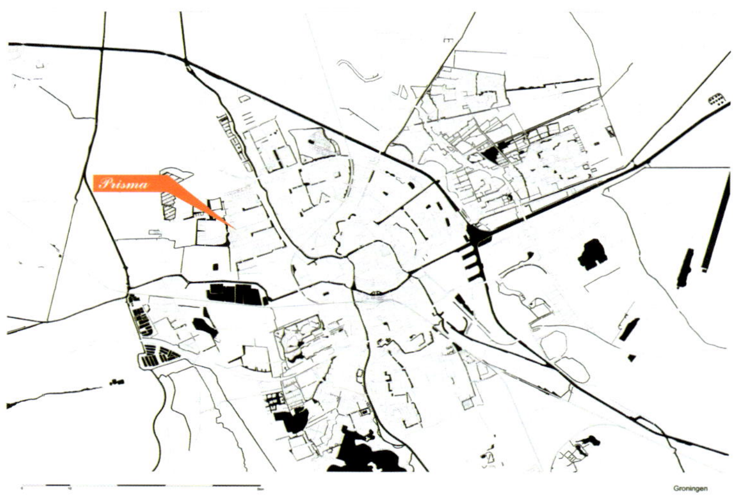

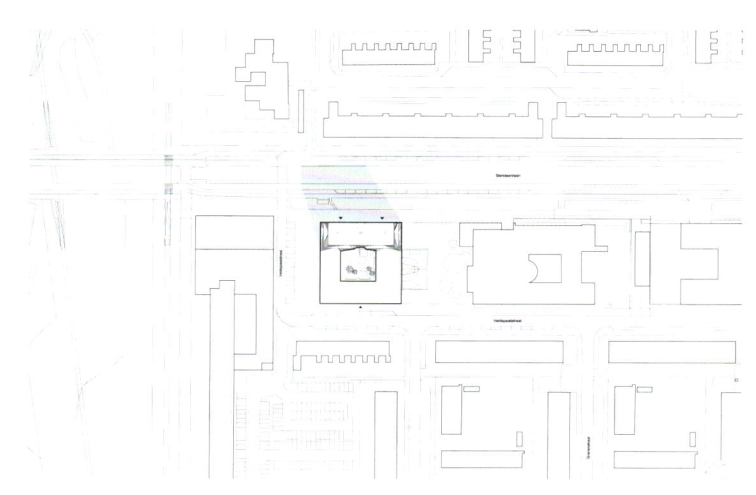

在荷兰这样地势极为平坦的国家却少见塔楼的身影是令人惊奇的。半无遮挡的视野——作为荷兰油画的主旋律，其潜能暗藏在正在开发的居住项目中。荷兰民众似乎都对电梯持有恐惧心理（这可以从荷兰电影《断头电梯》的成功中略见端倪）。Prisma作为最富魅力的住宅项目之一，将把景观视野作为将建筑内外空间相结合的起点。

Prisma是一个拥有52所公寓的住宅项目，总共有16层，并在格罗宁根有一些额外的设施。50米的高度在荷兰已经算是罕见的高楼了。格罗宁根以其比较有激进性的建筑著称。西北部的区域被称为Vinkhuizen，这是一种典型的基于CIAM的战后发展：一种开放式框架的中高层公寓楼与半高层的结合，结合了许多绿化以及公共空间与停车空间。现在这部分区域仍然并不是十分受欢迎。大多数的住宅都不再适合二十一世纪的需求。Prisma是

全国性的大型改造项目的一部分，现在已经在荷兰全面开展了。

这栋大楼由一个十分简单的叠加起来的公寓类型组成。最大的在下面，小一点的则在顶部；就如同一个造型奇特的玛雅金字塔建筑的轮廓让人联想起了高层的原型，从一定意义上可与曼哈顿相媲美。

大楼将会吸引一些年长者在此度过他们的余生。

一楼设有托儿所，儿童游乐场和医务室。而这些设施的入口与公寓入口相结合，从而增强对有限的规定空间的利用。

It is surprising that in a miraculously flat country like the Netherlands only so few towers emerge. The potential of the unobstructed view—as one of the main topics in Dutch painting—is in housing under developed. There seems to be a collective fear of elevators. (Which maybe explains the success of the Dutch movie Down—originally 'De Lift'). Prisma takes the view—one of the most glamorous properties of dwelling—as starting point in combination with outdoor space.

Prisma is a housing project for 52 apartments in a total of 16 stories and some additional facilities in the city of Groningen. A block of 50 meters high can in the Netherlands already be considered high. Groningen is renowned for its progressive architecture. The North West area is called Vinkhuizen. It is a typical CIAM based post war development; a mixture of middle high apartments in an open layout and some semi-high rises with a lot of greenery and a lot of public space and a lot of parking. Still the area is not particular popular at the moment. Much of the housing is no longer considered fit for the 21st century. Prisma is part of the nationwide, large-scale renovation operation that now is underway in the Netherlands.

The building consists of a simple stacking of the desired apartment types. The largest ones are below, the smaller ones on top: a weirdly proportioned Maya pyramid. The contour reminiscent of the archetype of the high rise: in a way comparable to the Manhattan zoning.

The building will basically attract active elderly that hopefully will be able to spend the rest of their lives there.

At ground level there is a nursery and children's playground and medical facility. The entrance of these facilities and the apartments are combined in an attempt to enhance the regular dimensions that are quite limited.

Begane grond
0 5 10

Begane grond
0 5 10

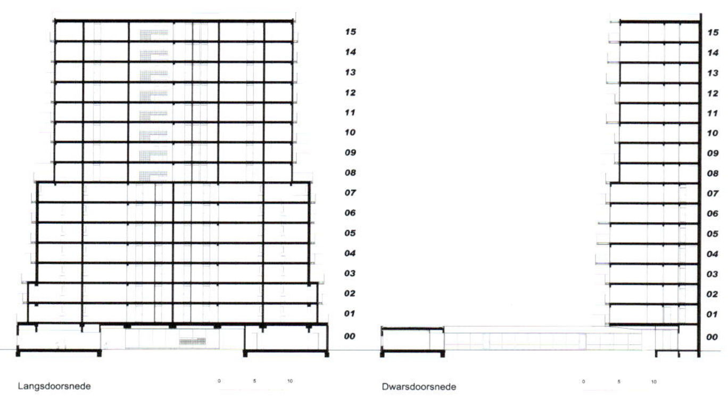
15
14
13
12
11
10
09
08
07
06
05
04
03
02
01
00
Langsdoorsnede
Dwarsdoorsnede

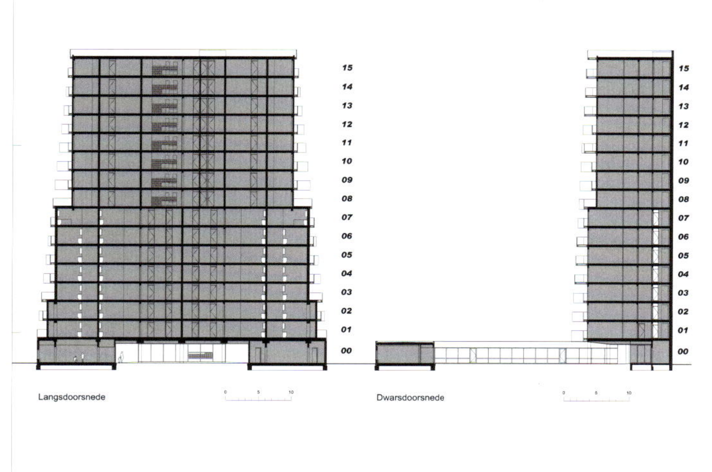
15
14
13
12
11
10
09
08
07
06
05
04
03
02
01
00
Langsdoorsnede
Dwarsdoorsnede

COP

ITALIA
9

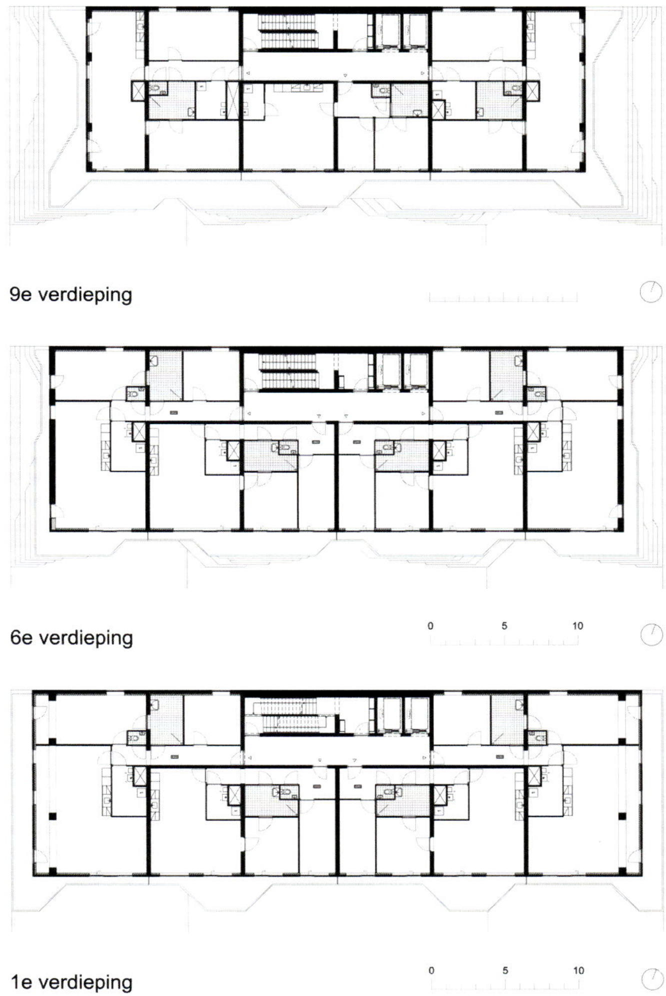
9e verdieping
0 5 10
6e verdieping
0 5 10
1e verdieping

Songzhuang Artist Residence

宋庄艺术公社

徐甜甜

项目地点	北京，通州区，宋庄镇
项目功能	住宅/艺术家工作室
客户	个人
设计时间	2007年8月~2008年5月
建造时间	2008年3月~2009年8月
场地面积	1400平方米
建筑面积	5300平方米
建筑师	徐甜甜
摄影师	Savoye/周若谷、Iwan baan

Project Location	Songzhuang town, Tongzhou district, Beijing
Program	Residence/art studios
Client	Private
Design Milestones	2007.08—2008.05
Construction Milestones	2008.03—2009.08
Site Area	1,400m²
Building Area	5,300m²
Architect	Tiantian Xu
Photographer	Savoye/Ruogu Zhou, Iwan baan

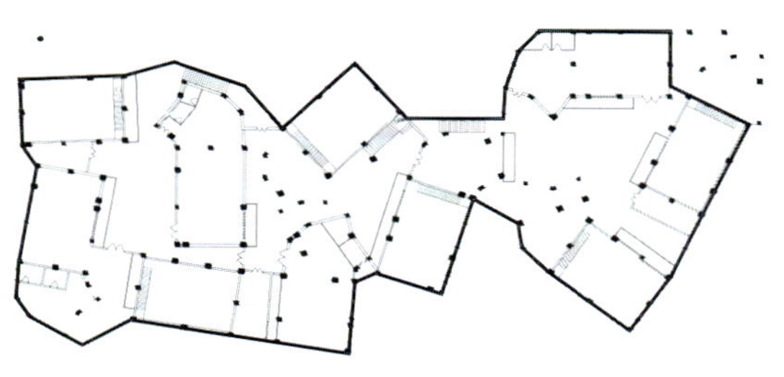

Basement Plan
地下层平面

Site Plan
基地平面

随着中国当代艺术家的激增，艺术家对居住和工作空间的需求也随之扩大。坐落在北京六环以东的宋庄艺术家村就正经历着由于这种增长所带来的变化，20个面向鱼塘的艺术家工作室在原室外储藏场地上应运而生。

按照工作室的居住和工作两种主要功能要求，每一个单元都包含两种不同的尺度与组合：工作部分为6米高的简洁的方盒子，居住部分高3米，有一系列几何形空间容纳起居厨厕功能．体量高差决定了每个单元的工作和居住功能或在同层联系或通过楼梯联系。整个建筑立面采用压型钢板，水平活动平面则是采用当地典型的红砖。

这20个工作室就像叠加在一起的集装箱，不仅和场地原有的工业气息相呼应，而且创造出富有张力的外形和独特的建筑空间。体量的虚与实，光与影，创造变化的室外交流空间，为艺术展示提供了更多的可能性。当地艺术节开放工作室时，这片区域可以为参观者带来多样的空间体验，同时成为丰富的创作展览现场。

换言之，这个带有艺术创作和居住功能的建筑何尝不是位于艺术村的一座另类美术馆。

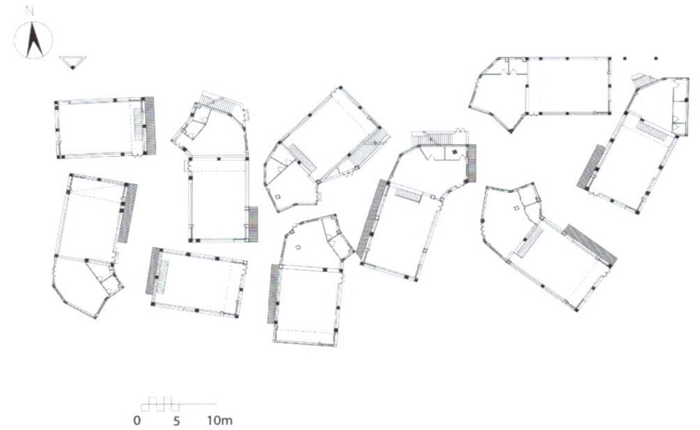
N
0 5 10m

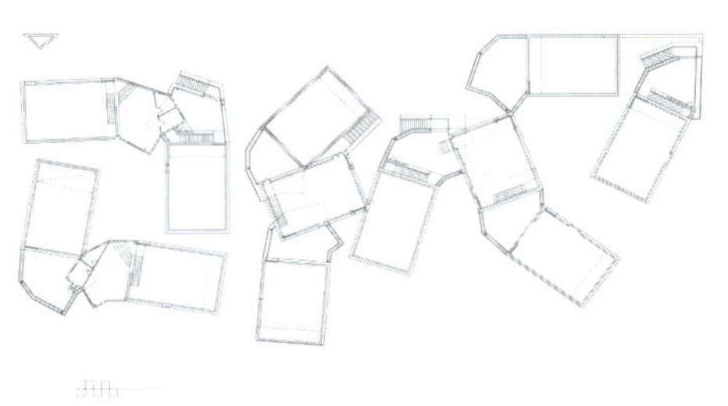

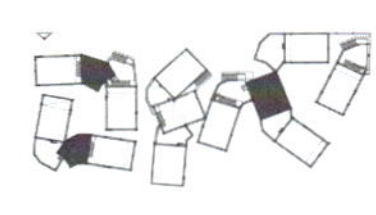

Located right next to east sixth ring road of Beijing city, Songzhuang Artist Village is undergoing a dramatic expansion of artist population and increasing demand of artist's working and living space. A 20-unit artist residence facing a fishpond at a former outdoor storage lot is one of the local development targeting such demand.

The programmatic requirement of working and living defines the height and geometry of both volumes: 6 m height for working and 3 m for living; a simple rectangular box for studio and a complex geometry for living indicating bedroom, kitchen and toilet. Living volume is plugged into working volume either on the same level or led by stair to upper level.

The roof material is the profiled steel sheet, and the ground surface is paved by the typical local red brick.

These 20 units are regarded as containers stacking up on this former industrial outdoor storage lot, creating an expressive configuration and spatial quality. The interplay of volume and void, light and shadow allows artists and visitors to constantly explore and experiment the outdoor community space, which could be the extension of art production and presentation as well as linking these 20 units as 20 individual showrooms on open studio days.

In other word, this complex becomes an alternative museum for living art creation and exhibitions.

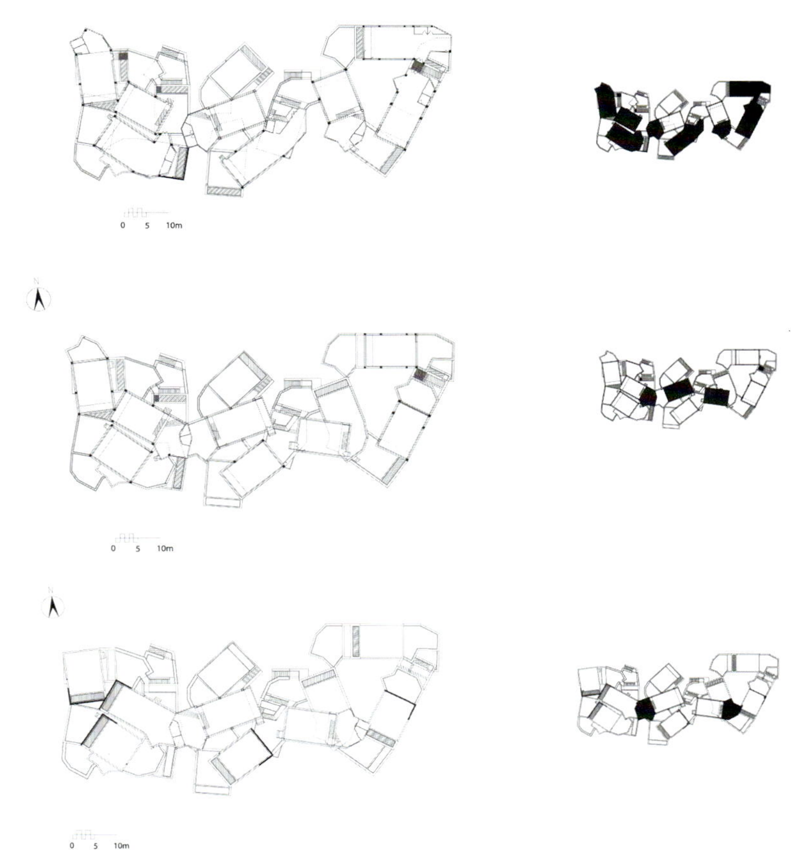
0 5 10m
0 5 10m
0 5 10m

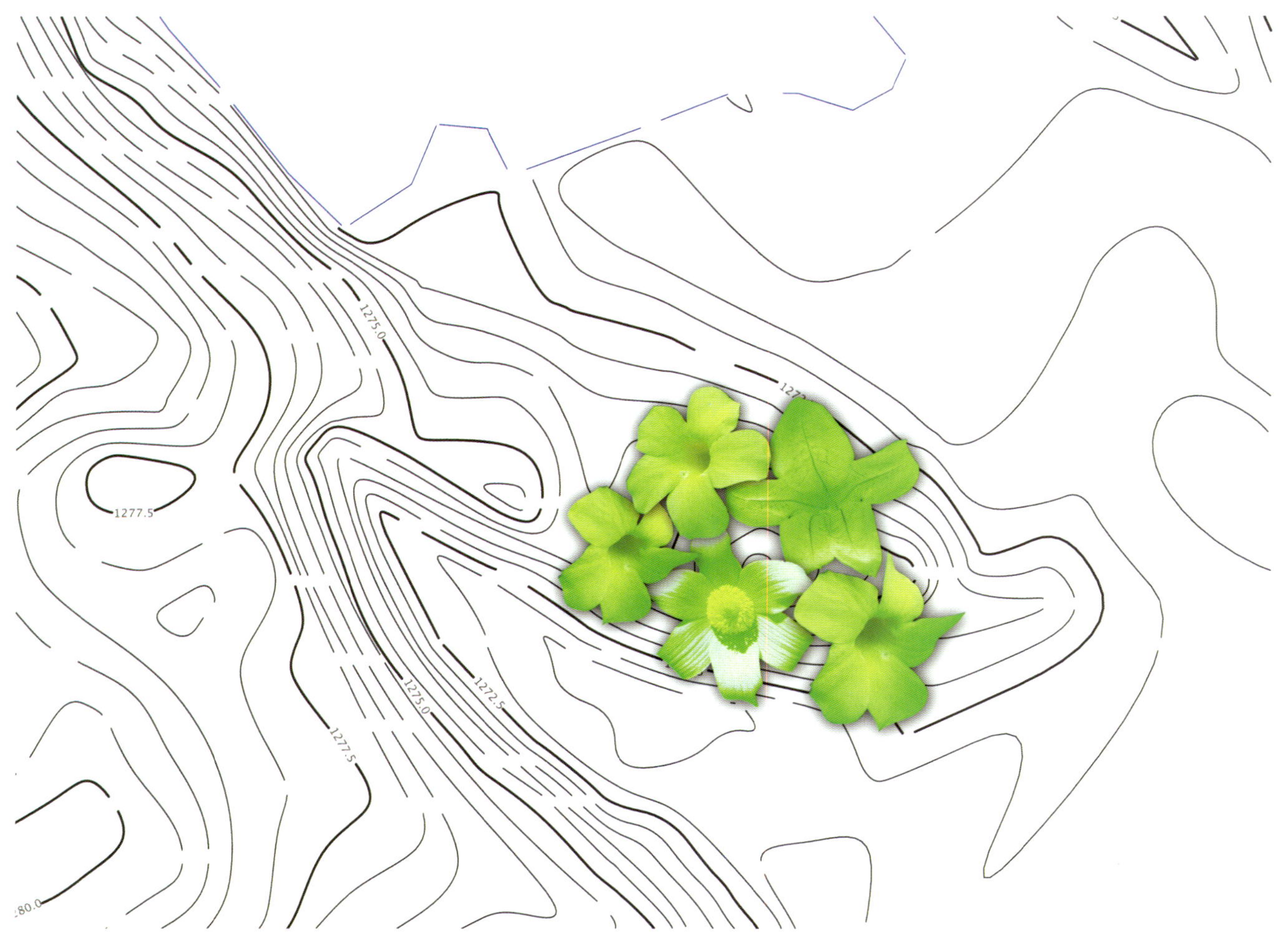

Villa by Water

水岸会所

徐甜甜

项目地点	内蒙古，鄂尔多斯
项目功能	住宅/艺术家工作室
客户	个人
场地面积	19800平方米
建筑面积	1200~1500平方米
建筑师	徐甜甜

Project Location	Ordos, Inner Mongolia
Program	Residence/art studios
Client	Private
Site Area	19,800㎡
Building Area	1,200–1,500㎡
Architect	Tiantian Xu

水岸会所位于鄂尔多斯市考考什那文化旅游区南端，面向考考什那水库。会所可被视为别墅和私人俱乐部，业主在此可以招待亲朋。其建筑面积为1 200平方米，功能要求包含5套私人套间，可以招待5个家庭，同时拥有客厅、饭厅、游泳池、图书馆和冬季花园等公共功能区。

这个唯一明确的功能要求——5套私人套间——给了我们最初的也是贯穿设计始末的灵感：5个单元像花蕊般向光和空气生长；其余的公共功能空间则如同花瓣在地面延伸交织。居住是私密的，脱离地面，朝向水景，不互相干扰；公共部分则是贯通的，花瓣的交接形式产生不同的空间和光影。

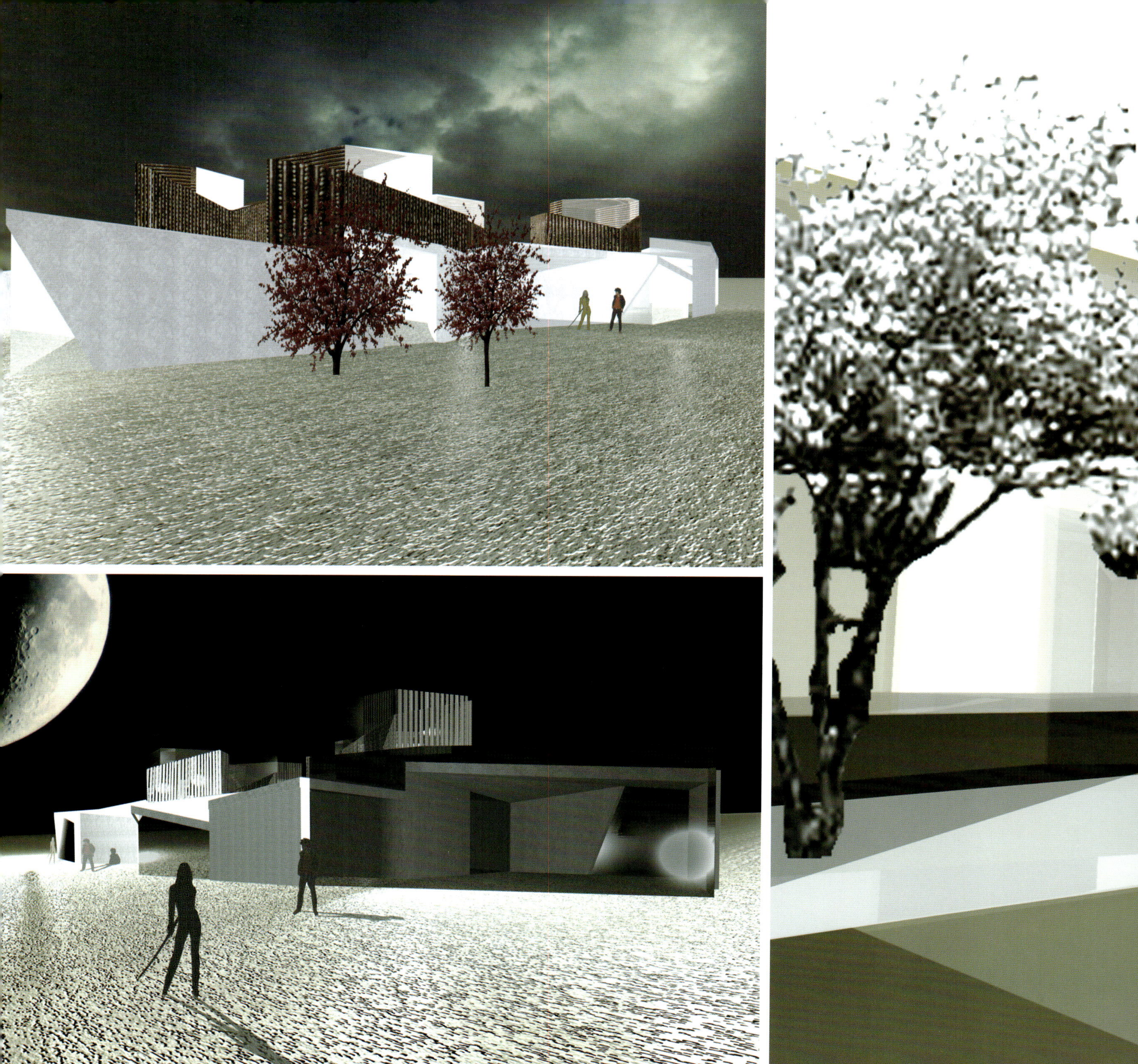

Villa by water is a private residence located on a vast mongolian land facing KaoKaoShiNa Lake in Ordos, Inner Mongolia. This will be a villa or private clubhouse for the owner to invite and host guests' families. In the program of 1,200 ㎡, there're 5 private suites which will accommodate 5 families, public functions including living room, dinning hall, swimming pool, library, winter garden, etc.

The natural Mongolian landscape, sky, sand dune, lake, the crispy air and breeze become key elements that define the atmosphere and inspire constantly that building becomes flowers blooming in Mongolian sky. The 5 private suites are flower cores striving up for light, air and view; each suite is a tower on second level starting with a double height living space, a stair leading to a bed mezzanine, and finally coming to a balcony looking over to the landscape beyond. The rest public programs are covered by petal-formed roof and walls merging and stretching on ground. The articulation of flower petals not only allows view on all directions and defines light and shadow, but also creates an organization of interior public spaces and a series of courtyards. Supporting functions (such as kitchen, wine cellar, bathroom, sauna, etc.) are embedded between petals under the bedrooms with setbacks allowing corridors and stairs becoming circulation for public spaces and bedrooms, and also create vertical shafts interfering with the continuous horizontal flow on ground level.

Materialization on each program further allows the differentiation between private or public: all the petals are painted white while bedroom suites are covered by ceramic tiles that shine under sunlight.

All-around Emerald Villa (One Phase)

绕翠（一期）

nikko建筑事务所

项目地点	中国，杭州
建筑师	王兴田，张峻，张凤春，苏晓宇
设计单位	nikko
设计时间	2010年
建筑面积	1603.74平方米
技术	混凝土结构

Project Location	Hangzhou, China
Architect	Xingtian Wang, Jun Zhang, Fengchun Zhang, Xiaoyu Su
Design Company	nikko
Design Milestones	2010
Building Area	1,603.74 m²
Technique	Concrete

江南富春，青山绿水，竹林溪流，古老中华文明的另一种风情。她传承着文明的激情，澎湃却又柔情似水。她就像一种气场，永远洋溢在空中，无需夸赞，无需改变，年复一年，日复一日，无声无息地自然衍生……

大美无言——东方美学思想的高度概括。它是东方人崇尚自然、顺应自然，并将自身融合于自然的世界观和生活态度。

而江南，自吴越以来文化发达，文人雅士云集，生成了与自然完美融合的独特的审美情趣和诗意品味。粉墙黛瓦的建筑在碧水蓝天、翠海山林中闪烁，仿佛一幅充溢着书卷气息的中国泼墨山水画，可谓"于朴实中显高雅"。它们在"天、地、人"三位合一的理念下，与自然和谐共生，让每一个生命感受到灵魂深处的宁静和在淡泊中饱含的憧憬……这，正是江南建筑空间意境——诗意栖居地的魅力所在。

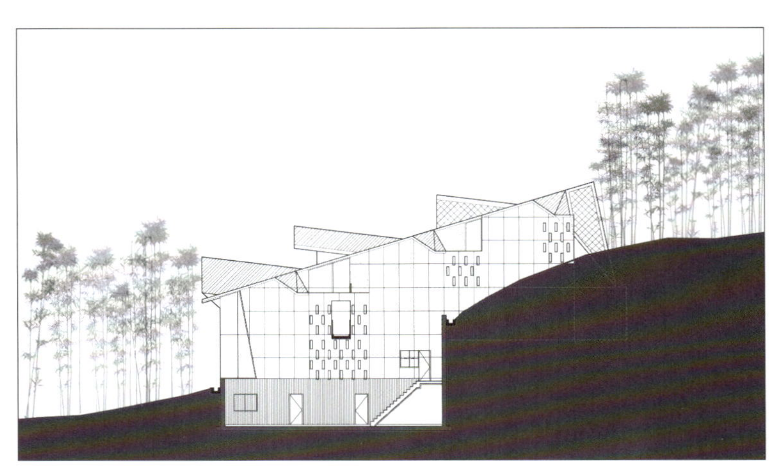

Fuchun in "Jiangnan" (the name for the south of the lower reaches of the Yangtze River), with fresh verdant hills and green waters, bamboo forests and clear streams, is another romantic charm of the millennia-old Chinese civilization. She inherits and conveys the passion of civilization, surging and yet tender like water. She is just like an atmospheric field, forever overflowing in the air, needless to praise, needless to change, year in year out, day in and day out, silently evolving in a natural course...

Nature's beauty is unspeakable — a succinct generalization of Oriental aesthetics. It is the world outlook and life attitude that Oriental people worship nature, comply with nature, and blend into nature.

While in "Jiangnan", culture has been advanced, and numerous literati and scholars have gathered here ever since the Wu-Yue period, so that the unique aesthetic sentiments and poetic tastes perfectly integrated with nature have been generated. The buildings of pink walls and black tiles shine

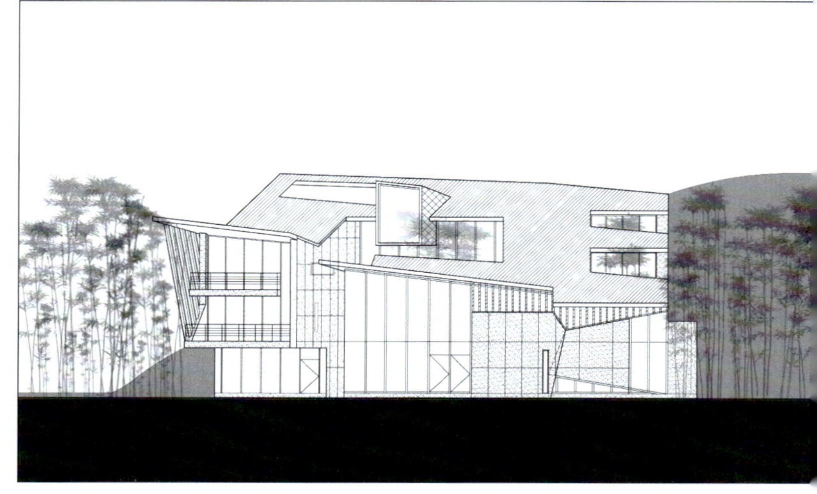

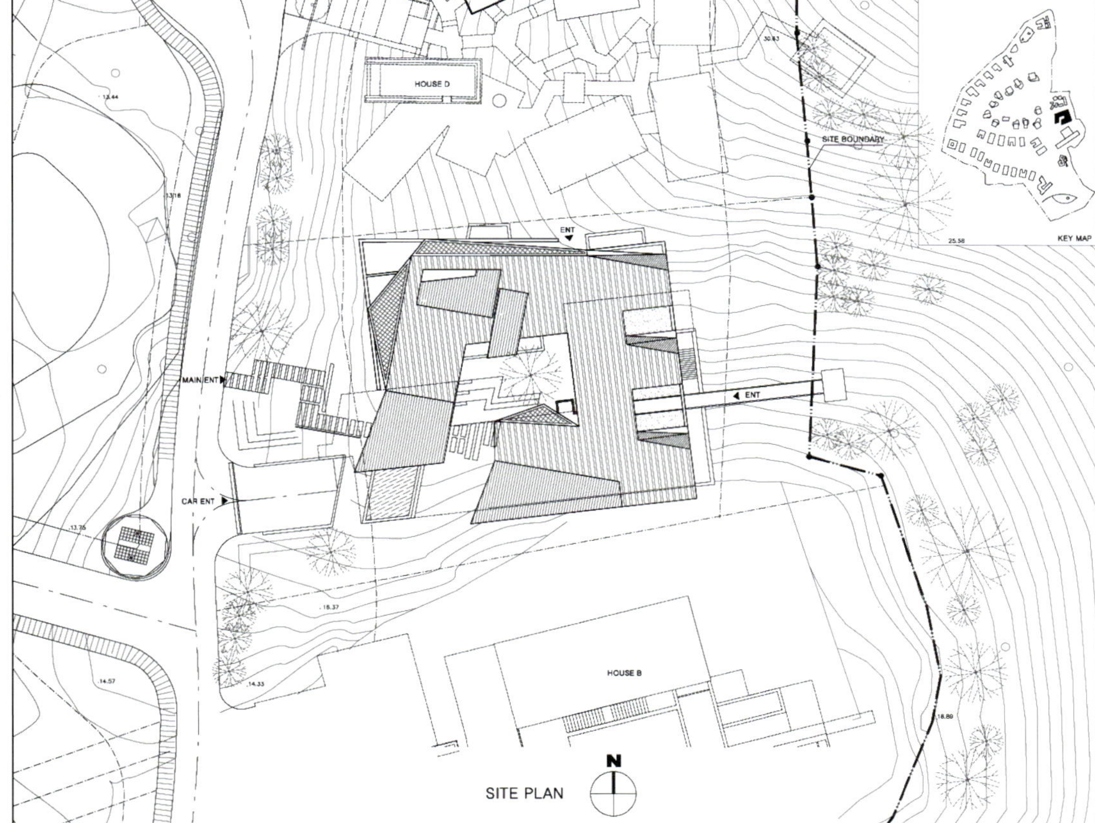

in green waters, blue skies, and hill forests of vast emerald, just like a Chinese splash-ink landscape painting, full of the fragrance of books and scrolls, which is just the "noble elegance shown in simple actuality". They according to the "heaven-earth-man" three-in-one concept, coexist with nature harmoniously, let every life feel and experience the tranquility in the depths of soul and the longing full in serene mood, and this is just the artistic conception of the "Jiangnan" architectural space — where the charm of a poetic dwelling place is.

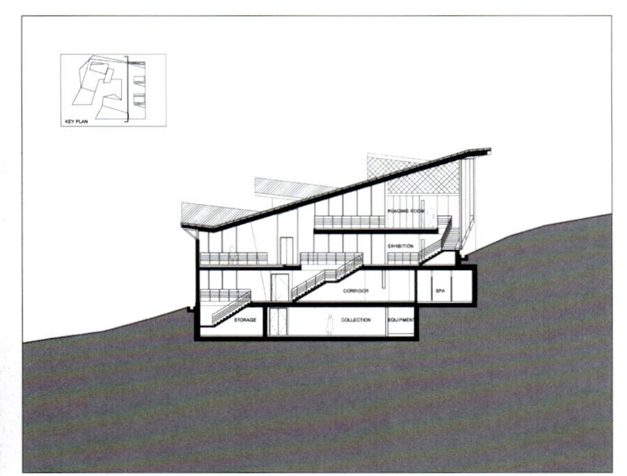

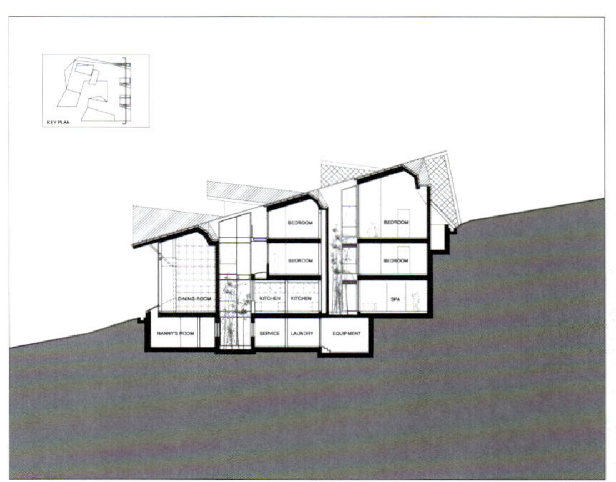

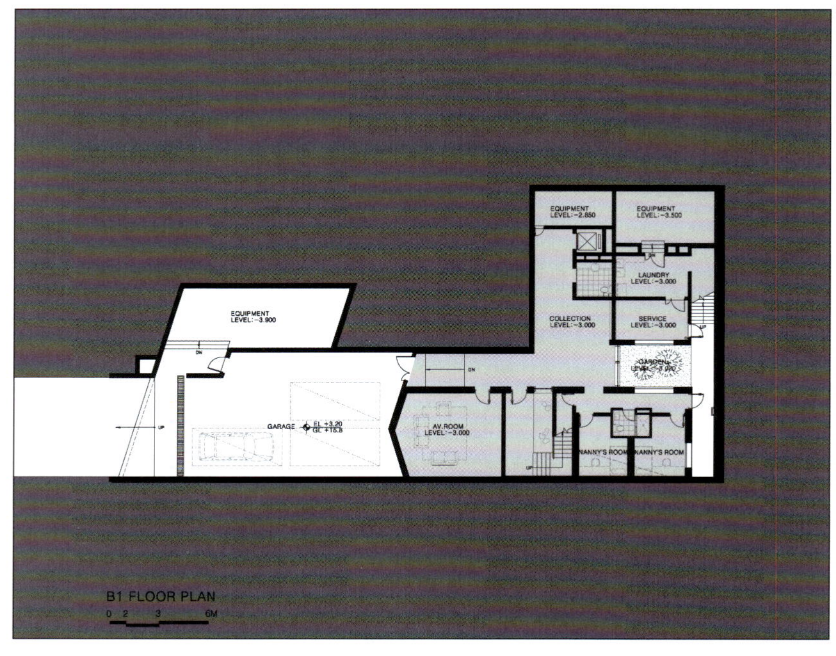
EQUIPMENT
LEVEL:-2.850
EQUIPMENT
LEVEL:-3.500
LAUNDRY
LEVEL:-3.000
EQUIPMENT
LEVEL:-3.900
COLLECTION
LEVEL:-3.000
SERVICE
LEVEL:-3.000
DN
GARDEN
LEVEL:-3.020
UP
GARAGE
EL +3.20
GL +15.8
AV.ROOM
LEVEL:-3.000
NANNY'S ROOM
NANNY'S ROOM
B1 FLOOR PLAN

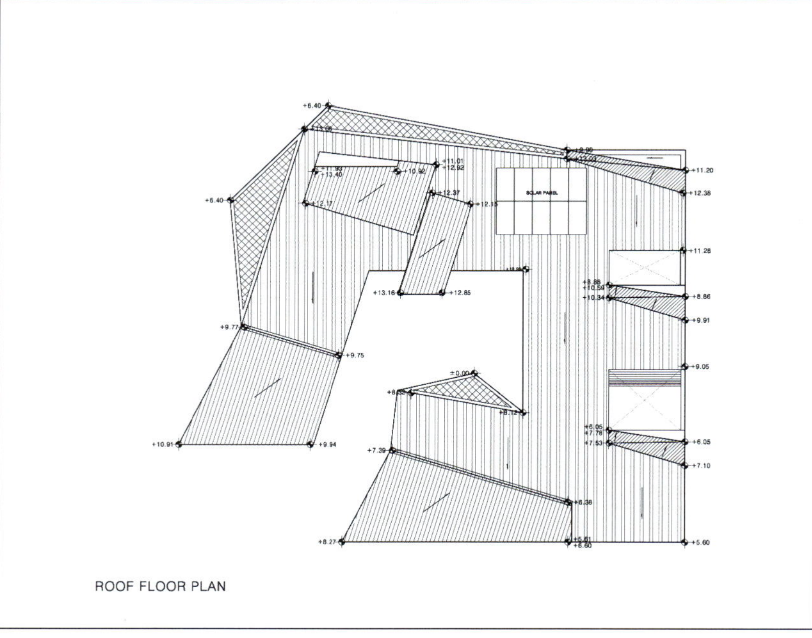
ROOF FLOOR PLAN

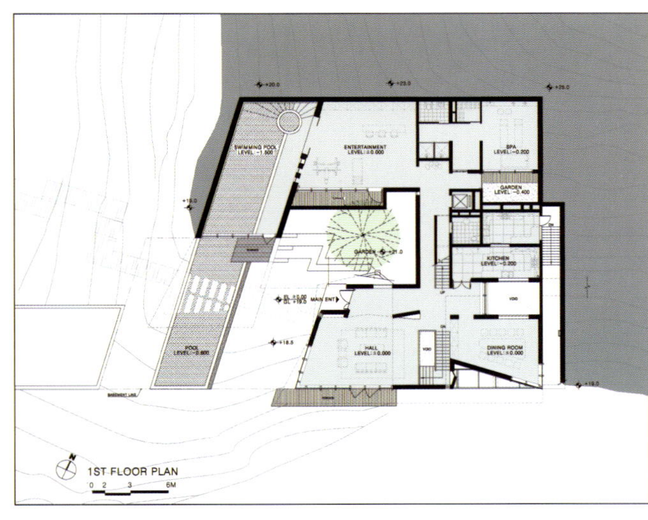
1ST FLOOR PLAN

2ND FLOOR PLAN

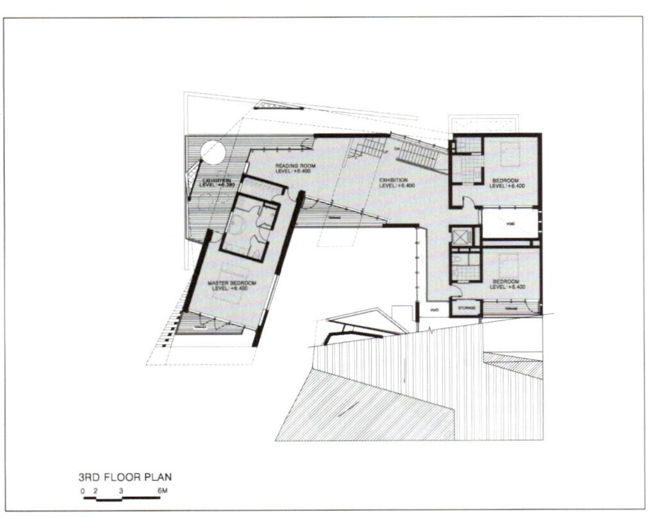
3RD FLOOR PLAN

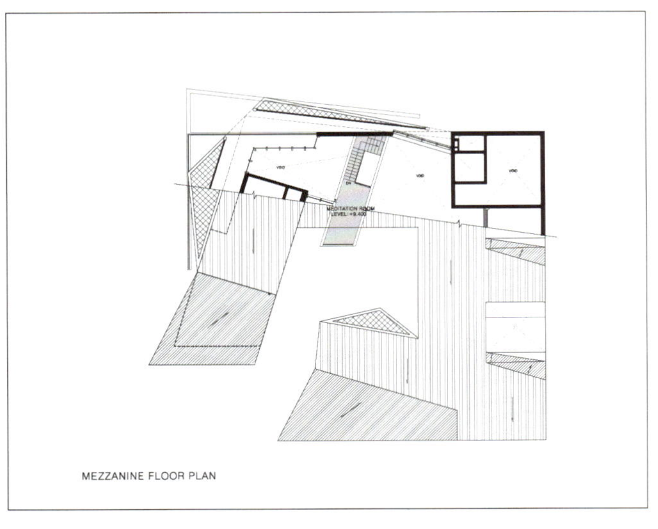
MEZZANINE FLOOR PLAN

"Diecui" — Overlapping Emerald

叠翠（二期Z12幢）

nikko建筑事务所

项目地点	中国，杭州
建筑师	王兴田，张峻，张凤春，苏晓宇
设计单位	nikko
设计时间	2010年
建筑面积	648平方米
技术	混凝土结构

Project Location	Hangzhou, China
Architect	Xingtian Wang, Jun Zhang, Fengchun Zhang, Xiaoyu Su
Design Company	nikko
Design Milestones	2010
Building Area	648㎡
Technique	Concrete

Z–12幢会所命名为“叠翠”，建筑以植物覆盖，依山顺势的层层叠落，通过台地和绿化，形成叠落的景观，庭园和山体自然融为一体。以清新幽雅为原则，建筑简洁、适用、大方。同时充分考虑了建筑的第五立面，使之更好地融入起伏的绿色环境。所选用的材料与色彩整体保持协调一致，将通透、轻巧的感觉贯穿其中。

底层为“工”形平面，两侧下沉内庭院不仅解决了地下层采光问题，还为各个功能空间创造了良好的景观视觉。另外，起居室屋面采用种植屋面使建筑与周围环境浑然一体，一方面建筑被周边自然所环抱，另一方面流动灵透的建筑空间也渗透着自然。景观设计总体上尊重原始地形，就地取材，依山成景。入口景观水池和中心水道遥相呼应，并增加了入户前的山林野趣。

Chamber Z-12 is named as “Diecui” (meaning Overlapping

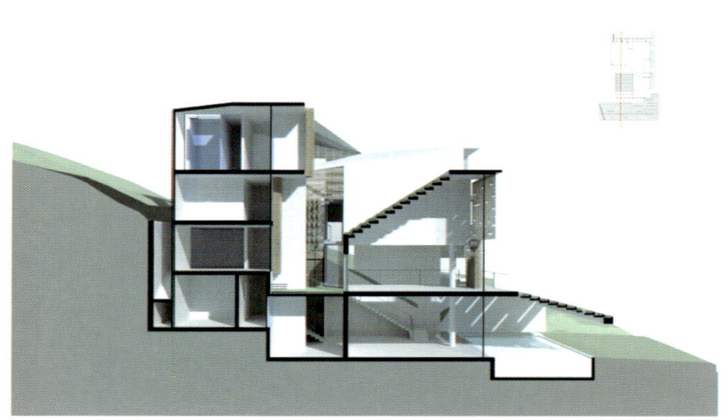

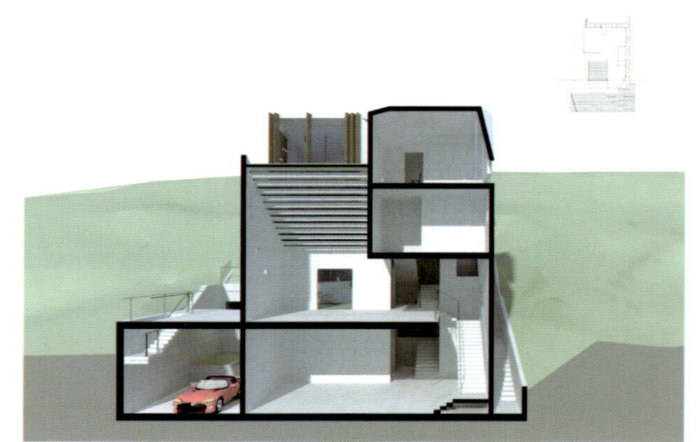

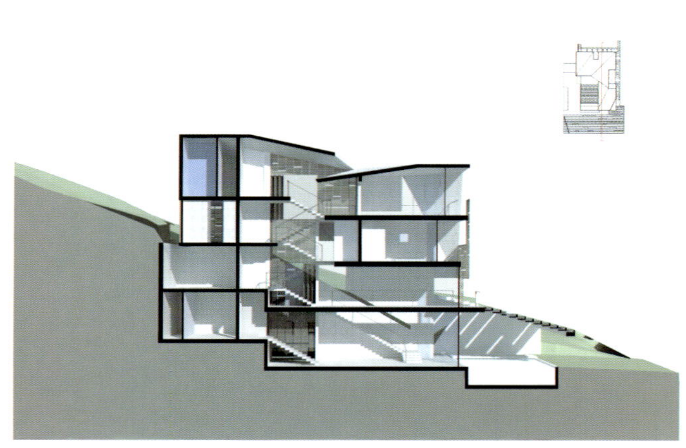

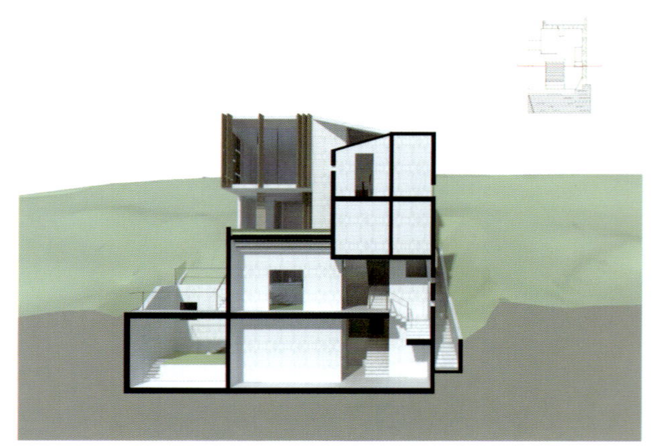

Emerald). The building is covered with plants, laminated along with the hill terrain, and a laminated landscape is formed via the terraced land and greening, and the courtyard is naturally integrated with the hill. Here the principle is freshness, tranquility and elegance, and the building is simple, usable and graceful. Meanwhile the fifth elevation of the building is fully considered, so that it better blends into the rolling green environment. The selected materials keep harmonious with the colors as a whole, so that a penetrating and light feeling runs through them.

The ground floor is a " I "-shaped plane, and the inner courtyard with the two sinking sides not only resolves the underground daylighting issue, but also creates a favorable landscape vision for every functional space. Besides, the living room takes a planting roof, so that the building is perfectly integrated with the surrounding environment. On one side the building is encircled with the circumjacent nature—on the other hand the flowing clear architectural space is also imbued with nature. The landscape design as a whole has respect for the original terrain, and locally available materials are used, and the scenery is formed along with the hill. The landscape pool at the entry and the central channel coordinate with each other over a distance, and a wilderness interest of hill forest previous to an indoor entry is added.

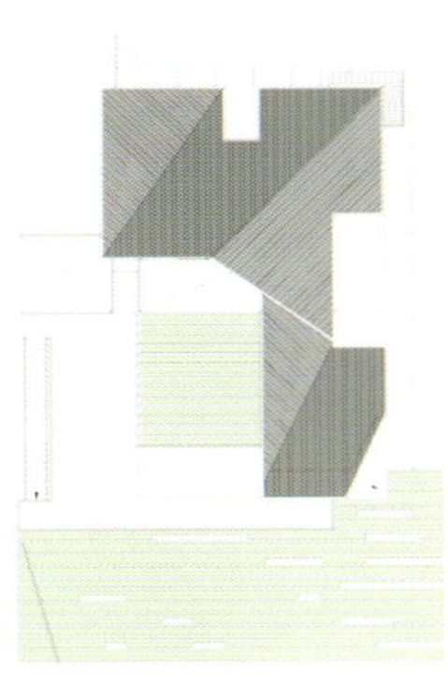

ROOF FLOOR PLAN

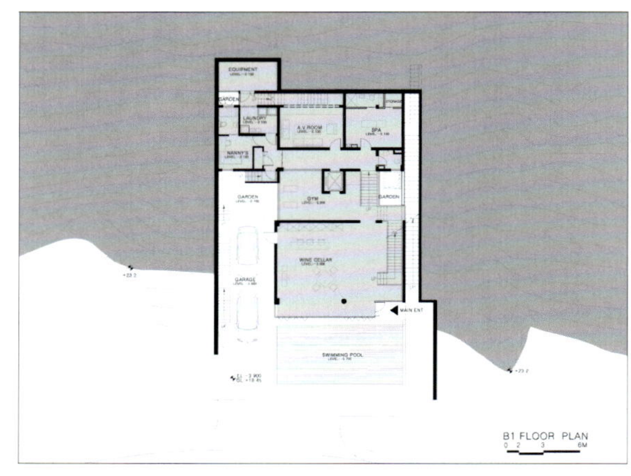
B1 FLOOR PLAN

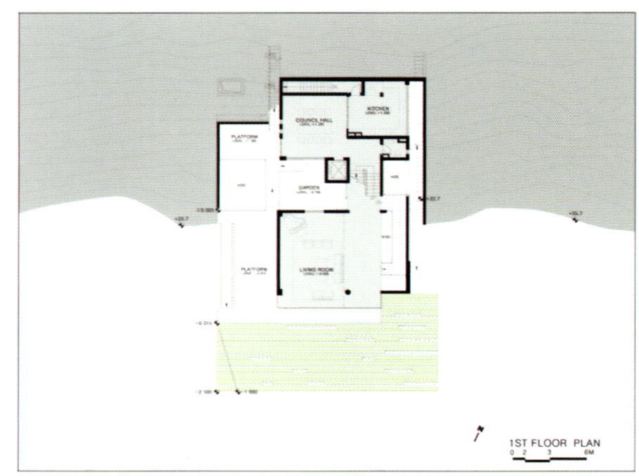
1ST FLOOR PLAN

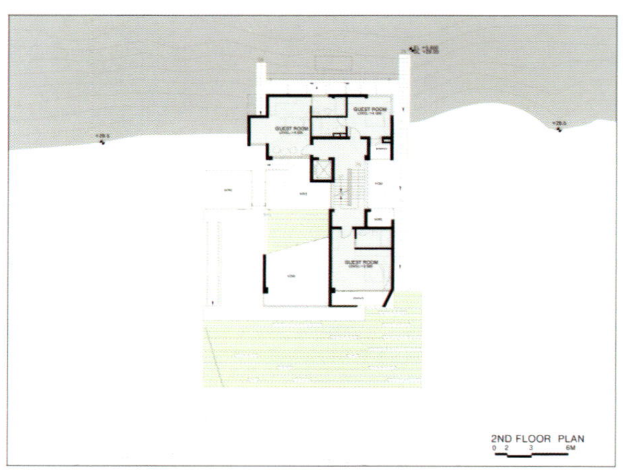
2ND FLOOR PLAN

3RD FLOOR PLAN

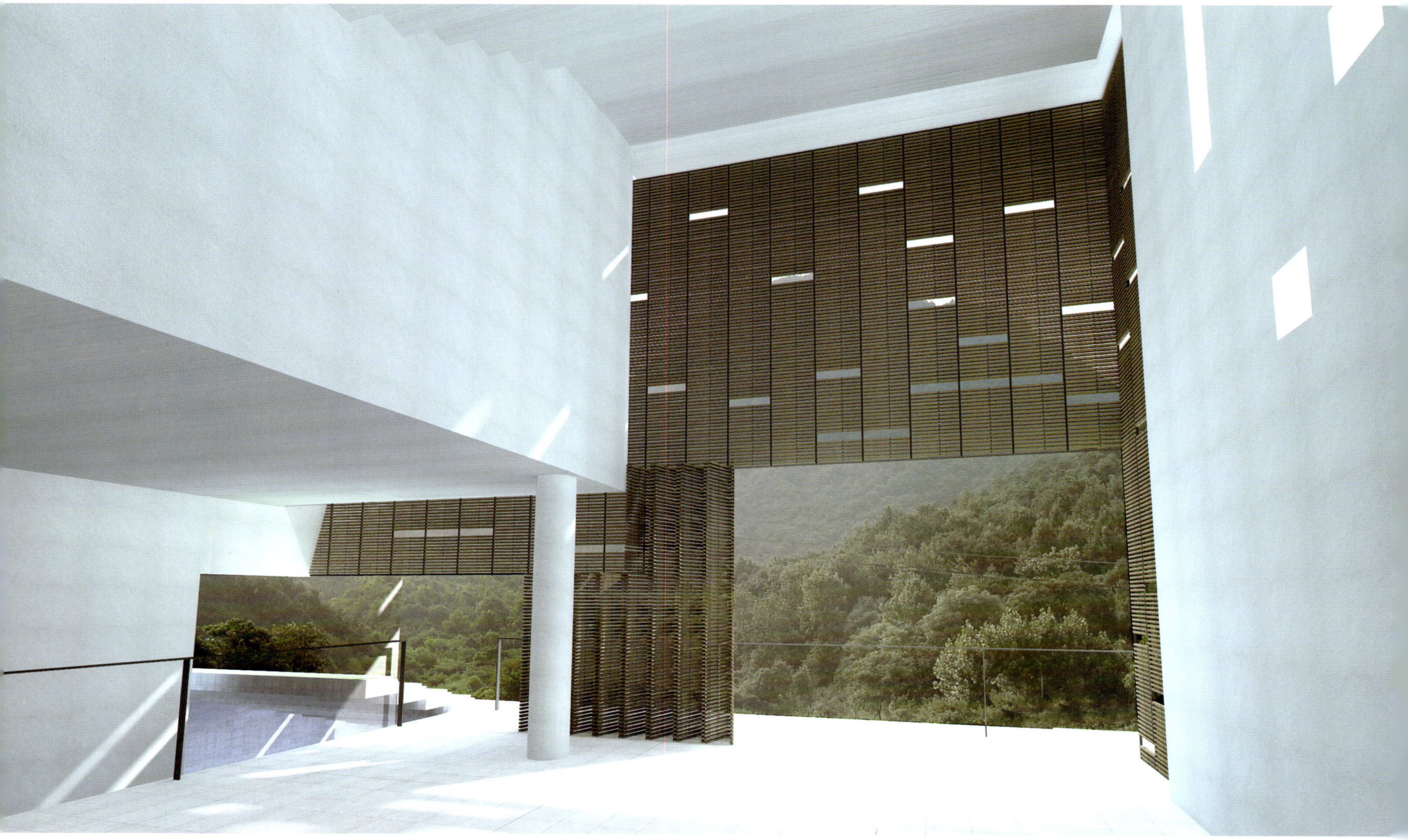

"Yingcui" (Two Phase Z1 Villa)
盈翠（二期Z1幢）

nikko建筑事务所

项目地点	中国，杭州	Project Location	Hangzhou, China
建筑师	王兴田，张峻，张凤春，苏晓宇	Architect	Xingtian Wang, Jun Zhang, Fengchun Zhang, Xiaoyu Su
设计单位	nikko	Design Company	nikko
设计时间	2009年	Design Milestones	2009
建筑面积	528.88平方米	Building Area	528.88 ㎡
技术	混凝土结构	Technique	Concrete

亚洲的生活态度——富春鸿茂　八墅系列之盈翠

闭门潇洒入窗明　一洗浮生梦幻情
夜月静移花下影　晚风轻度竹间声
天空寥廓人孤坐　地耸高寒鹤不惊
满眼青山云散尽　恍疑身世在蓬瀛

——《清虚堂》

诗中唯美的意境将我们从纷繁嘈杂的现代城市生活中剥离出来，带入宁静安逸的世外桃源，静心安坐，感悟人生。

尊重自然 依附自然

建筑依山傍水，坡屋顶的形象与山川水体的天然形成亲和关系，点缀山水，使建筑与自然环境和谐共处，达到你中有我，我中有你的共融情景。

屋面以干净利落的折线横跨整个建筑，连续折变的坡屋面由上至下，直至地面，以现代艺术中“拼贴”的手法打破了传统的三段式划分。屋面、台阶、窗洞选用高度抽象的几何形，每个块面都进行了光影处理以强化体积感。维持自然形态的山水和浮云，主张中西融合的设计理念。

有宅必有院

中间留出的院落既符合中国传统的建筑格局，又巧妙地解决了建筑北侧日照。有意模糊内外的定义，使居住者自融其中。

重构秩序是一种诗意：和谐、内敛、自省、自律

盈翠的意境呈现离不开材料的选择与应用。设计中分析研究不同材料的搭配及在不同角度下呈现的效果，进行推敲。屋面的结构形式和铺设纹理经看样分析确定。

Asia's Attitude towards Life— Fuchun Hongmao• All-around Emerald of the Eight Villas Series

Light-hearted I come to the shining window with the door closed behind, all illusory sentiments of the earthly life are cleared away;

The shadow behind the flower silently moves under the serene moon, the dawn breeze gives rise to gentle noises in bamboos;

Under the vast sky I sit alone, the ground towers high and the chilled crane stays motionless;

All clouds disappear from the lush hills extensive in the sight, and I feel like to live in a fairyland.

——Chamber of Pure Tranquility

The artistic state of pure aesthetics in the poem frees us from the noisy sophistication of urban life, leads us into a tranquil and cozy Arcadia, where we sit peacefully with a serene heart, and contemplate our life.

Respect nature and comply with nature

The buildings are embraced by hills and waters, and the hills and waters are adorned with the sloping roof image and the natural appetency between hills and water bodies. The buildings are harmonized with the natural environment in a coexistence, so that the scene of the building intermingling with nature is achieved.

The roof spans the whole building with clear and brisk bent lines. Continuously bent roof runs from high to low as far as to the ground, and the traditional three-sectioned division is broken by the "patch-up" manner in modern art. The roof, stairs and windows have a highly abstract geometric form, and every block face has been optically treated for a heightened sense of volume. The hills, streams and floating clouds are maintained in their original natural form, and the design conception of Chinese and Western integration is advocated.

A dwelling house must have a courtyard

The yard left in the middle conforms to Chinese traditional architectural pattern and also deals with the sunshine in the north side of the building artfully. The sense of being indoors

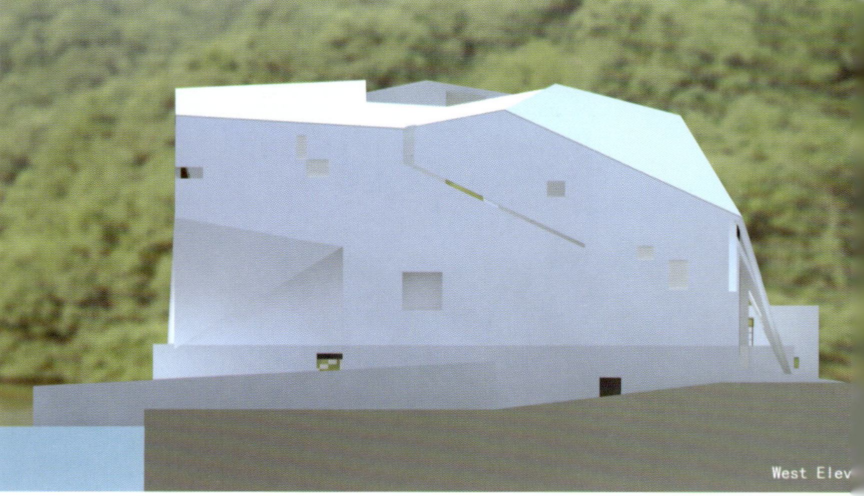

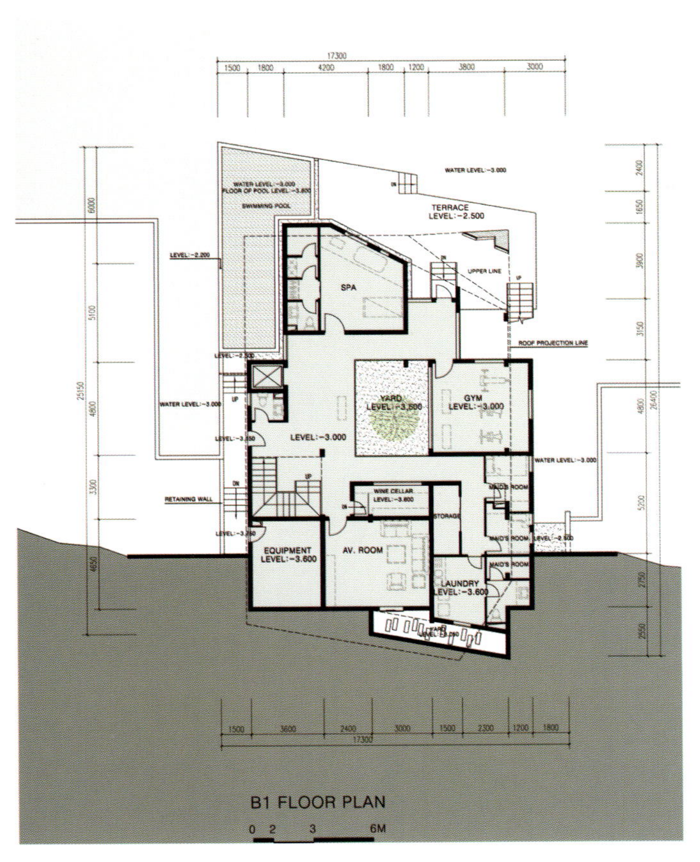

B1 FLOOR PLAN

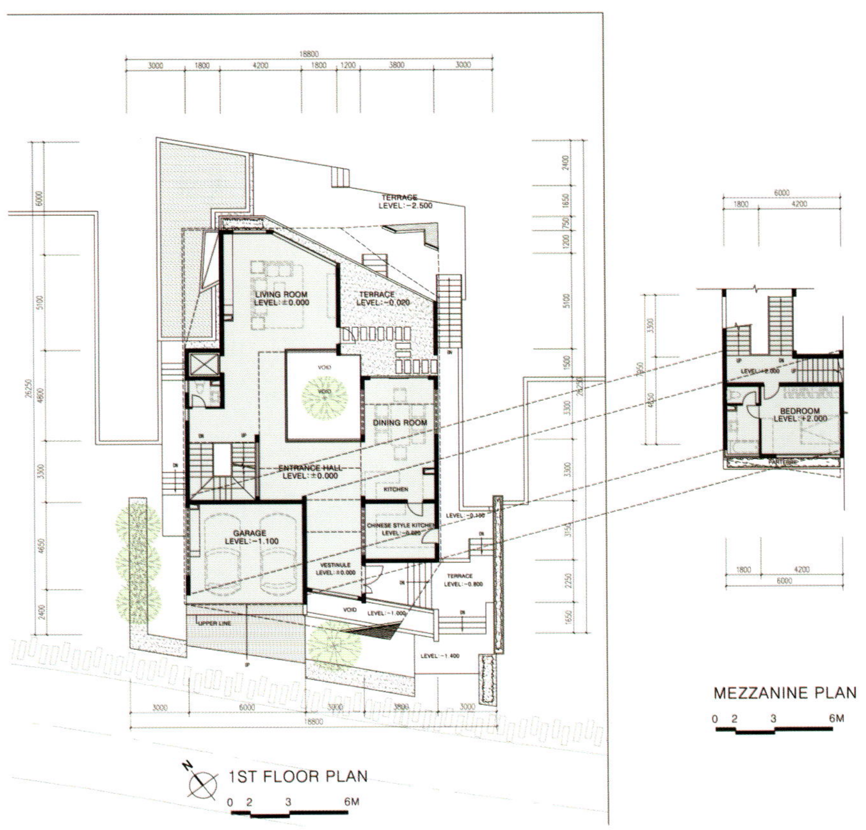

MEZZANINE PLAN

1ST FLOOR PLAN

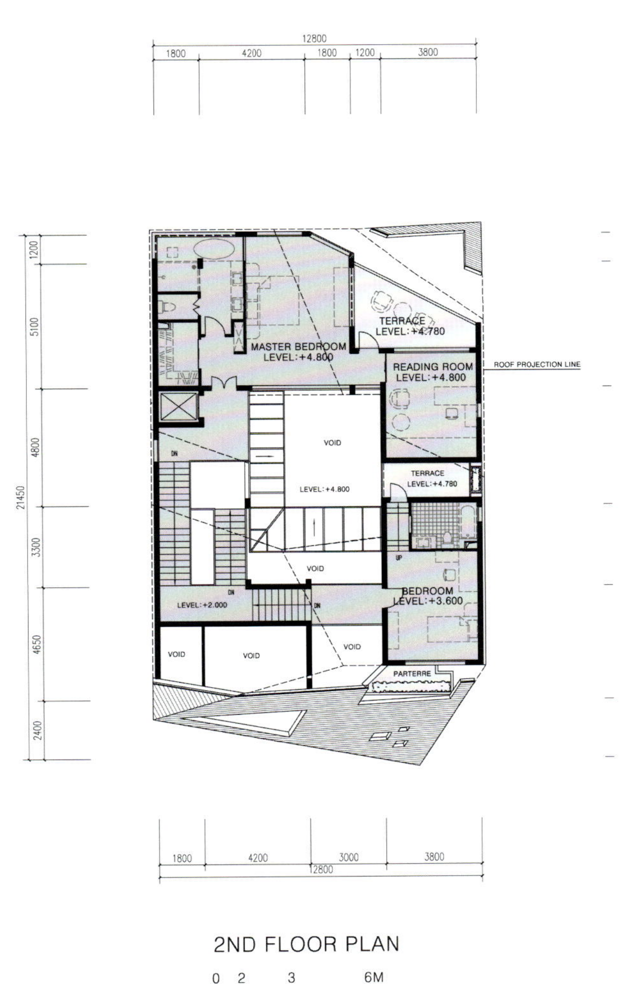

2ND FLOOR PLAN

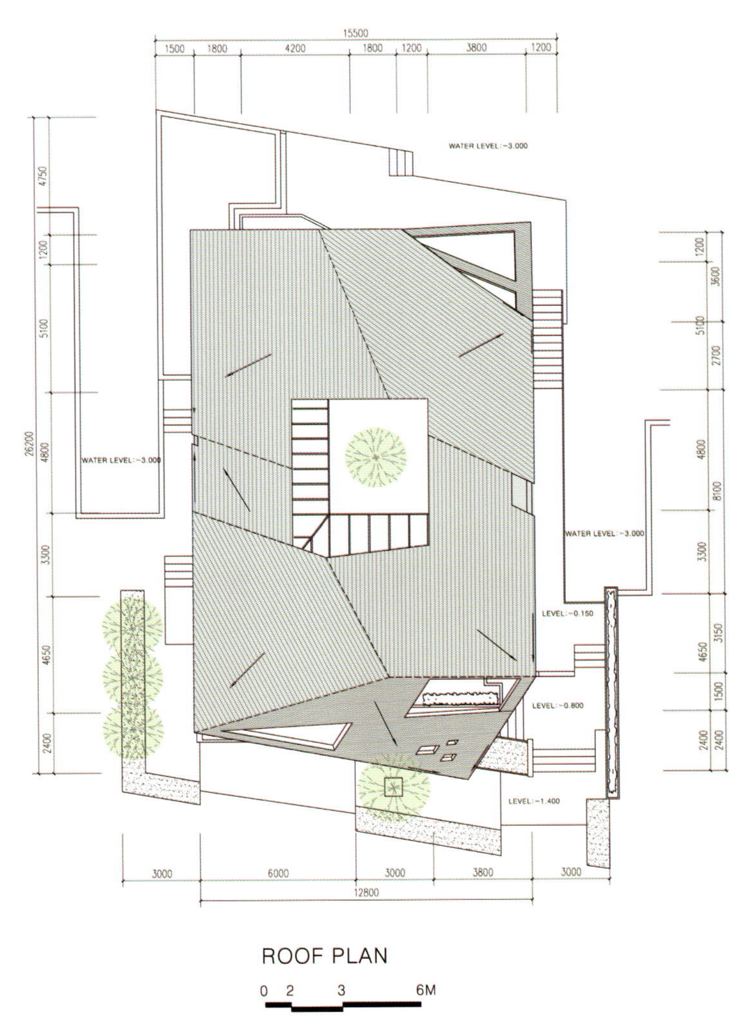

ROOF PLAN

or outdoors is intentionally blurred here, so that the dwellers naturally mingle there harmoniously.

"Yecui"—Bright Emerald

烨翠（二期Y5幢）

nikko建筑事务所

项目地点	中国，杭州	Project Location	Hangzhou, China
建筑师	王兴田，张峻，张凤春，苏晓宇	Architect	Xingtian Wang, Jun Zhang, Fengchun Zhang, Xiaoyu Su
设计单位	nikko	Design Company	nikko
设计时间	2009年	Design Milestones	2009
建筑面积	666.6平方米	Building Area	666.6㎡
技术	混凝土结构	Technique	Concrete

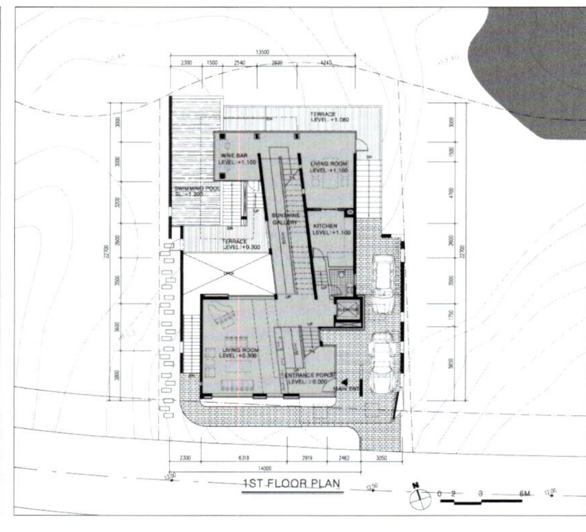

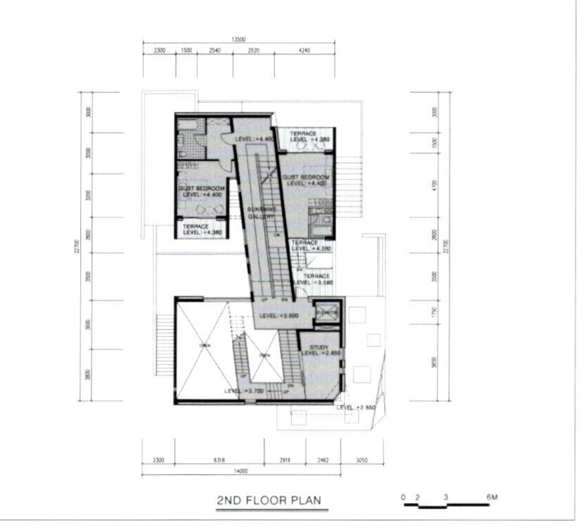

Y5幢会所命名为“烨翠”，顺应自然、布局上适应地形地势，将建筑适当地植入自然中，“空间和光影的渗透”是Y5幢会所的设计主题，核心的无障碍交通廊道将南北面的空间连接，各个空间自然地错落在不同的标高之上，产生相互渗透的空间关系。中央光廊两侧大小不同的方形窗洞闪烁着周围的光影和绿色意象，也是本案“烨翠”主题的由来。

Y5幢会所底层为“U”形平面，围合下沉内庭院开口向西，同时西北侧底层局部架空，将北侧核心景观引入建筑之中，与周围环境浑然一体，一方面建筑被周边自然所环抱，另一方面流动灵透的建筑空间也渗透着自然。景观设计总体上尊重原始地形，并在架空区域下方设置了室外泳池。

Chamber Y5 is named as "Yecui" (meaning Bright Emerald). It complies with nature, and is well adapted to the topographic terrain in its layout. The building is modestly planted in nature. The design subject for Chamber Y5 is "the infiltration

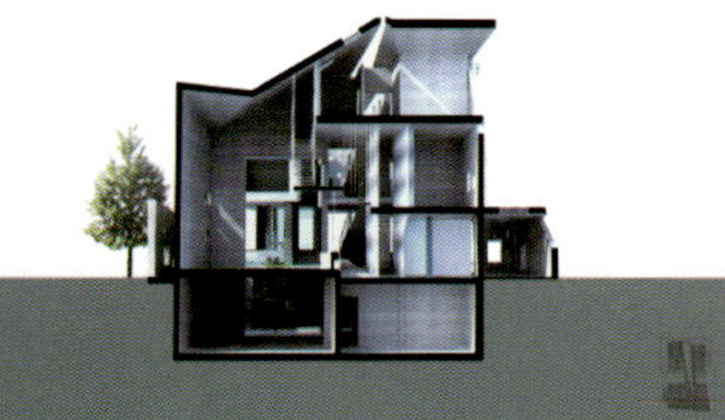

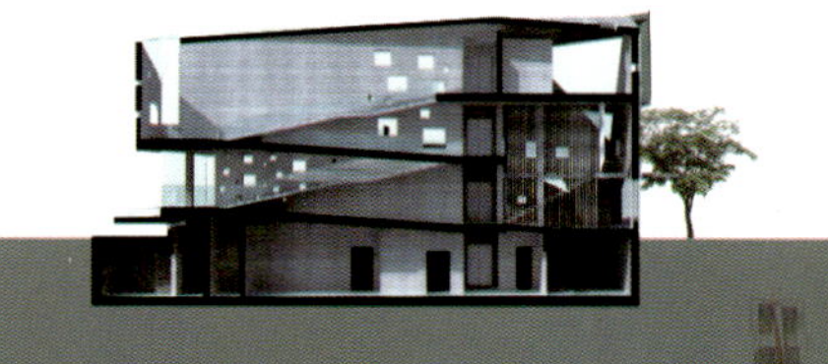

of space, light and shadow". The unobstructed come-and-go galleries at the core connect the spaces on the south and north sides; various spaces naturally are strewed at different elevations, so that the spatial relationship of mutual infiltration is generated. The different-sized square windows on the two sides of the central light gallery flickeringly reflect the surrounding light, shadows and green color, and this is also the reason for the "Yecui" subject of this case.

Chamber Y5 has its base floor in a "U"-shaped plane. The opening of the enclosed sinking inner courtyard faces westward, meanwhile the base floor on the northwest side is partially built on stilts, so that the core landscape on the north side is introduced into the building, and perfectly integrated with the circumjacent environment—on the one hand the building is encircled by the surrounding nature, on the other hand the flowing clear architectural space is also imbued with nature. The landscape design as a whole has respect to the original topography, and an outdoor swimming pool is provided below the area built on stilts.

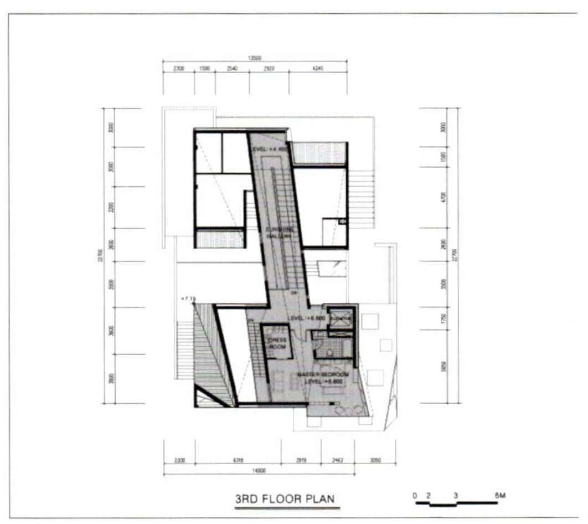

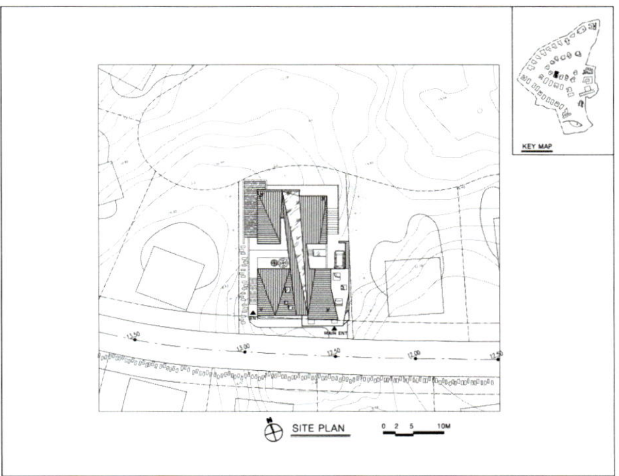

House in Bizan

眉山小屋

吉田周一郎建筑设计

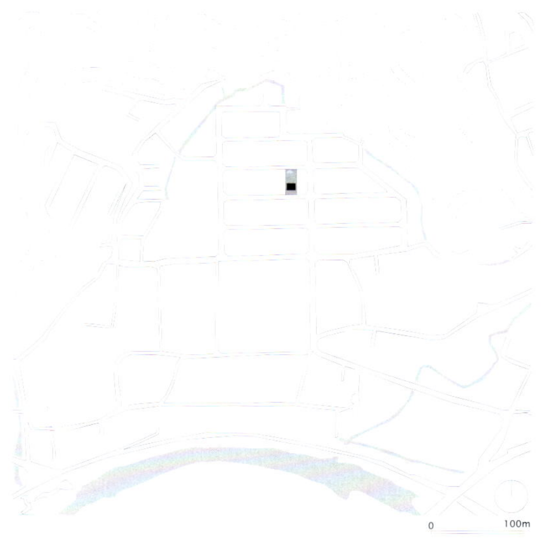

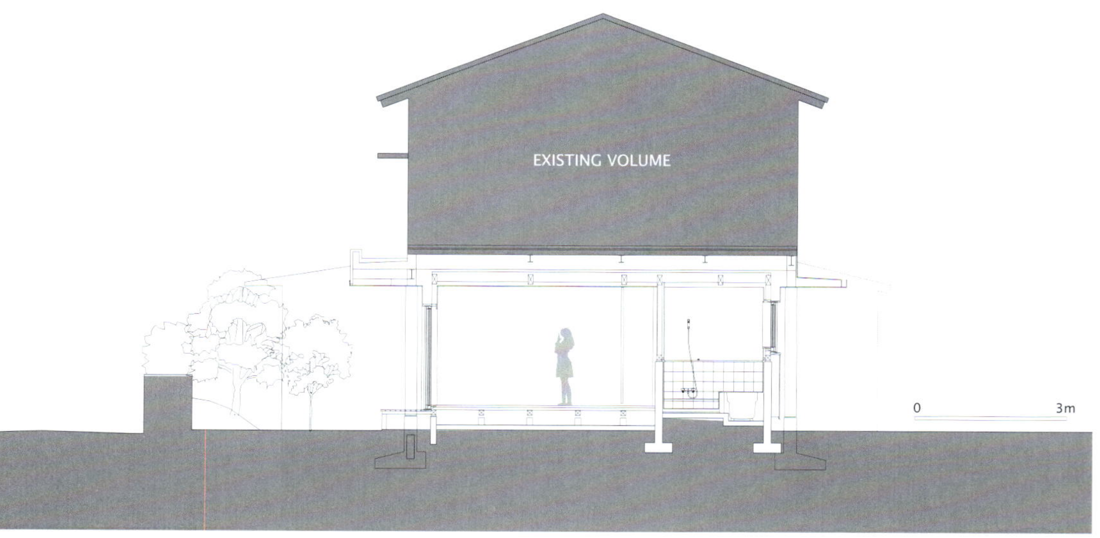

项目地点	日本，德岛
项目类型	住宅
面积	80平方米
开始时间	2008.12
竣工时间	2009.7
设计单位	吉田周一郎建筑设计
设计师	吉田周一郎
主承包商	Homare Kensetsu
摄影师	Akira Yonezu

Project Location	Tokushima, Japan
Type of Project	House for single family
Area	80㎡
Constuction Start	December 2008
Completion Date	July 2009
Design Company	Shuichiro Yoshida Architects
Architect	Shuichiro Yoshida
Main Contractor	Homare Kensetsu
Photographer	Akira Yonezu

材料描述

外墙、屋檐、天花板以及内部配件：日本雪松木

内墙装饰：白色水泥混以传统的日式材料“shoji”

地板：白色橡木地板、日式“榻榻米”（秸秆芦苇垫）

窗户、外部配件：黄色雪松木

家具、门：美国白蜡树

滑门：由日式纸张“shoji”覆盖的黄色雪松木网格门

业主于1974年购得了这所一层高的住宅，两年后以钢结构加盖了房屋的第二层。接着，业主在1999年扩建了厨房，并将一楼分成了许多小房间，但这使得业主和他的家人没有足够的空间款待宾客。所以在2008年的时候，业主决定对房屋进行改造，使其变得更加舒适。

除了柱子、房梁和地基等基本结构外，房屋的一楼被整个拆除了。墙壁、屋檐、地板和天花板都被重新装修，并在现有的花园中添加了拥有崭新的日式榻榻米的更大的空间。

尽管这片住址被眉山——德岛市的象征——所环绕，但原来屋子的一楼并没有能够一睹美丽山景的良好视野，所以在改建计划中，设计师精心安置了窗户和屋檐，使漂亮的眉山即使越过邻居的房屋和花园也能一览无遗。

在日本，该地区盛产雪松。但由于进口木材价格低廉并被广泛运用，许多二战后种植的日本雪松并没有得到充分利用。在这项改建计划中，日本雪松被大量地运用在外墙、天花板和饰面材料上。

内墙饰面“漆”有着极佳的防潮防火效果。日本的许多传统建筑，例如寺庙、神社和阁楼都在饰面上选用了这种材料。

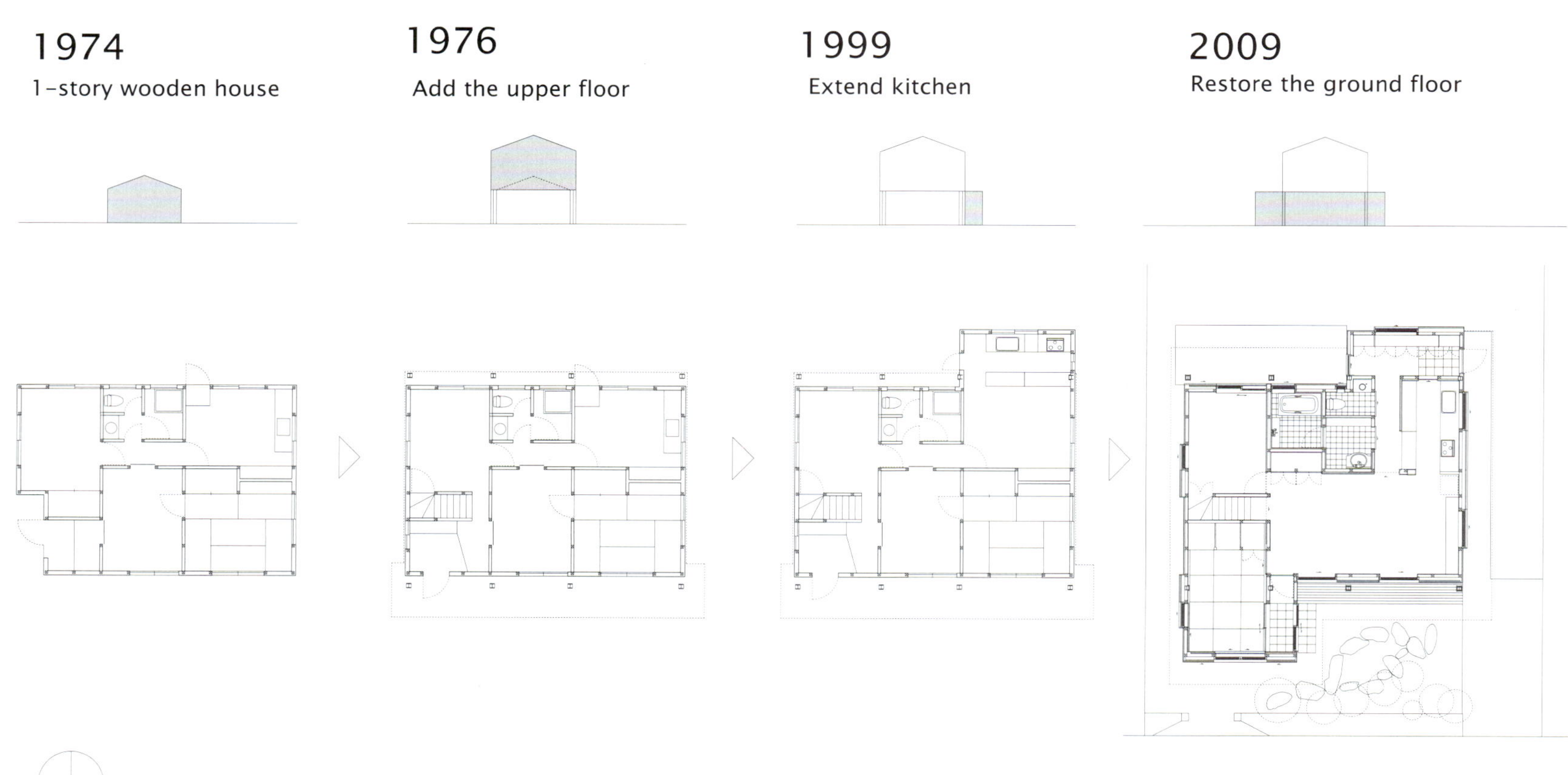
1974
1-story wooden house
1976
Add the upper floor
1999
Extend kitchen
2009
Restore the ground floor
0
5m

Material Description

Exterior wall, roof overhangs, ceiling and inside fittings: Japan cedar

Interior walls: white cement mixed with the traditional Janpanese material: shoji

Floor: white oak, Japanese tatami (reed mat)

Window and exterior fittings: yellow cedar

Furniture and door: white ash

Sliding door: yellow cedar grid door covered by the Japanese paper: shoji

The owner obtained the one-storyed house in 1974, and they added the upper floor of steel structure after 2 years. And moreover, they extended kitchen space in 1999. The former ground floor was divided into some small rooms and there was no space large enough for them to enjoy their time with their guests. In 2008 the owner decided to reform the ground floor to be more comfortable.

The ground floor is demolished except its structure such as columns, beams and foundations. The walls, eaves, floors,

ceilings are all renovated. A larger room is planned with a new Japanese "tatami" room added into the existing garden.

Although this site is surrounded by Mt. Bizan, symbol of Tokushima city, the former house isn't in good condition for looking at the mountain from the ground floor. In this project the windows and eaves are carefully placed so that they could enjoy the view of Mt. Bizan beyond neighbor houses and their gardens.

This area is rich in Japanese cedar trees. But a lot of Japanese cedars planted after the World War II have been left unused, because various kinds of imported timbers have been cheap and popular. In this project Japanese cedars are richly used as exterior walls, ceilings, and finish materials.

Interior wall finish "Shikkui" is effective for humid conditioning and fire prevention. A lot of walls of traditional Japanese architecture such as temples, shrines and castles are finished with this material.

The Luukku House

Luukku别墅

阿尔托大学——来自芬兰赫尔辛基的跨学科团队

项目地点	竞赛地点位于西班牙，马德里 最终地点位于芬兰
项目面积	25米x20米
主要材料	芬兰木 Kerto-Q结构、绝缘木材纤维、松木材内部镶板、桦木地板、杉木层&热处理松木板露台
设计单位	阿尔托大学 —— 来自芬兰赫尔辛基的跨学科团队
主负责人	Pekka Heikkinen

Project Location	Competition held in Madrid, Spain. Final location, Mäntyharju, Finland
Project Area	25mx20m
Main Materials Used	Finnforest Kerto-Q structure, Cellulose wood fibre insulation, Pine sapwood interior panelling, Birch lamella flooring, Spruce cladding & Heat treated Pine decking, terrace
Design Company	Aalto University — an interdisciplinary team from the Helsinki, Finland
Principal-in-Charge	Pekka Heikkinen

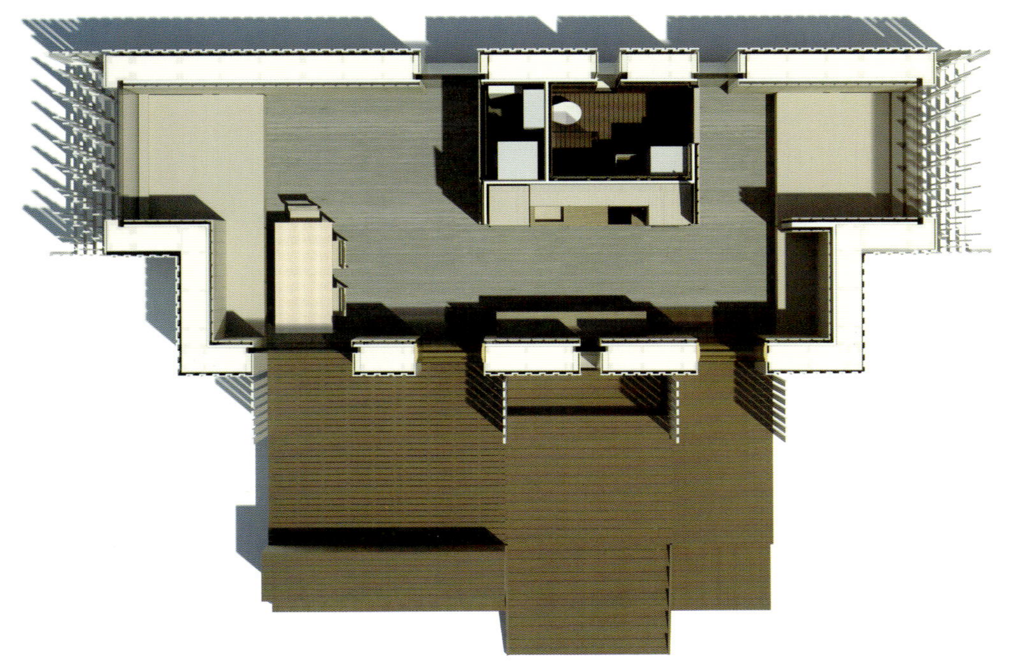

项目描述

房屋的设计与建造同时考虑到了马德里与赫尔辛基的气候环境。本项目是为2010太阳能大赛所设计，在17名参赛团队中名列前五并在建筑领域处于第一名。

背景

本项目的设计展示了木材以及与木材相关的材料如何被多用途地用到房屋的结构、绝缘材料、包层、内部楼层、墙壁以及天花板装饰中，在像卫生间这类潮湿的地区使用到湿度和热缓冲质量以及其实用性。本项目的目标是建造芬兰的第一座零碳建筑，成为马德里的第一家零碳房屋。

灵感

本项目是对于芬兰木结构房屋建造的一种探索与发展，这是一个对现有住房质量的提高的关键步骤，特别是对木结构度假型房屋意义重大。

挑战

强调了木材以及木材结构的产品可以在建造中运用

低碳排放（事实上在建造中达到了负的碳排放）

建造芬兰第一所零碳房屋

促进芬兰的设计、建造以及制造业

主要理念

通过这个项目，我们创造了一个舒适的生存环境，一个可以放松的居住空间：一个家。在本项目的每一个步骤我们都努力避免采用任何技术性的机器来体现功能性。以这样的方式，我们的房屋旨在达到天然建造方法的一种契约。在其他方法都无效的前提下，我们才会考虑使用机器技术。任何技术我们都是为适应马德里和芬兰的气候来进行运用的。

生态概念

建造：本项目的目标之一是验证木材可作为一种生态建造材料。房屋的碳平衡是经过计算的：大约26 500千克的碳在建造过程中被储存，仅仅15 000千克的碳足迹就使得房屋呈碳负极状态。85%的建造材料都是可再利用的。

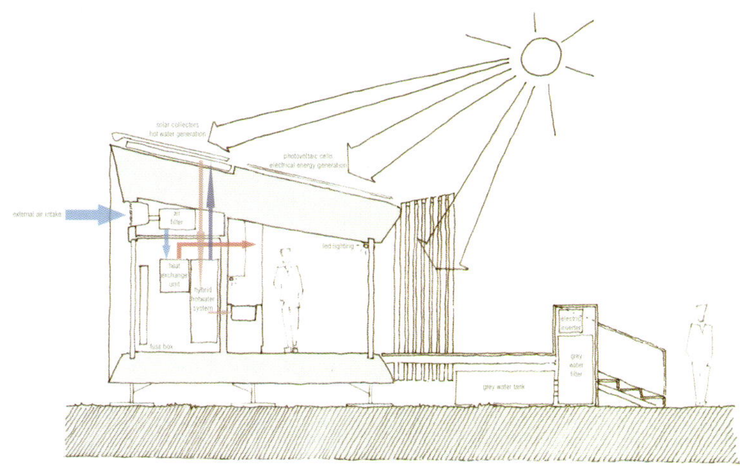

除了在材料上的运用，设计师还力图在设计中体现高效的节能作用。第一要务是通过减少建造过程中的损耗和隔热来实现能耗最小化。我们运用来自LVL（复合木板）以及木质纤维隔热层的U-Values值为0.1,0.08,0.08的气密层分别用在墙壁，屋顶以及楼梯上（达到被动式住宅的标准）。

窗户的U-value值为0.37（由无框架单元的四层釉面氩组成），门为0.52（三层带有软木热断层）。

水：我们还运用了拥有无触摸系统的低流量充气信号器来节约水流失，以及废水过滤器用于收集来自洗衣机、洗碗机、淋浴和自来水龙头的水。

Project description

The house has been constructed for both the climates of Madrid and Helsinki. Built for the Solar Decathlon competition 2010 it achieved overall 5th place out of the 17 contestants whilst taking first prize in the architecture contest.

Background

The house displays how timber and timber based materials can be used in a variety of applications from structure, insulation, cladding, interior floor, wall and ceiling finishes, for moisture and heat buffering qualities and applications in very wet areas such as bathrooms. The house has the goal of being Finland's first zero energy house, whilst being a plus energy house in Madrid.

Inspirations

The house is a step in the development of timber frame housing in Finland and can be seen as a crucial step in the improvement of existing housing stock and, in particular vacation housing from which the concept has spanned.

Challenges

To demonstrate the applications that timber and timber based products can be used in construction

To build with very low carbon emissions (actually achieved negative carbon emissions in construction)

To build Finland's first zero energy house

To promote Finland's design, construction and manufacturing

Key concepts

Throughout this project we have aimed to create a comfortable living environment, a house for living in and a place to feel at ease; a home. At every stage we have endeavoured to avoid any sense that one lives within some technological machine constructed to perform a function. In this way the aim of our house is to be a testament to honest and natural construction methods. Technology has been utilised only as a final resort before other methods have been exhausted and in order to achieve the goals that we have set for the climates of both Madrid and Finland.

Ecological concept

Construction: One of the design goals of the house is to demonstrate timber as an ecological construction material. The carbon balance of the house is calculated: approximately 26,500kg of carbon is stored in the construction whilst a carbon footprint of just 15,000kg makes the house firmly carbon negative. 85% of all consumption materials come from renewable sources.

Beyond material use, we want to show energy efficiency of our construction. The first principal here is to minimise energy loss through leakage in the construction and losses through insulation. We have an air tight layer constructed from the LVL (Laminated Veneer Lumber) and wood based cellulose fibre insulation with U-values of 0.1, 0.08, 0.08 in the wall, roof and floor respectively (Reaching passivehouse standard).

The windows have a U-value of 0.37 (made from quadruple glazed argon filled frameless units) and doors 0.52 (triple glazed with a cork thermal-break layer).

Water: We also use low flow aerated taps with a touchless system to reduce water loss, a greywater filter and tank for collecting water from washing machine, dishwasher, shower and tap .

Aptos Retreat Residence

Aptos小屋住宅

Neutelings Riedijk Architects、Rotterdam、The Netherlands

项目地点	加利福尼亚，阿普托斯，3600弗恩拉特路
面积	新型住宅：2800平方英尺 谷仓：1600平方英尺
竣工时间	2009年7月
业主	斯图尔特•加内和凯特•迪茨勒
建筑团队	首席设计师：卡斯•科尔德•史密斯 项目建筑师：蒂姆•奎尔
室内设计师	琳•罗斯
景观设计师	娜塔莉•施瓦茨
结构工程师	罗恩•贝尔纳普
土木工程师	大卫•多芬
摄影师	保罗•代尔
Project Location	3600 Fern Flat Road, Aptos CA
Area	New Residence: 2,800 square feet Barn: 1,600 square feet
Completion Date	July 2009
Owners	Stuart Gasner & Kate Ditzler
Architecture Team	Design Principal: Cass Calder Smith Project Architect: Tim Quayle
Interior Designer	Lynn Ross
Landscape Designer	Natalain Schwartz
Structural Engineer	Ron Belknap
Civil Engineer	David Dauphin
Photographer	Paul Dyer

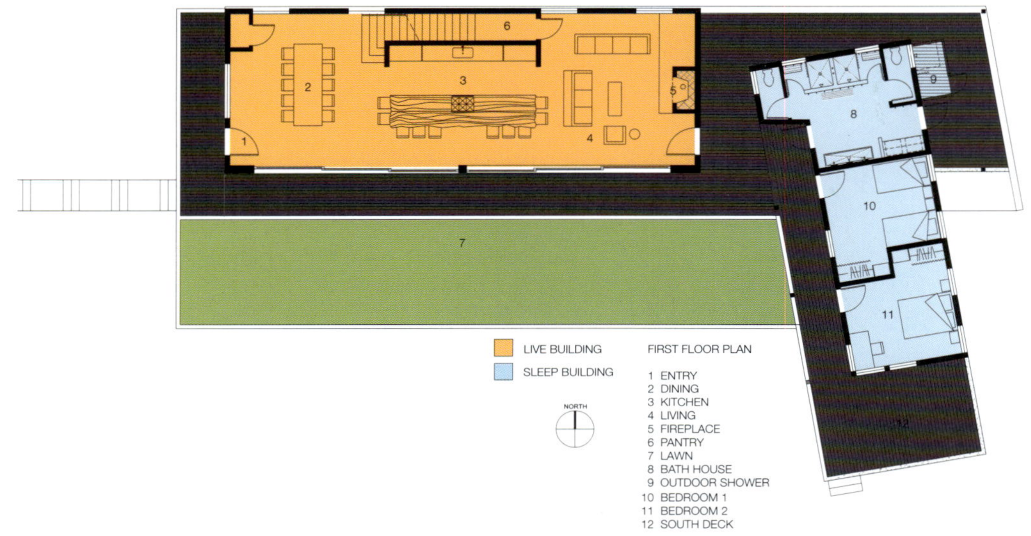

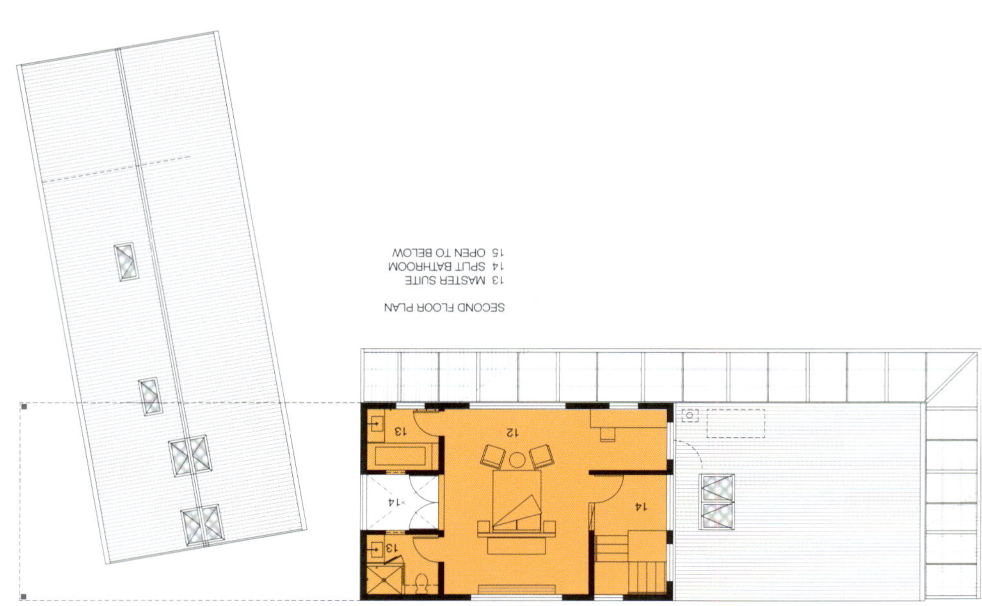

本案是为圣弗朗西斯科的一对夫妇设计的，他们有6个孩子，年龄跨度从高中到大学不等。本案占地20英亩，坐落于加利福尼亚圣克鲁斯山脉附近的圣克鲁斯市，靠近海滩小镇阿普托斯，距离太平洋约5英里，拥有美丽的山景和海景。

业主希望营造一个轻松惬意的乡土环境，并希望加入可持续元素来降低家庭的碳排放，此外，还要求房屋具有各种各样的功能——举办聚会、烹饪、健身、游泳、射箭、掷马蹄铁、园艺制作和劈砍木柴。

本案有两栋主屋，附楼和娱乐场所跟充满了田园气息的院子结合到了一起。

2 800平方英尺的主屋由生活区和休息区组成，两个区域的屋顶重叠在一起，外观上形成了一个相互连通又相互遮蔽的空间。生活区的屋顶比休息区的稍高一些，里面设有餐厅、起居室、厨房和通往主卧室的楼梯。生活区的中央位置是岛式厨房，它占据了生活区中的主要空间，两侧分别是餐厅和起居室。岛式厨房里有一块胡桃木质地的料理台，长18英尺，厚3英寸，由一棵自然倒下的大树制成，边缘有许多弯曲减损的部分。两三组8到10英尺长的滑动玻璃门成了通往院子的通道，当这些滑动玻璃门完全打开后，窗体将会形成一个32英尺宽的全景框。滑动玻璃门的设计也为这幢别墅带来了充足的自然光线。在这里，灯具都只在夜晚使用。

本案原有两幢主楼和附带的小楼以及娱乐设施共同组成了这个带有两间卧房、一间公用浴室、一个L形院子的乡村谷仓别墅。再生木材及合金耐腐蚀钢搭成的屋顶遮蔽了房屋的外墙，内部材料则运用了实木地板、木材、石料和钢材。

除主楼外别墅还有一个1 600平方英尺的谷仓，它被

用作聚会活动室，外墙材料选用合金耐腐蚀钢，一楼可以打乒乓球，看电视，冲浪板和其他沙滩娱乐工具也放在这里，二层的阁楼有台球桌和休息用的沙发床。谷仓的内部结构是一个钢壳，外部则是木质的。

这栋别墅和周围环境融为一体。别墅的主楼和谷仓都坐落在红杉木环绕的草地斜坡上，南面不远处是辽阔的海景。围绕着主楼和谷仓外面的区域是游戏场。较主楼地势相对低一些的地方有游泳池，游泳池旁边是一个火坑。射箭场在别墅的正上方，玩掷马蹄铁游戏时则可以用主屋和谷仓之间的空地。红杉木中隐藏着的两间帐篷小屋被用作客房，桑拿房设在客房和主屋休息区之间的通道上。

本项目具有如下可持续性特征，即太阳能热系统所提供的家庭热水供应、水池加热、及水力辐射地板。

太阳能热水不仅作为生活用水使用，地板辐射供暖系统和游泳池也需要它。太阳能供热系统由院子里一系列真空太阳集热管组成，它通过收集太阳能来为生活、游泳池及地板辐射供暖系统提供热水。地板辐射是整幢房子里除壁炉以外的唯一热源。

加利福尼亚北部是一片林地，那里的住宅通常用木材建造，特别是那些乡村小屋。别墅的外墙其实是用谷仓的木材做的，这些材料都具有可持续性，别墅内部木材的使用使其与周围田园景致遥相呼应。

别墅的屋面、谷仓的屋面及外墙主要采用可高度回收的合金耐腐蚀钢板。为了达成别墅业主和建筑设计师致力于保护自然资源的愿望，别墅的外部材料选用了金属质地的屋顶和墙面系统，它们不仅能够长期使用，而且环保，能够循环再造。用商品金属板材建造的屋墙具有热偏转涂层和隔热涂层，能帮助减少40%的加热和制冷花费。

高性能窗户让整栋别墅冬暖夏凉。

朝南的窗户带有檐篷，檐篷制造出的阴影能够避免阳光直射。通风井则帮助散热。

This project is designed for a San Francisco couple with six children—with ages ranging from high school to college. The property is located inland from the beach town of Aptos, California in the Santa Cruz Mountains, near the city of Santa Cruz. The 20-acre site has ocean and mountain views and is about five miles inland from the Pacific Ocean.

The family desires a setting that would be casual and rustic, and that would incorporate sustainable features to minimize the home's carbon footprint. Diverse activities are part of the design program, including: partying, cooking, tanning, swimming, archery, horseshoes, gardening, and wood-splitting.

The project has 2 primary buildings, plus accessory buildings and recreational components that are designed to work together as a country compound.

The 2,800-square-foot Main House is composed of a live building and a sleep building that overlap at their roofs to create a linkage and sheltered outdoor space. The sleep building is slid under the higher roof of the live building. The live building contains the dining, living, and kitchen areas, plus a master suite upstairs. The central kitchen anchors the main space, which is flanked by dining and living. The kitchen island has an eighteen-foot long, three-inch thick single walnut slab countertop that has many edges and is cut from a large fallen tree. Two, triple sets of eight by ten feet sliding glass doors open this main living space to the yard and the views, creating a panoramic, 32-foot-wide clear opening when fully deployed. The design concentrates a lot on bringing in natural light. Only artificial lighting is needed at night.

The project has 2 primary buildings, plus accessory buildings and recreational components that are designed to work together as a country compound maller building, with two bedrooms and a shared bathhouse, angling out to form an L-shaped yard. Reclaimed barn wood and Corten rusted steel roofing covers the exterior of the buildings. The interior is a composition of concrete floors, wood, stone, and steel.

The barn is a 1,600-square-foot, Corten rusted steel warehouse outfitted for use as the property's "clubhouse." The first floor is set up for ping pong and large-screen TV watching, and houses surfboards and other recreational beach equipment. The loft level is set up with a billiard table and sofa beds all around for additional slumber party needs.

The interior is a combination of the structural steel shell and the exposed wood loft.

The property itself pulls it all together. The house and barn sit on the sloped meadow surrounded by redwood trees and with distant ocean views to the south. The activity areas are arranged around the outside of the house and barn. The swimming pool is below the main house, with a fire-pit to the side. The archery range is above the house, and the horseshoe pit is between the house and the barn. Two tent cabins are set among the redwoods and serve as guest houses. A sauna is located between the sleep building and the tent cabins.

The project incorporates the following sustainable features: Solar thermal system for domestic hot water, the pool, and the hydronic radiant floor.

Solar hot water is used for all the home's hot water, radiant floor heating system and for pool heating. Solar thermal system is a series of tubes in yard that is called the "Evacuated Tube Solar collector", which collects solar energy, and in turn provides the hot water for the entire house, the pool, and the radiant floor. Radiant floors both upstairs and down heat the entire house, and are the only source of heat other than the fireplace.

Northern california is a wood region, and residential buildings are usually built of wood—especially when they are country houses. In addition to those reasons, the exterior is reclaimed barn wood which is a sustainable product and adds to the "Barn Look." The use of wood inside the house is to make it feel more connected to its rural setting.

Corten rusted steel, a high-content recycled product, is used for the roof of the main house, and for the walls and roof of the barn.

For building owners and architects committed to preserving natural resources, metal roof and wall systems offer an environmentally responsible solution to their building's exterior requirements. Metal roof and wall panels are long-lasting, environmentally friendly and highly recyclable. Commercial metal roofs with heat-deflecting coatings and finishes can save building owners up to 40% in heating and cooling costs.

High performance windows keep heat inside in winter, cool air in summer.

South facing orientation of windows is for passive solar gain, with deep canopies for seasonal shading. Natural ventilation shafts are built into design for passive cooling.

The Homestead House
霍姆斯特德家园

Michael Jantzen

该设计的灵感来自于二十世纪六十年代后期Michael Jantzen的一个实验性作品，那会儿他还只是南伊利诺斯州爱德华兹维尔大学的本科生。Michael Jantzen决定对某些现成的农业建筑原件进行复审，看它们在选择性建筑制度的建立中是否具有利用潜力。

霍姆斯特德家园是一项针对选择性建筑的概念设计，用以探索农业用途上的市售钢材、预制构件、模块化、高强度、低成本、拱形建筑系统的潜在利用价值。材料方面，该建筑选用了可再生钢质薄板制成的拱形和直线形翼片，哪怕是技术不娴熟的工人，也只需利用简单的工具便可将它们拼接到一起。值得一提的是，拼接到一起的拱形翼片通常都不需要额外的辅助支撑，因此，只需很少的材料便能搭建出坚固的房屋外壁，同时，又可以采取同样的组装方式将其拆分开来。由此，整个建筑可以回收，并在不同的地点根据不同的用途重建使用。

霍姆斯特德家园的极端模块化设计允许模块灵活地组合以适应不同的需求。整个建筑的大小和形状可通过添加和删减完整的模块组件或者单独的拱形结构来改变。

隔热方面，霍姆斯特德家园也有很多途径。目前的设计中，用于架设整个二级结构（由大量的轻量材料制成）的纤维素绝缘材料（比如碾碎的报纸），按照不同的厚度要求被熔断在了房屋内墙和外壁之间。

霍姆斯特德家园的另一些功能通过公共电网起作用。它凭借光伏电池和一个小型垂直轴风力发电机来产生电力。这些设备运用被动式太阳能加热冷却机制，此外，霍姆斯特德家园里的生活用水也是通过太阳能来加热的。

通过拱形屋顶，雨水能够直接落入安置在地面表层和地面里层的雨水收容器中。其他替代能源和存储系统，包括利用太阳能和风能来供电的氢气制造机，都将应用到整个霍姆斯特德家园中。

对于致力于环保建筑系统中替代性建筑设计的Michael Jantzen来说，设计时最重要的是要优先考虑那些已建成的部分，即使它们和最终用途相悖。在这个过程中，通常能够发现意想不到的方式来为这个世界对可持续、低成本住宅的不断增长的需求提供解决方案。此外，如果替代性建筑元件已经存在，那就没有必要为了再建一个新的生产基地去消耗更多的资源。

Inspired by his experimental design work in the late 1960s as an undergraduate at Southern Illinois University, Edwardsville, Michael Jantzen decided to re-examine the potential use of certain readily available agricultural building components in the creation of alternative housing systems.

The Homestead House is a conceptual design for alternative housing that explores the potential use of a commercially available steel, prefabricated, modular, high strength, low cost, arch building system normally used for agricultural purposes. The steel arches and straight panels used in this building system are formed from thin recyclable steel sheets, which can easily be bolted together with simple tools, and with unskilled labor. Once the arches are bolted together, they normally do not require an additional secondary support structure. As a result, very little material is required to form an extremely strong envelope that can be taken apart in the same manner in which it is assembled. In this way, the entire structure can be recycled by erecting it again in a different location for a different function.

The extreme modularity of the Homestead House design allows for a great degree of flexibility in the way in which the modules can be clustered together to accommodate different needs. The size and shape of the entire structure can easily be altered over time by adding or subtracting complete modules, or by adding or subtracting one arch at a time.

There are various ways to insulate the Homestead House. In the present design, an entire second structure (made of much lighter gauge material) is erected inside of the outer shell and cellulose insulation (ground up newspaper) is blown in-between the two structures in any thickness needed.

This Homestead House is designed to function off of the standard utility grid. It would be able to generate its own electricity with photovoltaic cells and with a small vertical axis wind turbine. The structure would be passively solar heated and cooled and the domestic water would also be heated by the sun.

Rainwater would be collected off some of the roof arches and directed to above or below ground storage containers. Many other alternative energy gathering and storage systems can be employed, including the possible use of solar and wind powered hydrogen manufacturing, for use throughout the house.

It has always been important to Michael Jantzen as a designer of alternative, eco-friendly building systems, to first look at what is already being manufactured, even for a different end use. In this way, one can often find unexpected ways in which to provide solutions to the growing world need for sustainable, low cost housing. In addition, if alternative construction components already exist, there is no need to consume even more resources in order to establish a manufacturing facility.

The Loq-kit Research Program

Loq-kit研究计划

Patrick A. Freet

Loq-kit是一项正在进行的研究和发展计划，由位于美国明尼苏达州明尼阿波利斯市的PAF建筑有限责任公司实施。Patrick Freet，AIA 和 Carl Olson负责领导这一计划。

计划的目标在于发展预制的、不太昂贵的、可持续利用的住所，它们可以被装运给有需要的人。

世界范围内大学的相关团体、团队成员已经在研究这项工作。PAF建筑将自己的创新之举命名为"服务式住所"，为Loq-kit房屋所有者开创出一个生命周期服务计划。由于房屋的部件可交换使用，使用过的部件可以在全世界房屋业主之间交换再用，这就可以降低房屋成本，减少浪费。

PAF建筑有限公司所设计的Loq-kit"摇篮到摇篮"之家，首次体现了这些新概念。它为弗吉尼亚州劳诺克一个腾空地盘（infill site）而设计，获得了2005年"摇篮到摇篮"世界可持续性房屋设计竞赛第二名。

寻找模本

大规模生产的过程为降低房屋成本提供了巨大的潜在空间。过去，建筑和设计师已经将工业化的房屋系统发展起来了。直到今天，我们仍然还没有一个具有市场可行性的模本，以便用于大规模地建造消费者支付得起的房子。对住宅建设来说，大规模的生产方式面临着一个独特的挑战，因为它的效率是通过生产的同一性来实现的。然而，我们的住所，乃是我们自身、家庭和认同感的延伸。多样化、独特性、空间的个人化，对我们都有着高度的吸引力。因此，那些大量地、重复地生产出来的房屋，很少能激起我们自我表达的愿望。出于这个原因，Loq-kit被发展为这样一个系统——各部件可以交换使用。

绿色Loq-kit

Loq-kit房屋能够被大规模生产，并且装配成功后，每个都能拥有自己的独特性。房屋拥有者可以更改布局，让家庭空间更加个人化。而且，由于各部件能够彼此分拆（仅从室内空间中进行分拆），它们可以被重新安排、一次又一次地再利用。Loq-kit易于维护，在它们的生命周期里能够持续更新，这样，房屋拥有者可以增加或减少空间，重新安排部件，改变它们的样子和布局。如果需要使用到再利用的部件，还能得到进一步的折扣价格。Loq-kit旨在表明，这个具有市场可行性、能够大规模生产的住宅模本，确切地说，是一个由可交换使用、可再利用、可维护的房屋部件所构成的系统。

Loq-kit是一个可以大规模生产的系统，它由可交换使用、可重复利用的房屋部件组成，容许人们作出独特的室内布局和外部设计。你可以增加或移除墙、门、窗，可以增建房屋额外的部分，也可以减去一部分——包括二楼。消费电子学的技术也得到了应用，通过线束或电气母线槽，完全被整合到结构框架里。可增减的电子技术包括光伏发电系统、安保系统、低压以及音频系统。这些房屋部件还包括一个屋顶雨水收集系统。Loq-kit房屋由三个要素组成：结构框架（再生利用的钢材），填料和易操作的覆盖面（使用天然的加强纤维树脂，这是一种理想的、非基于石油制成的材料），各种各样的结构成员由弹簧锁安装就位——所有的填料和覆盖面都利用弹簧锁来进行连接。由于使用了弹簧锁，各部件能够在不同家庭的房屋间交换使用，而那些没有被损坏的部件则能够一次次地重复利用。普通的木制房屋拆解起来既费时又费钱，它们在安装时使用钉子和胶水，这是单向的组装行为，很难逆向操作或进行拆解。因此，建筑材料很少能够再利用，而是被送到了垃圾堆填区。比如说，“在1998年，美国因建造和拆除建筑产生了1.36亿吨垃圾，只有20%到30%得到了再循环或再利用。”（《一种描述：美国在建筑相关行业建造、拆除行为中产生的瓦砾堆》，1998）而且，Loq-kit并不是一项基于建筑结构的技术。为了实现房屋的个人化，建造过程中，必然要通过改变已经生产出来的原材料来进行，制造出大量垃圾。训练有素的木工在工厂或工地进行切割，去改动那些已经被生产出来、包装好、作为成品运送过来的材料。建筑结构的形成，是使用材料来进行的二次制造，会产生大量垃圾。“每年，美国的建筑商生产出约3.15千万吨建筑垃圾，几乎占据了这个国家市政固体垃圾的24%之多。”（全美住宅建筑商协会，《建立一种平衡：清除固体垃圾环境教育情况说明书》2004）与之相反，Loq-kit是一项基于组装范式的房屋建造技术。组装的过程和部件的生产都是一次完成的，房屋最终的布局在生产时就能够定下。在组装的过程中，几乎不产生垃圾。

可持续的维护技术

Loq-kit 推介这样一种观念：房屋不是作为销售的产品，而是一种对房屋所有者及其家庭的维护服务。PAF建筑有限公司将它命名为“服务式住所”。房屋系统的设计和布局具有灵活性，采用了“可交换使用部

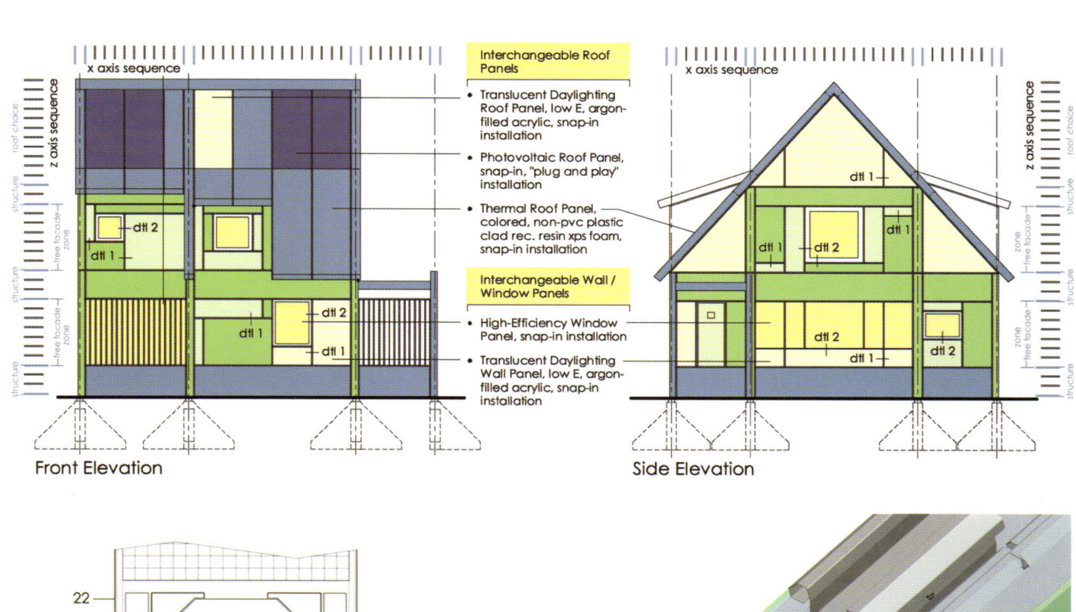

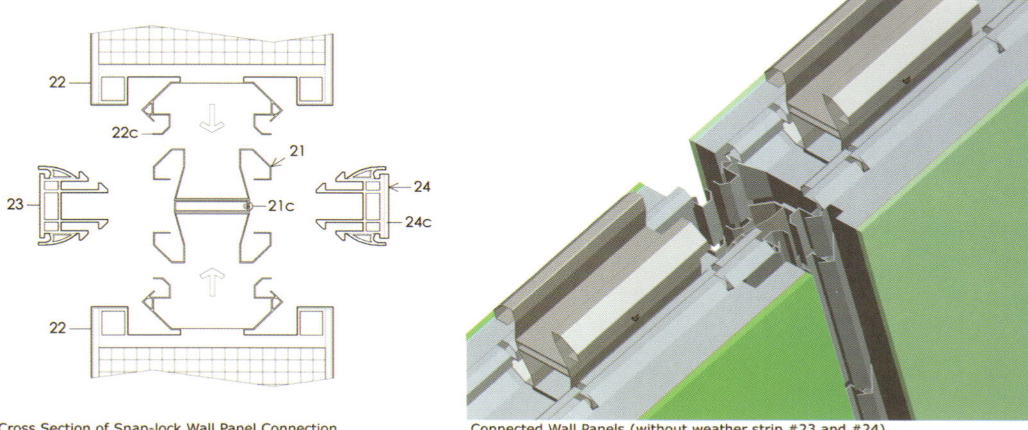

Cross Section of Snap-lock Wall Panel Connection

Connected Wall Panels (without weather strip #23 and #24)

Wall Panel Connections showing 6-Way Connector (Part #21)

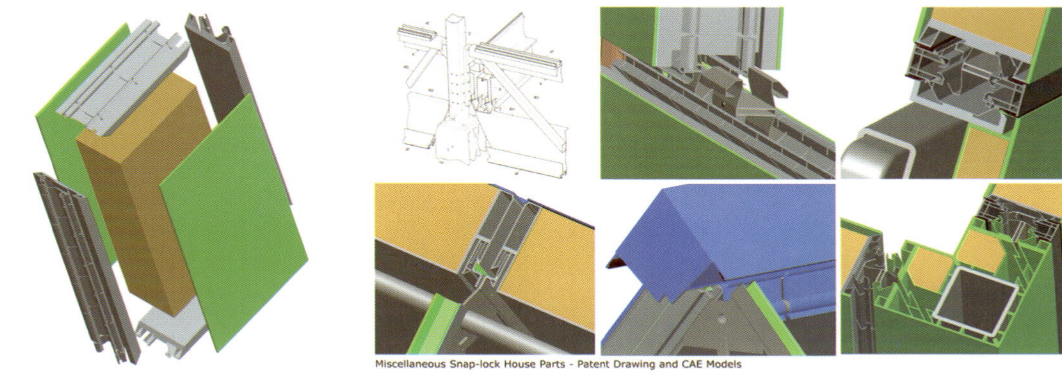
Miscellaneous Snap-lock House Parts - Patent Drawing and CAE Models

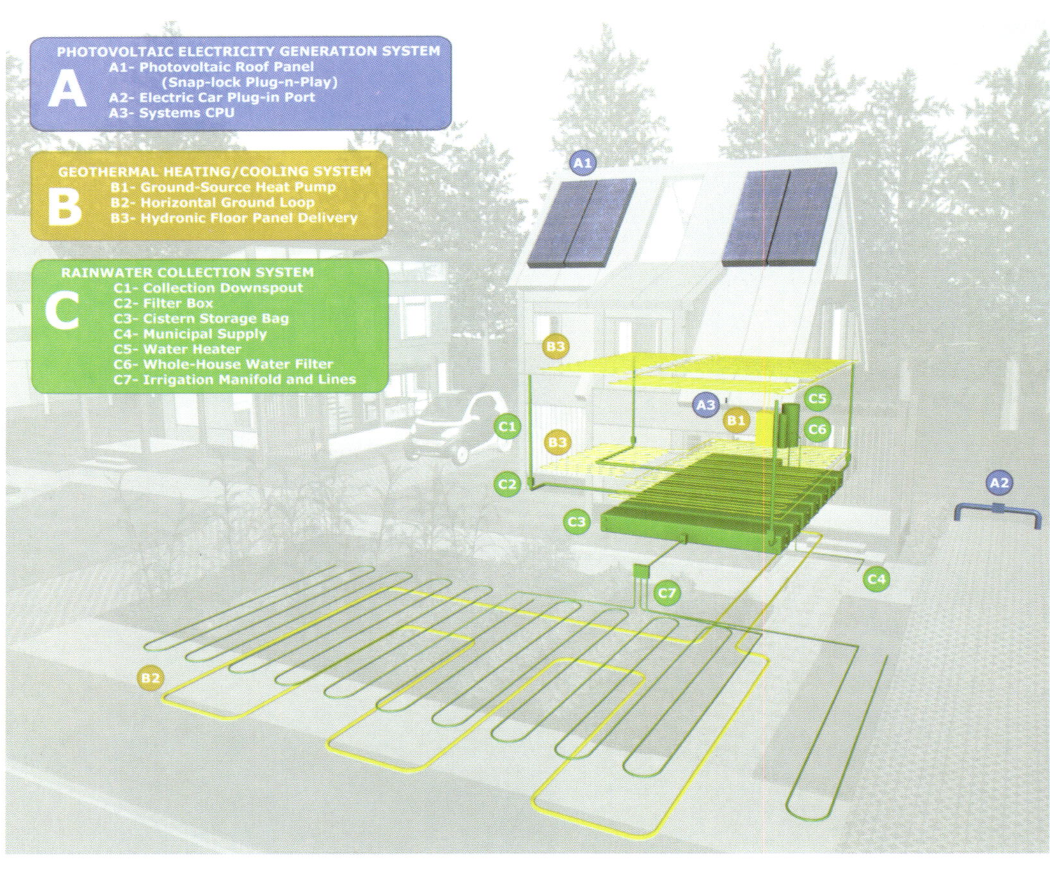

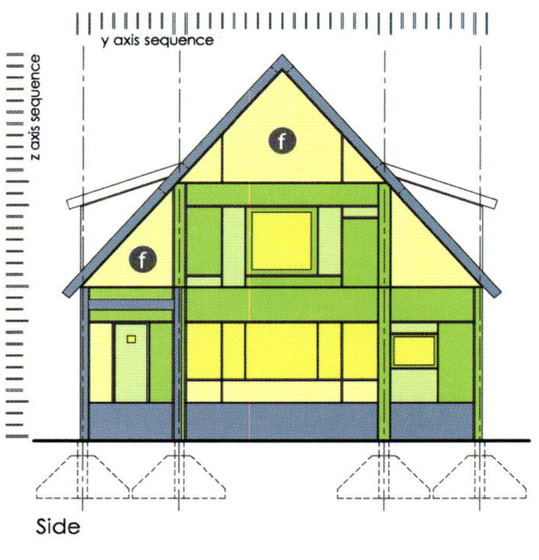

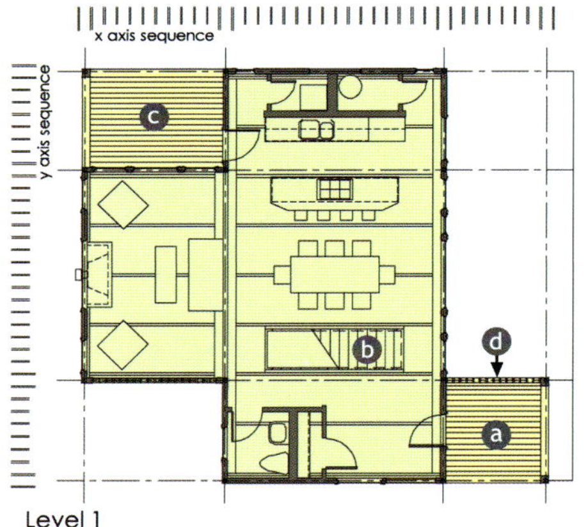

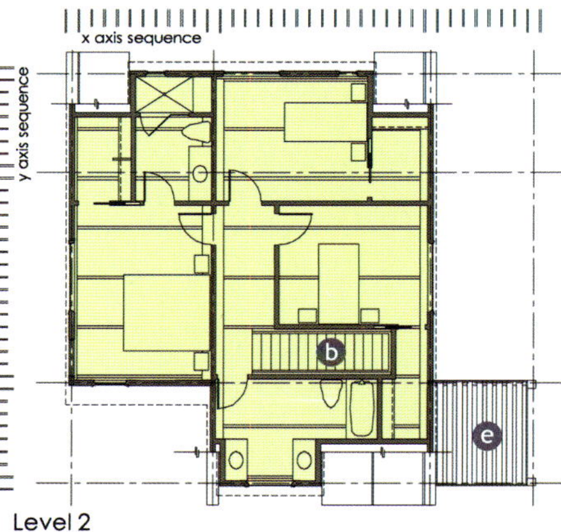

HOUSE 5

3 Bedrooms, 2½ Baths

Indoor Living - 1,702 SF

Outdoor Living - 169 SF

a - entrance porch
b - stair carriage
c - back-side porch
d - privacy screen
e - sunshade
f - translucent daylighting wall panel
g - translucent daylighting roof panel
h - photovoltaic roof solar panel

件”的概念。当消费者购买了Loq-kit的房屋，他们的家将被纳入到终身维护计划中。Loq-kit将为房屋所有者提供该栋房屋整个使用周期内的维护服务。如果，未来房屋所有者不再想要其中的一部分——不管是出于部件失效，还是业主个人的选择，这一部分都可以免费更换、或得到一个折扣价。比如说，当一个家庭人口增加，变成更大的家庭，Loq-kit可以为现有的房屋增加卧室。对于那些不再需要的部件，也可以在重建过程中被移除。由于组件在移除时不会受到破坏，Loq-kit可以把它们重新包装好，将来折价提供给其他有需要的房屋所有者。这是一项经济合理的再利用计划，能够保证这些部件不会被扔到堆填区去。Loq-kit对房屋的终身维护，确保了再利用的部件在公司内部的流通，它们将会被分出等级，以进行后续的维护服务。如果一个部件被判定为不合标准，Loq-kit会继续保留它，把它分解成单体的材料进行循环利用，用到新的部件上去。经由有效的循环利用计划，对于未经使用的原材料的需求能够得到控制。这项“摇篮到摇篮”的房屋维护服务计划，作为一个能够大规模生产、交换使用部件、消费者支付得起的系统，不是一次性的、会制造垃圾的待售商品，而是Loq-kit一个新颖的概念。在概念的背后，Loq-kit房屋已经在“摇篮到摇篮”世界可持续性房屋设计大赛获得过第二名。Loq-kit将“可支付”的概念具体化，通过大规模生产、可交换使用的部件、个人化、终身维护服务和重复利用的办法来降低成本。本房屋系统使用绿色材料、新技术，并推介了这样一个计划，用来提供人们支付得起的房屋，同时消除建筑垃圾。

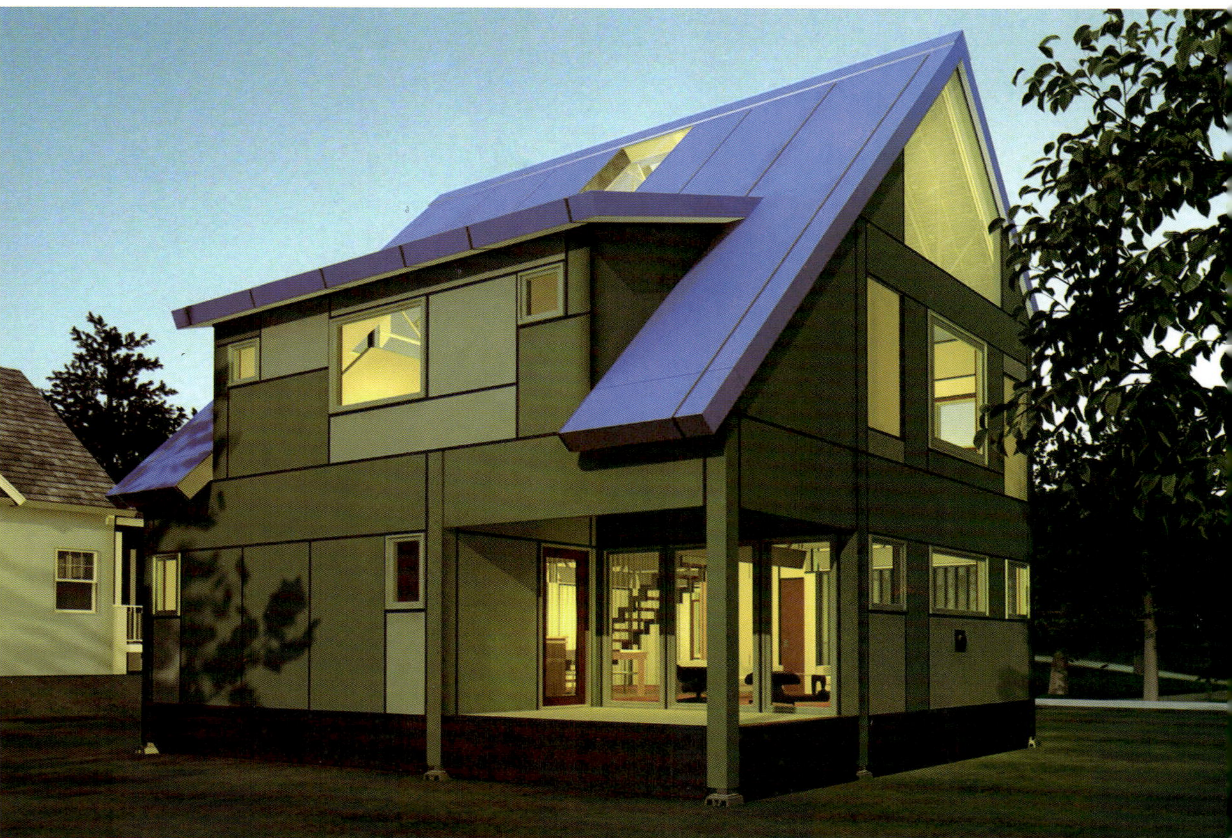

Loq-kit is an ongoing research and development program by PAF Architecture, LLC in Minneapolis, MN and is headed by Patrick Freet, AIA and Carl Olson.

The program is aimed at the development of ready-made, affordable, and sustainable homes that may be shipped to those in need.

The work has been studied by team members and groups at universities around the world. The innovative approach that PAF Architecture has named Shelter with Service initiates a lifecycle service program for Loq-kit homeowners where used interchangeable house parts are exchanged between homeowners around the world to reduce housing costs and waste.

PAF Architecture's Loq-kit C2C Home is the first home to demonstrate these new concepts. Designed for an infill site in Roanoke, Virginia, the house won 2nd Prize in the International Cradle-to-Cradle Home Sustainable Design Competition, 2005.

Pursuing the prototype

Mass production processes offer enormous potential for reducing housing costs. Industrialized housing systems have been developed by architects and designers in the past, with mixed results. And today we still do not have a market-viable prototypical mass-produced affordable house that is available to consumers. Mass production presents a unique challenge for housing, because its efficiencies are realized by producing sameness. Yet, our homes are an extension of ourselves, our families, and our identity. Variety, uniqueness, and the ability to personalize our space are highly desirable—while mass-produced repetition in housing does little to inspire our desire for self expression. It is for that reason, that Loq-kit is developed as a system of interchangeable building components.

Loq-kit green

Loq-kit homes can be mass-produced and assembled to be unique from one another. They may be modified by the owner, enabling homeowners to personalize their space. Furthermore, because the components can be released from each other (from the interior of the home only), they may be rearranged and reused over and over. Easy to service, Loq-kit homes can be updated continuously over their lifetime. This enables homeowners to add or subtract to their homes, rearrange components, change their look and layout, and obtain reused parts at a further-reduced price. Loq-kit aims to demonstrate that the market-viable prototypical mass-produced house is rather a system of interchangeable, reusable and servicable house parts.

Loq-kit is a mass-produced system of interchangeable and reusable house parts that allows unique interior layouts and exterior designs. Walls, doors, and windows can be added or removed. Home additions can be added or subtracted — including second stories. And consumer electronics technologies are incorporated via wire harnesses or electrical Buss duct — fully integrated within the structural frame. Integrated electronics technologies that may be added or subtracted include photovoltaic electricity-generation systems, security systems, low-voltage, and audio systems. The building components also incorporate a roof rainwater collection system. Loq-kit building components are of three varieties: structural frame (recycled steel), infill and snap-cladding (natural fiber-reinforced resins — ideally, non-petroleum based). A variety of structural members snap-lock into place — while all infill and snap-cladding components utilize snap-lock connections. With snap-lock parts, components may be interchanged between homes, and the undamaged parts may be reused over and over. For conventional wood-based assemblies, deconstruction is very time-consuming and expensive. Nails and glues are one-way assembly practices and are difficult to reverse or separate. Consequently, construction materials are seldom reused, and are lost to landfills. For example, "in 1998 the U.S. generated 136 million tons of waste from construction and demolition. Only 20 to 30 percent was recycled or reused." (*A Characterization of Building-Related Construction and Demolition Debris in the United States*, 1998.) Furthermore, Loq-kit is not a construction-based technology at all. Construction is a waste-producing process of modifying already produced raw materials to take on custom installations. Skilled carpenters cut and modify (in a factory or on site) materials that have already been manufactured, packaged and delivered as a finished product. Construction is a second fabrication of materials that produces vast amounts of waste. "Annually, builders in the United States generate approximately 31.5 million tons of construction waste, almost 24 percent of the total municipal solid waste stream in this country." (National Association of Home Builders, "Building a Balance: Solid Waste Disposal Environmental Education Fact Sheet," 2004.) In contrast, Loq-kit is a home-building technology based on the paradigm of assembly. Assembly is the process of manufacturing components once, in their final configuration – and ready for installation. Assembly produces almost no waste.

Sustainable service technology

Loq-kit introduces the concept of housing not as a product for sale, but as a service for homeowners and their families. PAF Architecture calls it Shelter with Service. The design and layout flexibility of the system employs the concept of interchangeable parts. When a customer purchases a Loq-kit house, their home will be included in a lifecycle maintenance program. Loq-kit will provide their homeowners with lifetime care of their home. Should any component become undesirable in the future, either by failure or choice, the part will be replaced for free, or at a reduced price by Loq-kit. For example, when a family grows, Loq-kit is available to add bedrooms to the house. The family will be credited for the no-longer needed components that are removed during the remodeling process. Because the components can be

removed undamaged, they can be repackaged by Loq-kit for further cost reduction to other homeowners. This is an economically-sustainable program of reuse which assures that components do not end up in landfills. The lifetime care of homes by Loq-kit ensures that reused components are continuously routed through the company, where they may be graded for continued service. At a point where a component is judged to be sub-standard, Loq-kit will be in possession of the part, where it can be broken down into singular materials for recycling into new components. Virgin material needs can be limited by an effective recycling program. This cradle to cradle service program of housing, as an affordable system of mass-produced and interchangeable parts — rather than a one-time, waste producing product for sale, is the novel concept behind the Loq-kit house part's 2nd place finish in the C2C Home international sustainable design competition. Loq-kit embodies the concepts of affordability through mass-production, interchangeable parts, personalization, lifecycle service, and reuse. It is a system that employs green materials, new technologies, and introduces a program for providing affordable homes while eliminating construction waste.

Oogst 1

Oogst 1 Solo住宅

Bio Frank Tjepkema / Tjep

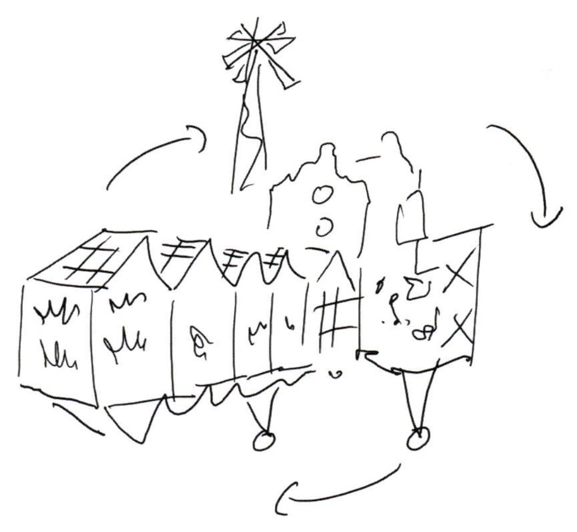

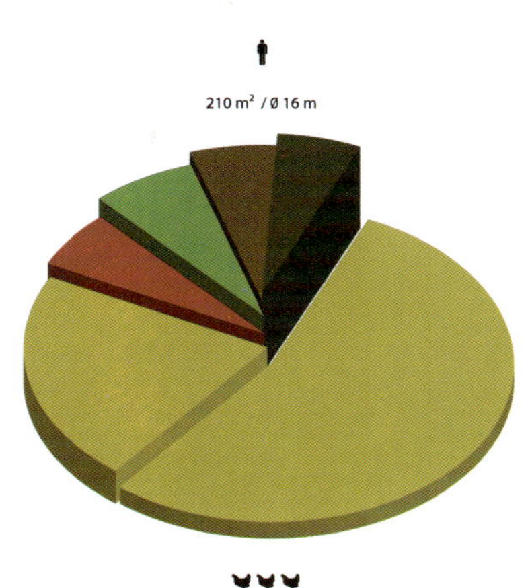

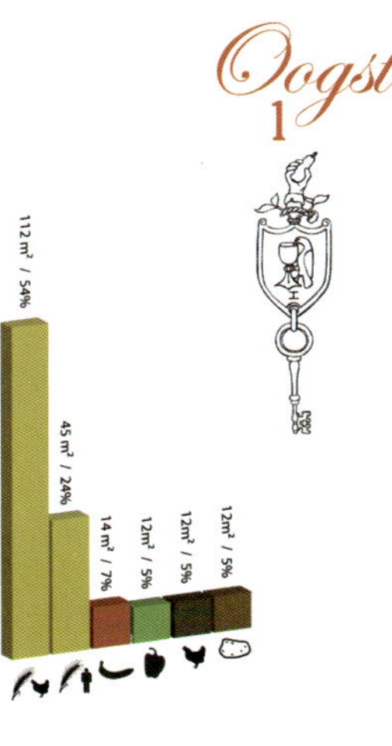

Oogst 1 Solo是一幢提供人们食物，能量，供暖和氧气的住宅。原则上，一个人居住在这里无特别情况都不用离开Oogst 1 Solo。

Oogst 1 Solo有一个十字形的楼层示意图。其中心区是居住区域，有厨房，浴室，储藏室和起居室，在最顶端，是卧室。其最大的部分是温室，其中一部分类似于温室村庄循环加工过程。

Oogst 1内唯一的动物是鸡，居民以鸡蛋为食，偶尔也食肉。在房子的最顶端可以看到发电风车，里面有太阳能控制板，里面拥有不同能源，得以最低限度地依赖一种资源。植物将二氧化碳转变为氧气，让氧气尽可能地保持新鲜，使其免受空气污染。

温室村庄模型贯彻一屋一主人的原则。其目标是实现图标结算，提供一个人可以舒适地居住在循环再加工的空间里的现象。可持续发展这一最终理念得以慢慢得明确且可预见。

模型屋可以在任何地方放置，就像在月球上一样，例如，可以放置在建筑物上，也可以放置在水上。模型屋从地面到顶层的三个点就好像在地球上逐步留下的小脚印。

模型屋被内外转动，所有的系统都可以进行明显的维护和修补。但是最重要的是展示运用于实际操作的技术原理。

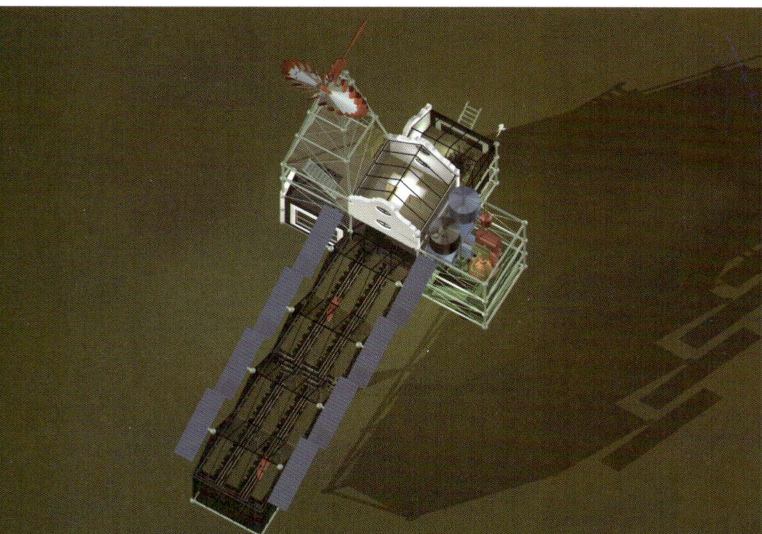

Oogst 1 Solo is a house for one person that provides its resident with food, energy, heat and oxygen. In principle, one could live in Oogst 1 Solo without ever having to leave the house.

Oogst 1 Solo has a cross-shaped floor plan. The heart is the living area, with the kitchen, bathroom, storage room, living-room and at the top, the bedroom. The biggest part is reserved for the greenhouse and one section is reserved for all the recycling process such as seen in the Greenhouse Village.

The only animals held at Oogst 1 are chickens, for their eggs and an occasional meat dish. On the top of the house one can find a windmill for electricity production. Also included are solar panels. Different energy sources are included to minimize dependence on one source. All CO_2 is turned into oxygen by the plants, making it possible to keep the house air-tight and air pollution proof.

The Greenhouse Village model has been implemented for a house with one occupant. The goal is the realization of an iconic statement and to provide evidence that one person can live comfortably within an identifiable recycling process. The ultimate idea of sustainability is made visible and tangible.

The house can be put almost anywhere, like a lunar module. For example, on top of a building or on water. The house stands from the ground on three points, literally leaving the smallest possible footprint on the earth.

The house has been turned inside out in the sense that all systems are visible for maintenance and repair. But more importantly, to show the technology that is put to work.

Oogst Community 100

Oogst社团 100

Bio Frank Tjepkema / Tjep

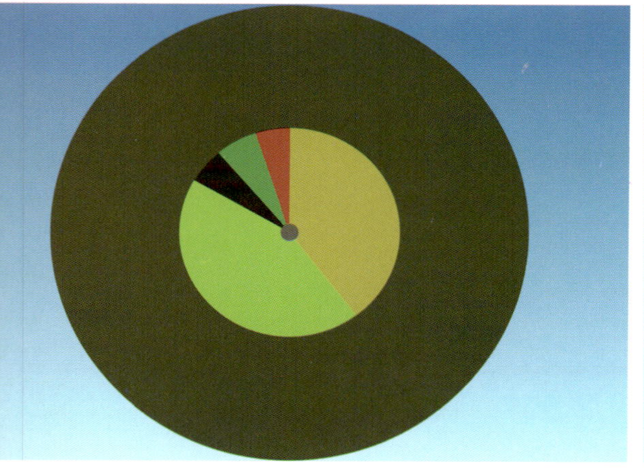

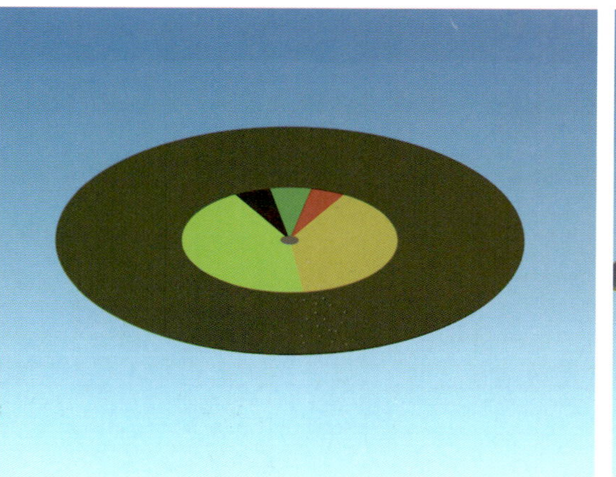

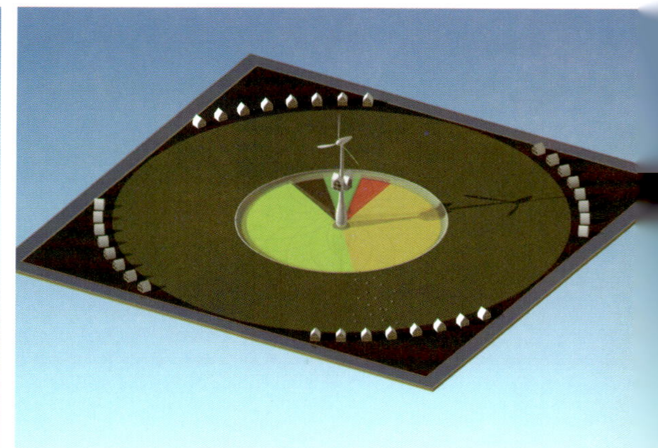

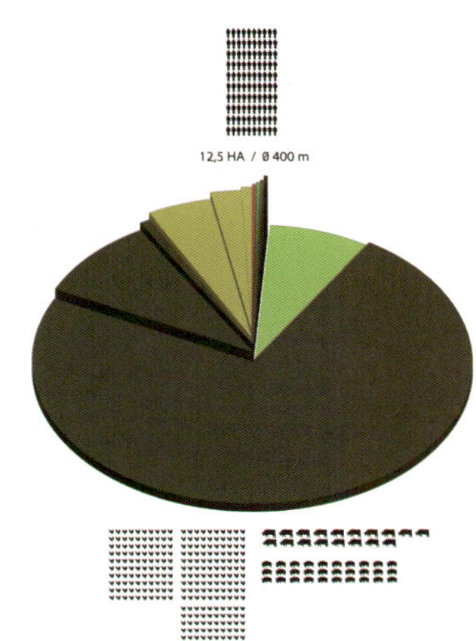

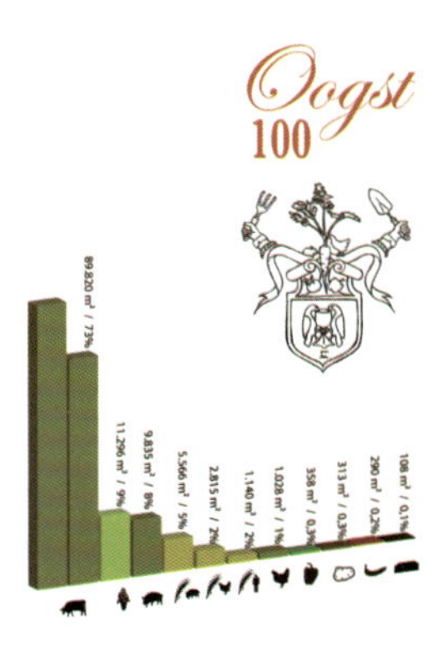

Oogst社团100是为100人提供自给自足的农场。在此居住的都是农民。在温室的中心，所有需要的庄稼都在生长，周围的田地都养着家畜。中央的风车提供所有必需的能源，在风车房的下面有个水井。但纯净的水资源主要由温室来提供。

虽然温室和Oogst 1所运用的技术相同，但有一点是隐蔽的，就是拥有100到150人的社区是最理想和最和谐的。因此，在设计概念上从来没有超过100人，但是它可以组成一个没有购物中心、银行、马路、汽车和飞机的社区。灌溉系统使水上运输成为可能（依靠电力）。

每一个社区需要一个聚集的地方，在建造风车房的时候有一个十字形的楼层图，在社区内有食物供应，进行婚礼，但要注意的是，这个社区的活动最多只限100个人参加。

Oogst Community 100 is a self-sufficient farm for 100 people. The residents are all farmers. In the central greenhouse, all necessary crops are grown, and the surrounding fields are for livestock. The central windmill provides all the necessary energy, and there is a water-well under the windmill. But clean water is mainly collected condensed water from the greenhouse.

The technology involved is the same as in the Greenhouse Village and Oogst 1, but only here it is hidden. There are studies that show that communities of between 100 and 150 people are the most optimal and harmonious. Therefore, the concept never grows beyond 100 people but it can be multiplied to form a society without shopping centres, banks, roads, cars or airplanes. An irrigation system makes navigation possible (electric of course).

Every community needs a space to gather. The construction halfway up the windmill has a cross (church like) floor-plan. This is where meals are served and marriages organized between communities, but be careful, never more than 100 people can join in.

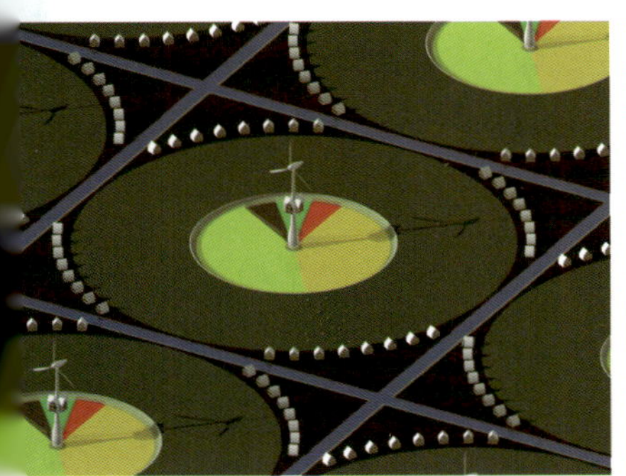

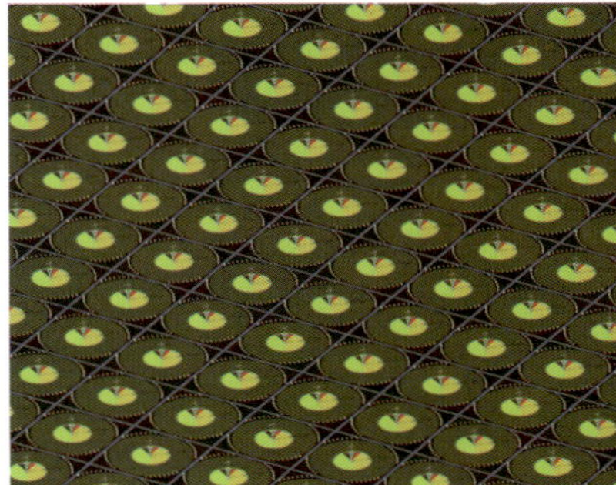

Oogst 1000

Oogst 1000

Bio Frank Tjepkema / Tjep

Oogst 1000仙境是一个每天供给一千人的农场、饭店、旅馆、娱乐场所。饭店所有的食物来自于建筑物中央和直接邻近的土地。Oogst 1000集合了许多有趣及有用的东西。你可以看见游乐场就好像一个巨大的人类数据处理机。旅馆的客人是在这里工作的农民，当他们工作他们可以免费地居住。

最初的荷兰农场建筑是贯彻比较传统的农场设计风格，现在已经全部被重新布局。在温室建筑物里所有的一切都和技术相连，为了实现一个自给自足的系统。

访客可以明显地看到最完整的进程，起到一个综合的教学功能并运用于新的农业发展。Oogst 1000仙境的洗手间与沼气能源系统相连，因此Oogst 1000提供全球领先的卫浴设备，每个参观者需要支付0.50欧元。

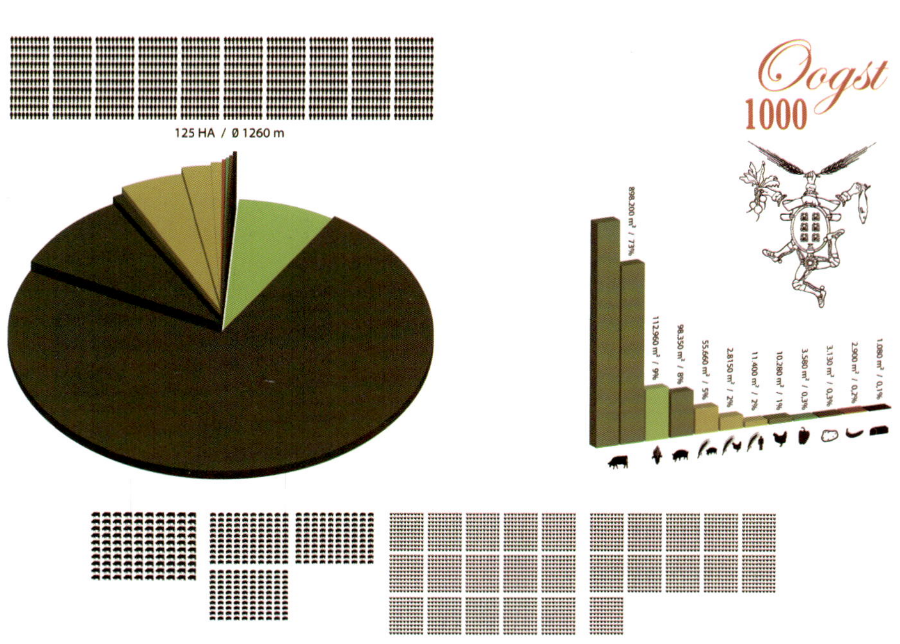

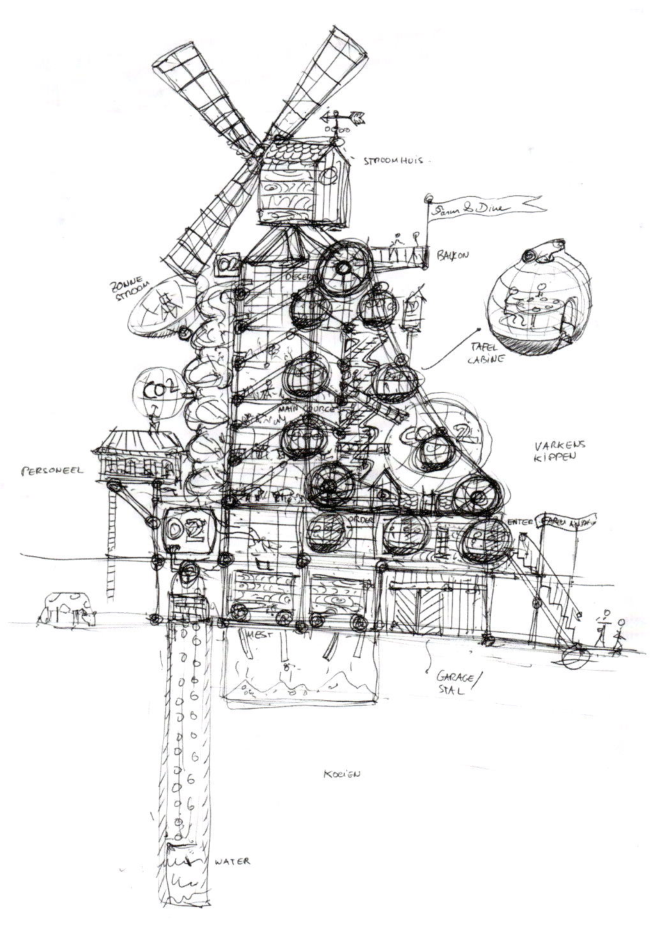
STROOMHUIS
BALKON
ZONNE STROOM
TAFEL CABINE
CO2
O2
VARKENS KIPPEN
PERSONEEL
MEST
GARAGE/ STAL
KOEIEN
WATER

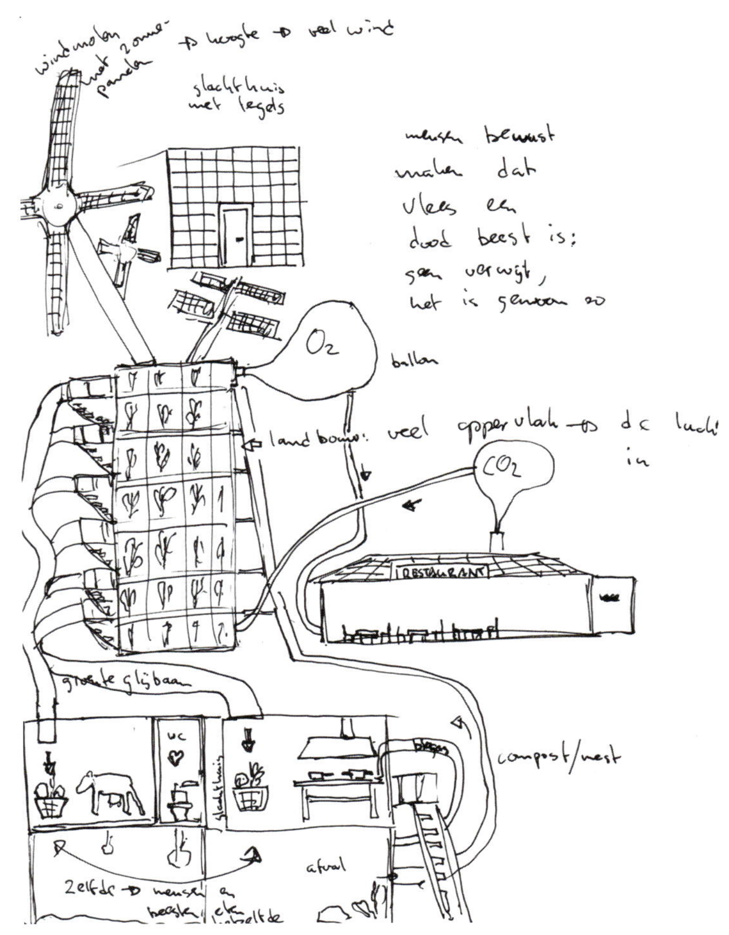
mensen bewust
maken dat
vlees een
dood beest is;
geen verwijt,
het is gewoon zo
O2
ballon
CO2
RESTAURANT
compost/mest
afval

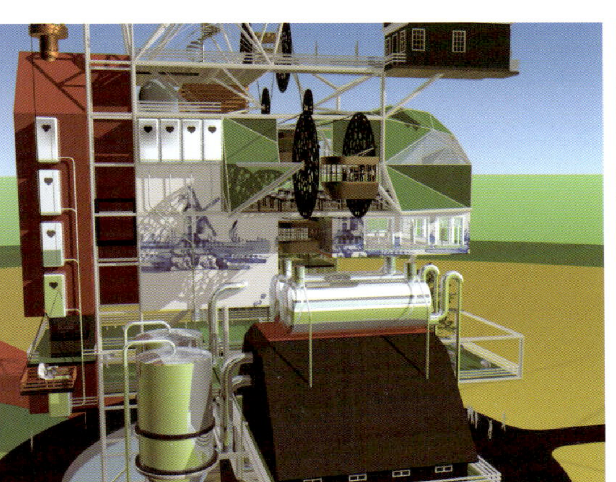

www.erwinvoogt.com

Oogst 1000 Wonderland is a self-sufficient farm, restaurant, hotel and amusement park for 1,000 people per day. All food for the restaurant comes from the central structure and directly adjacent fields. Oogst 1000 combines extreme fun with extreme usefulness. One can see this amusement park as a huge people processor. Hotel guests are also the farmers. When you work, you can stay for free.

Original Dutch farm buildings have been implemented but the traditional layout of the farm has been completely re-arranged. Everything is linked by technology proposed in the Greenhouse Village in order to create a self-sustainable system.

The entire process is visible to the visitor, giving the complex a didactic function as to new agricultural developments. Oogst 1000 Wonderland toilets are also linked to a bio-gas energy system, so Oogst 1000 offers the world's first toilets where you actually get paid Euro 0.50 per visitor.

图书在版编目（CIP）数据

生态建筑实验与实践 /《设计家》编. — 天津:天津
大学出版社，2012.2
ISBN 978-7-5618-4218-8

Ⅰ. ①生···Ⅱ. ①设···Ⅲ. ①生态建筑—建筑设计—
作品集—中国 Ⅳ.①TU206

中国版本图书馆CIP数据核字（2011）第237808号

主　　编　许晓东
编　　辑　宣　文
版式设计　徐晓霞
责任编辑　油俊伟

出版发行　天津大学出版社
出 版 人　杨　欢
地　　址　天津市卫津路92号天津大学内（邮编：300072）
电　　话　发行部：022—27403647　邮购部：022—27402742
网　　址　publish.tju.edu.cn
印　　刷　上海瑞时印刷有限公司
经　　销　全国各地新华书店
开　　本　240mm×320mm
印　　张　21
字　　数　296千
版　　次　2012年2月第1版
印　　次　2012年2月第1次
定　　价　328.00元